The Design of Microprocessor, Sensor, and Control Systems

The Design of Microprocessor, Sensor, and Control Systems

Michael F. Hordeski

RESTON PUBLISHING COMPANY, INC.
A Prentice-Hall Company
Reston, Virginia

Library of Congress Cataloging in Publication Data

Hordeski, Michael F.
  The design of microprocessor, sensor, and control
  systems.

  1. Microprocessors.   2. Process control.   3. Transducers.
  I. Title.
  TK7895.M5H66   1985   629.8'95   84-6897
  ISBN 0-8359-1269-8

© 1985 by Reston Publishing Company, Inc.
A Prentice-Hall Company
Reston, Virginia 22090

3 5 7 9 10 8 6 4 2

Printed in the United States of America

To the spirit of freedom typified by the Solidarity movement in Poland
and
To my mother, Julia Marion.

Contents

Preface

The Design of Microprocessor Sensor and Control Systems is based on the author's experience in applying microprocessors to the automation of industrial and commercial products and systems. Both manufacturing and process control today require the development of microprocessor systems for data acquisition and automatic control applications. This book provides the design information required for a successful and efficient microprocessor implementation.

From the initial specification to system testing and integration, the design must be based on a knowledge of microprocessors as well as the design techniques required to apply the devices. A broad array of applications for today's modern microprocessors exists for data acquisition and automatic control tasks when interfaced with various types of sensors and control elements. This book is intended as a reference source with chapters on the various system elements which meet the broad application spectrum.

Included are discussions on the basic structure and technology of microprocessors, industrial sensors, data acquisition, data conversion, microprocessor programming, control system elements, and analysis and system design techniques. These topics are necessary to achieve the objective of not only understanding the required technology, but also to apply the technology to actual microprocessor products and systems.

In chapter 1 we view the microprocessor as the main element in any microcomputer system. The recent significant strides in technology now offer more cost-effective solutions for many sensor and control systems. The user of these systems must understand the interaction of these components to effectively adapt and apply the new technology. This chapter considers the operation of microprocessors, microprocessor instructions, technology considerations, and semiconductor memories.

The users of microprocessor systems have a wide range of options by which to implement systems, from fully integrated networks to individual components for specialized product applications. Chapter two treats the hardware and software techniques now available. Topics include the proper use of specification definition, documentation concepts, flowcharting, algorithm development, programming techniques, program structures, hardware/software trade-offs, the use of single-board microcomputers, the selection of a programming language, the correct use of interpreters and compilers, editors, programming style, program optimization techniques, system testing and debugging techniques, and the use of operating systems and other software aids.

The collection, capture, and presentation of data from sensors is the basis of performance monitoring in applications ranging from industrial plants to test stations to probes deep into space. Recent years have seen great strides in the technology available for these tasks. We now have multiplexing methods available which can reduce the cost of communication channels, digital techniques to improve the transmission integrity, low cost computation methods which allow error checking and correction, and high frequency recording techniques as well as new graphic methods for the user. Some of these issues are considered in chapter three, including sensor scaling; signal conditioning techniques; instrumentation error analysis; filtering techniques; multiplex switching; the use of sample hold devices, digital storage techniques, and correction methods; distributed systems, and plantwide techniques.

The use of microprocessor systems in many production operations may require capabilities for sequencing, timing, speed, direction, and logic functions. The proper control techniques can produce strategies for energy management that minimize operating costs. Opportunities exist for optimization in these areas with the use of modern sensors and microprocessor controls. Topics in chapter four include speed control, motor regulation, the use of stepping motor control, dynamic considerations, energy considerations, drive circuits, open vs. closed loop control, the use of variable ratio systems, digital controllers, position control, acceleration control, and motor feedback techniques.

The incentives that now exist to raise the production capacity of installed capital equipment create the need to use new techniques with variable ranges of control. Combustion control systems are prime examples, since they allow boilers and heaters to operate at maximum efficiency. Other examples include systems for monitoring and controlling energy conversion and delivery using compressors, pumps, and other conversion devices. Chapter five examines feedback compensation techniques, gain control, stability, dynamic loop response, process lags, deadtime, the use of corrective networks, sampled systems, temperature control systems, nonlinear feedforward systems, and pulse frequency modulation systems.

Computerization and automation are becoming a major part of all manufacturing. Transducer design and application are an integral part of these systems.

Temperature is the most widely measured variable. It is important as a direct indication of the state of a process as well as a means of inferring other conditions that cannot be determined by specific measurement. The instrumentation is diverse, reflecting the long period of development and the broad range of applications. Chapter six covers resistance techniques, semiconductor methods, thermocouples, and optical radiation pyrometers.

Many materials in process, finished products, and energy transfer media are in liquid or gaseous form for all or part of production. The measurement of pressure and flow are fundamental to these operations. The traditional sensors in some applications continue to be in demand. Other applications require new techniques with higher accuracy, faster response, and the greater stability required for computer control. Chapters seven and eight treat both classes of sensors while concentrating on the advanced analog and digital signal types. The following types of sensors are considered: potentiometric, strain gage, inductive, piezoelectric, capacitive, encoder head meters, variable-area meters, turbine meters, anemometers, sonic and ultrasonic meters.

Many applications require determining the viscosity, moisture, or humidity of materials or objects. The capabilities of the available sensors are reviewed to aid the user in selecting units with the proper features, functions, and specifications for the application. Chapter nine considers pressure drop techniques, capillary methods, float techniques, ultrasonics, torque methods, resistance techniques, hygroscopic films, oscillating-crystals, spectroscopics, psychrometrics, dew-point sensing, and the use of lithium-chloride salts.

Strides have been made in recent years in the measurement of displacement, velocity, and acceleration. Optical techniques can allow the rapid transmission of signals or complex computations which are needed in many digital systems. The user must be familiar with these techniques as well as the traditional methods for many specialized applications. Chapter ten covers resistance techniques, capacitive sensors, inductive methods, digital techniques, optical encoders, laser interferometers, electromagnetic methods, tachometer techniques, seismic masses, piezoelectric accelerometers, and capacitive charge measuring systems.

Because stress and strain are generated in parts and systems, the proper instrumentation is the key to safety precautions in almost all applications. Their use involves monitors and controls to detect incipient or existing hazards and issue warnings. Force and torque measurements can also be used to ensure safety as the vessels and machinery used in many applications are capable of releasing large amounts of power. Chapter 11 discusses strain gage techniques, flame spraying, photoelasticity, Moire fringe analysis, proving rings, piezoelectric techniques, capacitive methods, vibrating wires encoder, and optical techniques.

This text is designed to provide the critical knowledge required in the fast growing microprocessor-oriented systems market. It relates the use of microprocessors to a broadening array of applications. The design techniques of microprocessor systems are presented to allow an easy transition into the new technology. The sensor and control techniques are illustrated with many application examples to provide the in-depth knowledge required to solve today's engineering and management problems in a cost-effective manner.

The book can function as a reference book as well as a text to be used for advanced undergraduate or graduate study. The exercises presented at the end of each chapter are designed to help the student better understand actual sensor and control system solutions.

The book emphasizes the systems concept, stressing that a device is almost always a part of a system. The devices of the systems—sensors or motors—have this concept in common. The operation of these devices in the system is highly dynamic, which it must be, for these systems to be effective. The microprocessor can optimize use of these systems for modern applications. It is no longer enough to consider these systems as constant processes or operations. We must include the dynamic operations, transient performance, and the possibility of unstable behavior in the modern microprocessor system.

One important area is the growth in robot use in the application of isolated processes or handling tasks. Much of the focus is now directed toward the integration of robots into the total manufacturing process, including computer aided design (CAD) and the creation of flexible manufacturing systems. This integration is possible due to the programming capability of some robots allowing access to the CAD/CAM data bases.

Flexible manufacturing systems incorporating robots offer small batch producers the economies which special-purpose automation brings to high-volume manufacture. The development of efficient software using the methods discussed in this text is required before significant integration with other manufacturing operations can be achieved. For example, the proper research and development have brought vision and other sensory capability as used in arc welding to widespread commercial use. It is hoped that the techniques discussed in this book will be used to achieve similar advances in productivity.

As I travel and advise various industries ranging from small start-up companies to major international corporations, I notice a concern but not a commitment to the technological progress required for success in today's world. I hope that this book will help reduce some of the barriers that exist in recognizing the power and usefulness of today's microprocessors, and will help to incorporate that power effectively into a product or system.

This book owes much to those in the many universities and industrial groups that have contributed the articles and books from which we build our technical knowledge. Some of these works are included in the bibliography, but I also thank those that are not included. I am indebted to the conscientious students at the California Polytechnic State University at San Luis Obispo for their interest and feedback. I am also grateful to those who have attended my seminars on the use and development of microprocessor products. Their stimulation and support and that of many in aerospace, automotive, communications, electronics, and process-related industries have also helped greatly. A special thanks to the red-haired lady who typed the manuscript and corrected many of the errors through the course of this project.

Atascadero, California   MICHAEL F. HORDESKI, P.E.

There is every reason to believe that the closing decades of the twentieth century will see changes as rapid as those that characterized the fifty years between 1860 and 1914, where a major new invention with new big businesses appeared on the scene every two to three years.

—Peter F. Drucker

The Design of Microprocessor, Sensor, and Control Systems

1

Microprocessor Characteristics

The central processing unit (CPU) for a computer can be constructed with medium scale integration (MSI) functions such as registers, arithmetic logic unit (ALUs), and decoders. A CPU made with MSI functions might require hundreds of integrated circuit (IC) packages, but it can be tailored for a particular application. When the CPU is in the form of large scale integration, it becomes a microprocessor (Figure 1-1). Unlike the custom designed CPU, the functional characteristics of the microprocessor are usually fixed by the internal architecture. The only access available to the microprocessor is through the terminals on the package.

An 8 bit microprocess such as the 6800 has 8 pins for the movement of data into and out of the device. These 8 pins are called the *data bus.* Information can flow in both directions along the bus, but during different times. This technique of multiplexing is different than the wiring in most logic circuit systems, which carry data in only one direction.

The 6800 microprocessor also has a group of 16 pins, which are used to move the binary words called *addresses;* together these are called the *address bus.* The address bus carries information outward from the microprocessor to memory and the input/output (I/O) chips. The signals on the address bus are used to select a certain part of the memory or I/O circuits.

At the core of the basic microcomputer system is a microprocessor with its arithmetic processing circuits and a control memory for its instruction set. Even a simple microcomputer system with its microprocessor requires some list of instructions to perform its task along with the control of input/output operations and communications.

Memories are used to hold programs and any data that must be manipulated by the instructions. The programs are stored in the memory as a series of binary

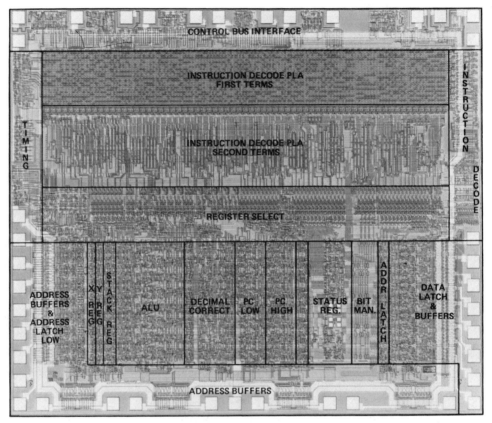

FIGURE 1-1   The chip layout of a CMOS microprocessor. (*Courtesy Rockwell International.*)

words. Each word may have from 4 to 64 bits, depending on the particular micro-processor used in the system and the accuracy required. Each word or series of words represents an instruction or data that the microprocessor will decode and act upon when that information is presented.

The microprocessor communicates with the outside world via three paths: the data bus, the address bus, and the control bus. The data bus is a group of parallel line signals that permit bidirectional digital data transfer. It may have 4, 8, 12, or 16 lines, depending on the microprocessor. Digital words transmitted over the data bus are either instructions for the processor, data to be manipulated, or the processed results of an operation.

The overall throughput and efficiency of the systems are directly dependent on the hardware and software interconnection mechanisms supported by the mi-croprocessor chips. Many different interconnection systems have evolved over the years, but the single time-shared bus offers advantages as an interconnection mechanism.

Arithmetic Operations

When an "add command" instruction occurs in the microprocessor, a binary adder operates on the data presented to the adder. In a similar way, subtraction is performed.

When subtraction is being performed, complementing one of the inputs yields the difference answer.

The microprocessor's ALU performs the arithmetic and logical operations as required by the instructions on the binary data that are stored in the registers. These operations are performed using the programmable logic of the ALU for adding, subtracting, multiplication, and division as well as Boolean algebra functions.

Most arithmetic logic units consist of a binary adder and other logical circuitry. A typical ALU has data inputs, function inputs, a carry input for performing multiple precision arithmetic, as well as data and status outputs for setting the various status flags. The function inputs determine which function the ALU performs. Typical functions other than the basic arithmetic operations may include:

1. Logical AND.
2. Logical (INCLUSIVE) OR.
3. Logical EXCLUSIVE OR.
4. Logical NOT (complements).
5. Increment (adds 1).
6. Decrement (subtracts 1).
7. Left shift (adds input to itself).
8. Clear (set to zero).

Since most chips operate in two's complement, subtraction is performed by taking the two's complement of one input before sending it to the adder. Multiplication or division can be performed by repeated additions or subtractions, respectively. Other circuitry can be used to form status outputs, such as a zero indicator.

The ALU for the 2900 bit slice microprocessor can perform five logic operations or three binary arithmetic operations on the two 4 bit input words via multiplexers. The multiplexers can be inhibited for a zero source operand. The 10, 11, and 12 microinstruction inputs to the ALU are used to select the ALU source operands for two 4-bit input words (R and S). The 13, 14, and 15 inputs are used to select the desired ALU function as shown next:

| Micro Code | | | | ALU |
15	14	13	Octal	Function
0	0	0	0	R + S
0	0	1	1	R − R
0	1	0	2	R − S
0	1	1	3	R OR S
1	0	0	4	R AND S
1	0	1	5	R AND S
1	1	0	6	R EX-OR S
1	1	1	7	R EX-NOR S

The adders and subtracters are the basis for many arithmetic circuits in microprocessors, but other special circuits are used also. The adder/subtracter circuit usually contains the necessary temporary storage registers to hold the individual operators and answer.

Often the addend register serves as both an addend and answer store. It is then called the *accumulator register.*

MICROPROCESSOR REGISTER CHARACTERISTICS

In microcomputers a majority of operations inside the computer are transfers of data from one register to another, sometimes between registers of different types, but still transfers of data. Memory can also be considered to be large groups of registers. Now that core memories have been replaced by chips, memory has become a large group of registers.

Registers are connected to other registers, to other parts of the control section, and to external buses by means of internal buses. The registers and internal buses usually have the same word length as the rest of the computer system; but registers and internal buses with half or double the system word length are also used in some microprocessors. The signals generated by the instruction handling section may place the contents of a register on a bus or the contents of a bus in the register.

The cost of registers and interconnections tends to limit the number found in a microcomputer. Registers were once so costly that most computer systems used less than ten. The rapid decrease in semiconductor chip prices and the increase of LSI technology caused these register costs to decrease. Newer computers use dozens of registers.

MICROPROCESSOR REGISTERS

A shift register in a microprocessor may be capable of shifting its binary information either to the right or to the left. An n bit shift register has n flip-flops along

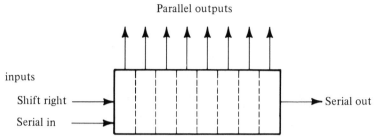

Parallel outputs

inputs

Shift right → Serial out

Serial in →

FIGURE 1-2   Shift Register with Shift Right Control.

with the gates to control the shift operation. The block diagram of a shift-right register is shown in Figure 1-2. It uses a shift-right control input to enable the shift. The serial input is a single line going to the left-most flip-flop of the register. The serial output is a single line from the output of the right-most flip-flop. The parallel output consists of n lines, one for each of the flip-flops.

The operation of a 4 bit shift-right register is shown in Table 1-1.

TABLE 1-1   SHIFT REGISTER OPERATION

Clock	Serial In	Parallel Outputs				Serial Out
0	1	0	1	1	1	1
1	1	1	0	1	1	1
2	0	1	1	0	1	1
3	1	0	1	1	0	0
4	Start of New Word	1	0	1	1	1

Register Operation

The initial binary information is stored in the register as 0111. The parallel outputs have one line to each flip-flop so the information stored can be inspected through these lines at the same time. The serial output is a single line from the output of the right-most flip-flop, so the 4 bits stored in the register come out through this line one at a time. The serial input is a single line and new information will enter the register 1 bit at a time during four consecutive clock pulses through this input. Thus information is transferred serially in and out as the register is shifted to the right.

We wish to transfer the binary word 1011 into the register in a serial fashion and at the same time to extract in serial fashion the stored 0111. Only one serial input line exists and the binary word must be applied at this input 1 bit at a time. The least significant bit (LSB) of the word is brought into the input line prior to the arrival of the clock pulse. At the same time, the bit in the right-most flip-flop of the register is extracted from the serial output line. When the input bit is transferred into the left-most flip-flop (with a clock pulse), the other bits are shifted to the right, bringing the second bit to the serial output line. The other input bits are shifted in a smaller manner. After the four shift pulses, the input word 1011 is shifted into the register with the lower order bit first and the previously stored word 0111 is shifted out, 1 bit at a time. At the same time, parallel output lines can provide the state of the register after each shift.

Registers may operate in a serial mode, a parallel mode, or a combination. The parallel outputs of registers are not needed if the CPU operates in the serial mode only. Words then are transferred into shift registers through serial input lines and taken out through serial output lines. Serial operations are slower because of the time it takes to transfer information in and out of the shift registers. Serial operations, however, require less hardware because one common circuit can be used over again to manipulate all bits sequentially.

Operation in the parallel mode is faster. In parallel computers, the output is taken from the parallel output lines. The serial input and serial output lines are used for cascading shift registers for increased bits.

Registers can have either shift-right or shift-left capabilities, or both. Some shift registers also use parallel input lines for parallel loading. An inhibit state may also leave the register unchanged, even though the clock pulses are applied. Chip sizes have tended to limit the number of registers in microprocessors, but this is changing with the newer, larger chips.

When the CPU has a large number of registers, programs do not require as many transfers of data to and from the memory. This can reduce the number of memory accesses along with the size of the memory.

Registers have various purposes; some microprocessors allow the programmer to assign registers for different functions. However, most registers are permanently assigned. Some common uses of registers in other microprocessors include:

1. Program counters.
2. Instruction holding.
3. Accumulators.
4. Index registers.
5. Condition code storage.
6. Stack pointers.
7. General purpose.

Program Counters

The program counters (PC) hold the address of the memory location for the next instruction. An instruction cycle starts with the microprocessor placing the con-

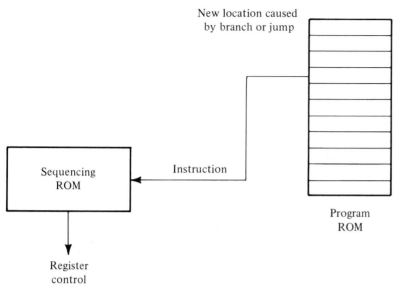

New location caused
by branch or jump

Sequencing
ROM

Instruction

Program
ROM

Register
control

FIGURE 1-3   Instruction Register Control.

tents of the PC on the address bus, thereby fetching the first word of the instruction from the memory.

Then the contents of the program counter are incremented so that the next instruction cycle fetches the next instruction from memory. The microprocessor executes instructions sequentially unless an instruction such as a jump or branch changes the program counter.

The program counter may have its contents changed, saved, or replaced by user instructions, thus causing the microprocessor to execute instructions from some part of the memory other than where it had been executing.

When a branch or jump instruction occurs, the contents of the program counter are changed to a value that is carried along with the instruction (Figure 1-3).

In a jump, the instruction may be anywhere in 16 bits of address, within memory. In the 6800 a branch instruction is limited to 8 bits, but this is not a general address. It is the distance—expressed as the number of locations, forward or backward—that we wish to jump over. This number is added algebraically to the program counter.

INSTRUCTION REGISTERS

The instruction register (IR) is used to hold the instruction until it can be decoded. The length of the instruction register is the same as the length of a basic instruction for the microprocessor. Some systems use two instruction registers, so they can fetch and save one instruction while executing the previous one. This tech-

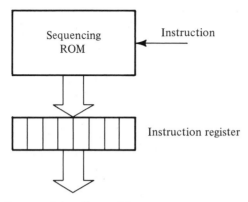

Commands to other registers

FIGURE 1-4   Basic Instruction Register Operation.

nique is called *pipelining.* The programmer cannot access the instruction register in most microprocessor systems.

The instruction "resides" in a read only memory, or ROM, and it is ready to be executed when the microprocessor reads it from the ROM, as part of a sequence of instructions from the program. If this particular instruction is 8 bits long—or 1 byte—then it occupies a single memory location in ROM. When it is read by the microprocessor, it acts as the signal for the microprocessor to go to a particular part of a smaller ROM which is part of the CPU chip. This on-chip ROM contains the microinstructions to control the microprocessor as shown in Figure 1-4.

As instructions are read from the program memory of the microcomputer, the first part of the instruction, called the *opcode* or operation code, is loaded into the instruction register (Figure 1-4).

The operation codes are addresses for the on-chip sequencer ROM which operates the control lines for register clears, gate transfers, and other signals such as commanding the arithmetic unit when to shift.

ACCUMULATOR AND REGISTER USAGE IN THE 8080

Accumulators are temporary storage registers for use during calculations. In most CPUs, the accumulator holds one of the operands in arithmetic operations. The CPU may also use the accumulator during logic operations, shifts, and other instructions. Accumulators are the most frequently used registers. Many microprocessors such as the 8080 used a single accumulator.

The 8080 also uses a bidirectional input/output data bus and a register array which allows registers to be addressed singly for single precision operations or in pairs for double precision operations. The H and L registers can be used for referencing memory locations. The H register will hold the 8 higher order bits of the memory address while the L register holds the lower 8 bits of the same address. The stack pointer holds the address of the next available location for a last-in, first-out pushdown stack. The actual stack remains in RAM for use in subroutine linking. This stack allows the use of programs with a large amount of nested subroutines with multiple priority interrupts. Since the stack is implemented in RAM, the size of the stack is limited only by the amount of memory that may be addressed by the stack pointer.

The 16 bit program counter contains the address of the next instruction to be executed by the processor. The program counter is updated and advanced automatically during memory cycles. An address latch is used to store addresses associated with the program counter updating requirements.

Temporary registers are used by the 8080 for arithmetic operations and data transfers. The temporary register connected directly to the ALU stores data during a machine cycle for execution by the ALU. The W and Z registers are for temporary storage of data during operations of the six general purpose registers, B, C, D, E, H, and L.

An address bus is used to address external memory in an 8080 microprocessor system. The 16 bit address bus can address up to 65,536 8-bit words in external memory or up to 256 input devices. The function of synchronizing the different elements of the processor is done by the timing and control block. The timing and control logic also controls the system functions based on inputs from external sources and provides inputs to external devices through the output pins.

The instruction decoder is connected to the internal data bus. At the proper time in the machine cycle, an instruction code appears on the data bus which indicates that the instruction should be fetched to the instruction decoder. When the instruction is in the decoder, signals are generated to decode the instruction and issue the correct control and timing. The 8080 is internally microprogrammed to translate the instruction code into sequences of microinstructions required to perform the arithmetic and logical functions requested by the user. The microinstructions are stored in a ROM on the chip and are not accessible to the user.

The accumulator is the main register used with the ALU for all arithmetic and logic operations in the 8080. An accumulator latch is used for the temporary storage of data before acceptance by the ALU.

In microprocessors with two accumulators such as the 6800, the contents of the A and B accumulators are fed into the arithmetic unit of the processor, and the arithmetic logic unit is instructed. The answer is then placed in the correct accumulator.

Index Registers

The index register (IR) is used to hold the addresses of important memory locations. This register may be incremented and loaded by various instructions. It can function as a pointer to direct the processor to the area of memory containing the required information. A program may be simplified by having the index register act as a pointer to the memory locations. This type of memory addressing is called *indirect.* In the most general form, the pointer register can be an internal register or any other memory location. This type of program addressing uses more memory space than direct addressing. However, for long lists of numbers indirect addressing combined with an increment and compare loop shortens the number of instructions needed. But, the instruction cycle for an indirect address command is longer than for a direct address instruction.

Another technique, used in more complex programs, is indexed addressing. Here, the correct data address is calculated by adding an offset value to a specified address. Usually, the offset is stored in the index register and the specific address is obtained by direct or indirect addressing. For example, if the IR contains 0005, and the instruction "Add (0A00) indexed" is used, the correct data address is obtained by adding the contents of the IR to 0A00 to give the correct address, 0A05.

Some microprocessors have an indexed autoincrement indirect addressing mode, where the correct address is obtained by adding the IR to the indirectly specified address; then the IR is automatically incremented. Other processors manipulate the indirect address rather than the IR.

Condition Code and Status Registers

Condition code or status registers hold 1 bit indicators or flags which represent the conditions inside the CPU or other external conditions. These flags are the basis for computer decision-making. Some older microprocessors have only one or two flags while newer devices may use several flags.

Some CPU's flags may be changed or observed externally as serial input or output lines. Among the common flags in use are the following:

CARRY indicates if the last operation generated a carry from the most significant bit. The CARRY flat retains this information and handles the carry from one word to the next in multiple precision arithmetic operations.
ZERO indicates if the result of the last operation was zero. It is useful in loop control or in searching for a particular data value in a set.
OVERFLOW indicates if the last operation resulted in an overflow. The difference between carry and overflow is that the OVERFLOW bit shows if the

result of an arithmetic operation has exceeded the capacity of a word or register.

SIGN indicates if the most significant bit of the result of the last operation was positive or negative. The SIGN bit is used in arithmetic and for examining bits within a data word.

PARITY indicates if the number of "1" bits in the result of the last operation had even or odd parity. PARITY is used for character manipulation in communications applications.

HALF-CARRY indicates if the last operation generated a carry from the lower half-word. This occurs in some 8 bit CPUs for BCD arithmetic.

INTERRUPT ENABLE is usually a "1" if an interrupt is allowed, a "0" if not. The CPU can disable interrupts during startup or service routines. The programmer can disable interrupts during critical timing loops or multiword operations. The CPU may have as many interrupt enable flags as it has interrupt levels.

Stack Pointers

The stack pointer register is a special register specifically used for keeping track of the next memory location available for the stack.

Stacks may be implemented two ways: with a shift register, where a push corresponds to a shift in one direction and a pop refers to a shift in the other; or with a RAM and the stack pointer register.

Every time a subroutine call is made, the current contents of the program counter are stored in the stack. The loaded value is always placed on top of the stack and, when a subroutine is completed, the top value from the stack is removed. Putting a word on the stack is called a *push operation* and removing a word from the stack is called a *pop operation.*

Stack Operation

A stack is similar to a springloaded dish storage device similar to those used in restaurants: place the dishes in one at a time, and weight makes the stack of dishes sink into the well, so that the last one in is the first one to come out. This is the way the computer stack works: as a word is written into the stack in the RAM (using the stack pointer contents as the address), the stack pointer is decremented to allow the next word to go into the next lower location or "above" it in the memory. Thus the stack pointer keeps track of the next location into which a word should go in the stack.

The shift register approach is used when the stack is part of the processor chip, and although it is faster than an off-chip register, it is restricted by the number of possible addresses it can hold. The stack can grow to almost any size when external memory is used since more can be added if necessary.

Push and pop instructions can be powerful, when they operate on registers other than the program counter. When the processor jumps to a subroutine, the contents of the PC should be stored as well as the contents of other registers. This is required if the processor is likely to be interrupted from its normal operation by a power outage or an interrupt signal from another device.

Stack Characteristics

The major advantage of the stack is that data can be added to it up to its capacity without disturbing the data already there. When data are stored in a memory location or register, we lose the previous contents of the storage place. The same memory location or register can be used again if its contents are saved elsewhere. The stack can be used over and over since its previous contents are automatically saved. The CPU can easily and quickly transfer data to or from the stack because the address is in the stack pointer and does not need to be part of the instruction. Thus, the stack instructions can be short.

The last-in, first-out nature of the stack is used to store subroutine return addresses. Each jump to subroutine instruction moves a return address from the program counter to the stack. Each return instruction fetches a return address from the stack and places it in the program counter. Now, the program can retrace its path from the subroutines using the stack.

The major disadvantage of stacks is the difficulty of debugging and documenting programs. Since the stack does not have a fixed address, its location and contents are difficult to remember. Lists of the current contents of the stack, called *stack dumps,* may be used for documentation. Errors in stack usage can be difficult to find; some typical problems are removing items from the stack in the wrong order, placing extra items in the stack, or removing extra items and overflowing or underflowing the stack capacity.

General purpose registers can have a variety of functions. Such registers can serve as temporary storage for data or addresses. The programmer may be able to assign them as accumulators or even as program counters.

MICROPROCESSOR CONTROL SIGNALS

A group of assorted control signals enter and leave the microprocessor. Some of these may carry control signals back and forth between the microprocessor and the memory, and I/O chips. They are usually grouped together and called the *control bus.* Other wires may go back and forth between the microprocessor and support chips. No connection is made directly to the registers, the ALU, or other internal components. Microprocessors have many of the features common to all computers. The characteristic that makes the microprocessor unique is that the CPU is contained in just a few integrated circuit packages which require a number of these control signals for system operation.

We will examine a few of these control signals for representative microprocessors starting with the 6800.

6800 Control Signals

In the 6800 the INTERRUPT REQUEST line is used to request interrupt sequences. The processor will wait until the current instruction is completed before it recognizes the request. If the interrupt mask bit in the condition code register is not set, the interrupt sequence begins. First, the contents of the index register, program counter, accumulators, and condition code register are stored in the stack. Then the CPU sets the interrupt mask bit high so that no additional interrupts can occur. Finally, an address is loaded which allows the CPU to branch to the interrupt routine in memory.

The NON-MASKABLE INTERRUPT allows a nonmask interrupt sequence to be generated in the 6800. The processor will complete the current instruction being executed before it recognizes this signal, and the interrupt mask bit in the condition code register has no effect on this sequence. The contents of the index register, program counter, accumulators, and condition code register are again stored in the stack and an address is loaded to allow the CPU to branch to a nonmaskable interrupt routine in memory.

The index register, both accumulators, the status register, and the program return address can be pushed to the stack by either interrupt or the HALT instruction. When the HALT input is in the low state, all activity in the 6800 is stopped. As the 6800 stops at the end of an instruction, BUS AVAILABLE will be a "1" and VALID MEMORY ADDRESS will be a "0".

The VALID MEMORY ADDRESS line is used to signal the peripheral devices that a valid address exists on the address bus.

The BUS AVAILABLE line is normally in the low state until the microprocessor stops and the address bus becomes available. The DATA BUS ENABLE line removes the data bus from its high impedance condition when activated. A RESET line, which is provided to reset and start the 6800 from a power down condition, is used during initial starting of the processor.

The THREE-STATE CONTROL sets all the address lines and the READ/WRITE line into an off or high impedance condition. The VALID MEMORY ADDRESS and BUS AVAILABLE are forced into a low condition.

The READ/WRITE line signals the peripherals and memory devices when the 6800 is in a read (high) or write (low) condition. The normal standby mode for this signal is read (high). In the halt state, this signal is low.

8086 Control Signals

Many of the lines of the 8086 control bus are similar to those of other microprocessors. The clock input (CLK) accepts a 5 V clock pulse train. The 8086, like the 280, has a nonmaskable interrupt request input line (NMI) and a maskable interrupt request line (INTR). The READY line is similar to the 8080 and is used to stop the processor during a machine cycle. The RESET line is different, following a logic "1" reset signal. The 8086 begins execution at location FFFF0H in memory rather than at location OOOOOH. $\overline{\text{TEST}}$ is a control input which can be tested by the wait for test instruction: If the $\overline{\text{TEST}}$ input is low, execution continues, otherwise the processor waits until the input does go low. The $\overline{\text{RD}}$ output is the same as in the Z80, indicating a memory read or an input port operation.

Several of the control lines of the 8086 are multiplexed. Since the 20 address lines are latched, address information appears only at the beginning of a machine cycle, curing the T1 state. One of the control output lines, $\overline{\text{BHE}}$, also appears only during T1. When $\overline{\text{BHE}}$ becomes low the high byte of the data bus can be used for a data transfer during the other states of that machine cycle. When address line AO is low during T1, the low order byte of the data bus can be used for data transfer during the other states of that machine cycle. The entire 16 bit data bus can be used for a data transfer in a machine cycle only when both $\overline{\text{BHE}}$ and AO are low during T1 of that machine cycle. The remaining eight lines of the control bus can serve one of two different functions depending on whether the 8086 is in minimum or maximum configuration. The MN/$\overline{\text{MX}}$ control input line selects one of the two modes. When the MN/$\overline{\text{MX}}$ line is high, minimum mode is selected and the eight lines function as follows: $\overline{\text{INTA}}$ is used to acknowledge an interrupt similar to the INTA status bit of the 8080. ALE appears during T1 and is used to strobe the address and $\overline{\text{BHE}}$ latches. When the 8286 bus driver is used on the data bus, $\overline{\text{DEN}}$ enables the outputs of the drivers and DT/$\overline{\text{R}}$ sets the direction of the bus drivers. M/$\overline{\text{IO}}$ separates a memory cycle from an I/O cycle, like $\overline{\text{IORQ}}$ in the Z80. $\overline{\text{WR}}$ indicates a write operation either to memory or an output port. HOLD and HLDA are used as in the 8080 to initiate and acknowledge DMA operations.

If MN/$\overline{\text{MX}}$ is low, the 8086 is in the maximum mode select. Then QSO and QS1 are status bits used to provide information on the status of the internal instruction queue. Status bits $\overline{\text{S0}}$, $\overline{\text{S1}}$, and $\overline{\text{S2}}$ are used to encode information that, in minimum mode, appeared on five pins. These status bits are decoded using the 8288 bus controller.

Three lines are used in microprocessor applications. The $\overline{\text{LOCK}}$ output is under control of the program and is used to prevent other processors from bus control. Request/grant pins are used to force the active processor onto a HOLD state to give up control of the bus.

Z8000 Control Signals

The 48 pin version of the Z8000 includes seven address lines (SN0 to SN6). These seven lines specify one of 128 different memory segments. Each segment can be 64K bytes in size, so this version of the Z8000 addresses up to 8 megabytes of memory. The segmented version also has a segment trap control input ($\overline{SEGT}$), which works like an interrupt input and is designed to trap specific operations.

The 16 low order address lines (A0 to A15) are multiplexed on the 16 bit data bus. The address strobe signal ($\overline{AS}$) will go low at the beginning of a machine cycle to indicate that the address information is valid. $\overline{DS}$ is a data strobe and $\overline{MREQ}$ is used for a memory request. Seven lines are used to indicate the processor status: ST0–ST3, WORD/$\overline{BYTE}$, NORMAL/$\overline{SYSTEM}$, and READ/$\overline{WRITE}$. $\overline{WAIT}$ is used to synchronize the processor to slower memory or I/O devices while the $\overline{STOP}$ line can be used to stop the processor when single stepping through a program. The $\overline{BUSRQ}$ and $\overline{BUSAK}$ lines are the same as on the Z80; a DMA request can be made by a logic 0 on the $\overline{BUSRQ}$ input and the processor will respond with a logic 0 on the $\overline{BUSAK}$ output line. The Z8000 also has two control lines, $\overline{\mu I}$ and $\overline{\mu O}$ (micro in and micro out), which are used to synchronize multiple microprocessors in a microprocessor system.

MICROPROCESSOR OPERATION

The operation of a typical microprocessor can be observed by following the flow of information between the CPU and other devices in a typical instruction sequence. The CPU structure of a typical microprocessor system contains the ALU, an accumulator, program counter, instruction counter, a set of registers, and a number of driver receiver circuits.

The section of the CPU that controls the sequence of instructions is the program counter. It may also be called the program address register, memory address register, or program register in other microprocessors. The program counter is set to zero by a power-on reset line when power is turned on. The PC always contains the address of the next instruction.

The stack pointer holds instruction addresses during subroutine branching operations. The stack allows a multilevel nesting capability for storing the return program subroutine addresses and contains segments which are sequentially shifted to and reloaded from the accumulator for stacking of subroutine branches or auxiliary storage.

Arithmetic and logic functions are performed and processed in the section of the CPU that contains the accumulator register, the ALU, and the carry register. The accumulator is the primary working register for the CPU and the results of all arithmetic and logical functions are transferred here. The accumulator, ALU, and carry register provide a parallel adder with carry-in and carry-out capabilities.

Other registers include a register for the storage of data addresses and a register for temporary storage. The accumulator provides data for directly modifying sections of the data address register and an indirect path for modifying this register through the temporary register. Control flip-flops are used as status indicators and they can be set, reset, and tested by the programmer. The carry register may also be independently set, reset, and tested, in addition to being used for arithmetic operations.

The CPU has a direct input/output capability using discrete input bits and discrete output bits. Inputs are loaded in groups of bits into the accumulator and an output buffer is loaded directly from the accumulator. The drivers and receivers provide an interface protecting the CPU while at the same time supplying drive to power additional devices.

INSTRUCTION CHARACTERISTICS

This chapter has discussed how a microprocessor might process an instruction. Now it considers how the instructions can provide the processing operations and tasks. The instructions used in microprocessors can be classified into seven basic functional groups:

1. *Data transfer* includes the methods of moving data from one point to another or organizing it for processing.
2. *Arithmetic operations* can include binary, binary coded decimal, or decimal manipulations.
3. *Logical operations* include the methods for manipulating the data as bits as well as the methods for obtaining logical functions of the bit combinations.
4. *Data address modifications* include the methods of modifying addresses in registers to facilitate data manipulation.
5. *Control transfers* include such techniques as the interrupts for modifying the normal sequence of the program to allow conditional and unconditional branching.
6. *Register manipulation* includes the methods for storing data along with the manipulation of registers or individual segments of data within registers.
7. *Input/output operations* are the methods of using the input/output instructions.

Subroutine usage such as the techniques for calling routines, setting up data addresses, returning from subroutines, and nesting of subroutines can be used in conjunction with other operations. Some typical instruction sets for 16 bit microprocessors will be discussed next.

8086 Instruction Characteristics

Using a design philosophy requiring storage efficiency, the 8086 special purpose registers use implied register addressing in most instructions and allow shortened instructions.

Intel has preserved compatibility between the 8086 and the 8080. The 8086 has 95 basic instructions, of which a number are only 8 bits long. In the 16 bit instructions, only the first 8 bits are used for operation codes; the additional byte specifies data displacement. Instructions longer than 2 bytes use the remaining bytes for specifying data. Arithmetic operations including 8 or 16 signed or unsigned multiply and conditional branching have been added. An added set of string manipulation instructions can be used for text or table processing. Relative addressing has been added to improve coding and program reduction. Source code for the 8080 may be translated into 8086 object code, thus allowing use of all the 8086 instructions.

68000 Instruction Characteristics

In the 68000 the status register contains a user byte and a system byte. The user byte contains five control bits. The system byte contains a trade mode bit, a supervisor state bit, and a 3 bit interrupt mask.

The 68000 instruction set emphasizes space efficient code through quick instructions and short jumps. Debug tools include single step execution and traps on illegal instructions. The instructions include multiple load and signed multiply and divide.

The 68000 supports 56 basic instructions with 14 addressing modes. The total number of instructions is greater, since many instructions perform multiple functions. The number 56 is a function of the assembler. The instruction sizes vary from one to five words.

INTEGRATED CIRCUITS

Digital integrated circuits may be classified not only by function but by also being members of a logic circuit family. Each logic family has its own basic electronic circuit upon which all circuits and functions are developed. This basic circuit is either a NOR or a NAND gate. The electronic components and their interconnection are used to derive the name of the family. Many different logic families of digital integrated circuits have been introduced. The logic families that have been used in microprocessors are:

1. TTL: Transistor Transistor Logic.
2. ECL: Emitter Coupled Logic.
3. I²L: Integrated Injection Logic.
4. MOS: Metal Oxide Semiconductor.
5. CMOS: Complementary Metal Oxide Semiconductor.

Two basic types of components can be made using integrated circuit technology: (1) the bipolar devices, which have two types of charge carriers, and (2) the unipolar devices, which only have one charge carrier. Bipolar technologies include: Transistor transistor logic (TTL), emitter coupled logic (ECL), and integrated injection logic (I²L) among those that are used in microprocessors. A cross-sectional view of a bipolar transistor is shown in Figure 1-5.

The contacts are usually made of aluminum using an evaporation process. The buried layer provides a more conductive path between the collector, base, and emitter regions. Figure 1-5 shows one transistor; a microprocessor or memory circuit would have thousands of these transistors connected to form the logic elements required.

Microprocessors use almost all digital integrated circuits except for some emerging microcomputer chips which contain analog-to-digital converter circuitry.

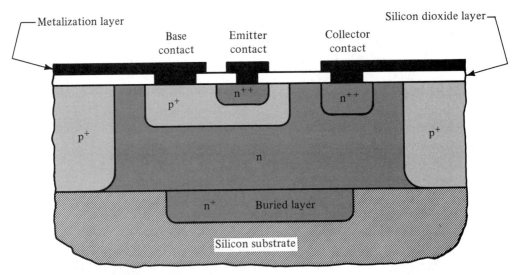

FIGURE 1-5   Bipolar Transistor Structure.

LOGIC CIRCUIT CHARACTERISTICS

The characteristics of digital logic circuits may be compared by analyzing the circuit of a basic gate for each type. The parameters that can be evaluated and compared are:

1. *Fan-out* defines the number of standard loads that the output of the standard gate can drive without impairing its normal operation. The standard load is defined as the amount of current needed by an input of another similar gate.

2. *Power dissipation* is the power consumed by the gate which must be available from the power supply.

3. *Propagation delay* is the average time for the signal to propagate from input to output when the binary signal switches from a high to a low state. The device operating speed is inversely proportional to the propagation delay.

4. *Noise margin* is the minimum noise voltage that can cause a change in the circuit output.

Transistor Transistor Logic (TTL)

The circuit for a TTL NAND gate is shown in Figure 1-6. This circuit uses four transistors and four resistors.

TTL is a saturated form of logic. During turn-on both the emitter base and collector base junctions are forward biased which causes an accumulation of charged carriers in the base region. As the device is turned off, this charge must be discharged through the collector. The time required for this discharge results in a delay in turning the transistor off. All saturated logic experiences this storage time delay.

Other variations of this TTL gate are listed here together with propagation delay, power dissipation, and the product of these two parameters which can be used as a figure of merit.

TTL Circuit Version	Standard Abbreviation	Propagation Delay (ns)	Power Dissipation (mW)	Speed Power Product (ns)×(mW)	Input Pull-Up Resistor (Ω)
Standard	TTL	10	10	100	4K
Low-power	LTTL	33	1	33	40K
High-speed	HTTL	6	22	132	2.8K
Schottky	STTL	3	19	57	2.8K
Low-power Schottky	LSTTL	9.5	2	19	25K

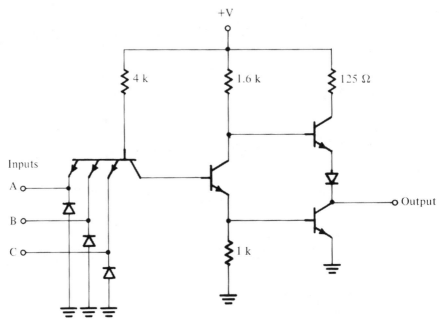

FIGURE 1-6   TTL NAND Circuit.

The standard TTL gate was the original version of the TTL family. Improvements occurred as the technology matured. In the lower power version, the propagation delay was increased to reduce the power dissipation. In the high speed version, power dissipation is increased to reduce the propagation delay. Schottky TTL increases the speed of operation without an excessive increase in dissipated power by the use of clamping diodes. The low power Schottky version sacrifices some speed for reduced power dissipation. It compares with the standard TTL in speed but requires less power. The fan-out of TTL gates is 10 when the standard loads of the same circuit version are used. The noise margin is greater than 0.4 volt with a typical value of about 1 volt.

ECL Technology

ECL is a nonsaturating technology and transistor storage time delays are avoided. In ECL circuits the transistors are biased to operate only in the linear region.

The basic ECL circuit consists of a differential amplifier input circuit. Emitter follower circuits are used to restore the dc levels and provide the buffering to drive output lines. High fan-out is possible because of the high input impedance of the differential amplifiers and the low output impedance of the emitter followers. The generated noise tends to be small because of the constant current drain of the differential amplifier circuit. Gate delays of less than 1 nanosecond are possible and operating frequencies for gates are close to 1 gigahertz.

The ECL family has a number of versions. Propagation delays range from 1 and 2 nsec, depending on the version used. Gate power dissipation is about 25 mW and the fan-out is greater than 25. Noise margin is the lowest of all logic families at about 0.2 volt.

The fastest digital circuits in use employ ECL, which has been the choice in many of the largest computers. ECL circuitry is also used in some microprocessors, notably the 10800.

I²L Technology

Integrated Injection Logic (I²L) is based on Direct Coupled Transistor Logic (DCTL). The problems of the older discrete component DCTL are eliminated with a single multiple electrode transistor using the structure shown in Figure 1-7a. An N substrate acts as the structure ground plane for connecting the grounded emitter transistor rates. An N type layer is grown on top of the substrate which acts as both the grounded emitter region for the vertical NPN transistor and the grounded base region for the lateral PNP transistor. The resulting structure offers improvements in speed, power dissipation, and density with a relatively simple structure. It is the newest bipolar technology to be applied to microprocessors. I²L uses saturated transistors, so the speed is lower than Schottky or ECL circuits.

The basic I²L circuit is shown in Figure 1-7b. The lateral PNP transistor acts as a current source for the multiple emitter transistor which performs the inverting function.

MOS Technology

Unipolar devices use field effect transistors made with MOS technology. The transistors are created on the surface of a small piece of silicon, called the *substrate.*

To make a MOS circuit, a single crystal of silicon is grown. This crystal is then cut into thin circular slices called wafers. The crystal must be cut in a specific direction of the crystal lattice. Several dozen chips can be made from a wafer. These chips might be microprocessors, memory, or other functions. Transistors are created on the wafer using a photolithographic process, similar to commercial printing. The positive and negative areas are made in the silicon with impurities using a process of masking and diffusion, called *doping.* After the functions have been processed on the wafer, it is tested using special test points which are on the

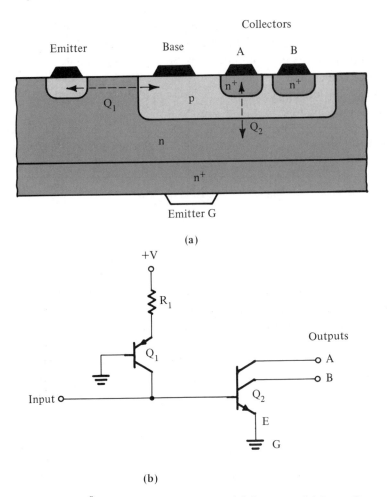

FIGURE 1-7 I^2L Integrated Circuit Logic. (a) Structure; (b) Basic Circuit.

edges and at the center. If the tests are successful, the wafer is scribed and cut into individual chips. Each chip is then mounted in a package and connected through wires to the pads of the package. The packaged IC is then given further visual, electrical, and environmental tests, then sealed, and given a final test. A modern MOS fabrication facility is shown in Figure 1-8.

A majority of the microprocessors being manufactured today use MOS (metal oxide semiconductor) rather than bipolar transistors. The main advantages of MOS technology over bipolar technology are higher density and lower production cost. The higher density of MOS allows more functions to be placed on a chip of a given size than can be attained with bipolar transistors. The unipolar devices

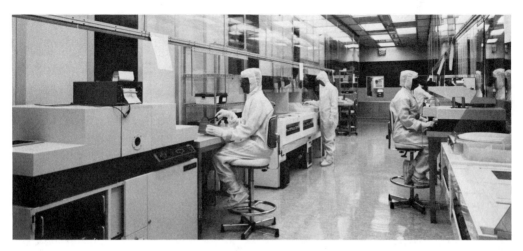

FIGURE 1-8   A modern CMOS fabrication facility for the production of CMOS micro-processors. (*Courtesy Rockwell International.*)

used to fabricate MOS microprocessors are field effect transistors (FETs) with P or N channels. A complementary type of MOS with both P and N channels (CMOS) is also used in some microprocessors.

The basic structure of the MOS field effect transistor is shown in Figure 1-9. Like the bipolar transistor, it is formed on a P or N silicon substrate which also serves as the supporting structure for the device. An oxide layer is formed on the silicon chip which serves as a protective layer against surface contamination. The source and drain are p or p+ regions, which are obtained by diffusing an impurity such as Boron into the desired sections. The gate is usually aluminum and serves as the control element, covering an area from the source and drain. The oxide under the gate can be much thinner than the oxide on the rest of the surface to

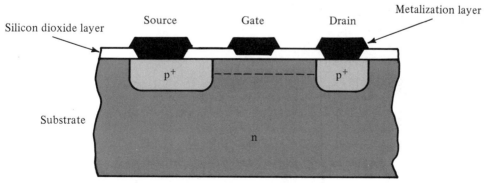

FIGURE 1-9   MOS-FET Structure.

allow close control of the conduction characteristics between the source and the drain.

The channel refers to an area directly below the gate which connects the source and drain. As a voltage is applied to the gate, the field created causes the channel to be converted into a P region which allows conduction. When a P substrate is used as the body or support, the channel becomes an N region when a voltage is applied to the gate and the device is called N type or NMOS as shown in Figure 1-9. Thus, the two basic types of MOS devices are the P-channel transistors and the N-channel transistors. In the P-channel devices, the electrical carriers are holes while in the N-channel devices, the carriers are electrons.

The earliest microprocessors used P-channel technology, which had two disadvantages: (1) Holes have a lower mobility in silicon than electrons, thus PMOS transistors are slower than NMOS devices, and (2) PMOS circuits provide an active pull-up but require a passive pull-down of external loads. NMOS amplifiers provide an active pull-down, which is more effective in driving the TTL interface circuits common in microprocessor systems.

MOS transistors can be also classed as depletion mode or enhancement mode devices. Depletion mode transistors are normally on and require a gate voltage to be turned off. Enhancement mode devices are normally off and require a gate voltage to be turned on. The early NMOS microprocessors (such as the 8080) used enhancement mode transistors for load resistors. This required a separate power supply to provide the gate voltage for the loads. The later MOS microprocessors starting with the 6800 use depletion mode loads which eliminated the need for the extra power supply voltage. These processors operate from a single 5 volt supply (Figure 1-10). Since NMOS is faster than PMOS, and newer versions such

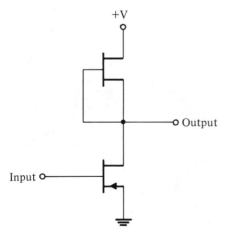

FIGURE 1-10   NMOS Amplifier
with Depletion Mode Load.

as H-MOS and V-MOS give excellent density, it is the most popular way to implement microprocessors today.

CMOS Technology

The CMOS (complementary MOS) circuit uses complementary transistors (P-channel and N-channel devices) for active pull-up and pull-down of the load. The complementary structure allows one transistor to be off when the other is on, resulting in a very low power dissipation. CMOS also allows high noise immunity, large fan-out, and the use of inexpensive power sources.

The CMOS structure requires a more complex fabrication process as shown in Figure 1-11. The P-channel device is formed directly from the N type substrate. A P doped tub must then be created for the N-channel device which adds to the processing steps. Channel stops are also required between the devices to prevent extraneous current flows between devices. In the standard MOS circuit like the one shown in Figure 1-10, the upper transistor acts as a load which accounts for a significant amount of the power dissipation. In a CMOS circuit with the P- and N-channel devices as shown in Figure 1-12, a complementary inverter is formed by applying an input to the gates of the two opposite polarity transistors. When a logical 1 is applied to this circuit, the upper P-channel transistor is off while the lower N-channel device conducts. The output is shorted to ground and current from the supply is only the leakage through the P-channel device after the initial transient has settled.

The characteristics of CMOS technology fall between those of NMOS and PMOS. CMOS is faster than PMOS, but slower than NMOS. The use of two transistors rather than one results in less density than standard MOS. The main

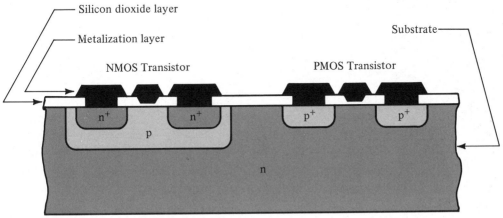

FIGURE 1-11   CMOS Structure.

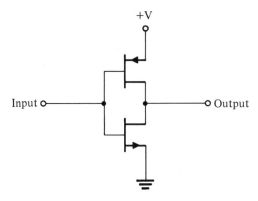

FIGURE 1-12   CMOS Circuit.

advantage is the very low power consumption. The noise immunity is about 1.5 volts and CMOS devices can operate between 2V and 12V. CMOS technology is well suited for systems that require portability or a very low power consumption. Several commercial microprocessors using this technology are available.

In addition to the low operating power, a low cost, unregulated supply is sufficient to operate most CMOS devices.

CMOS speed is limited by substrate capacitances. To reduce the effect of this capacitance, the CMOS devices can use an insulating substrate such as sapphire. This technology is called silicon on sapphire (SOS). This extends the speed capability of CMOS into the lower range of transistor transistor logic (TTL). The starting silicon layer is grown on a single crystalline wafer of sapphire and by a repeated process of producing silicon dioxide, etching, and diffusing, a CMOS configuation is obtained. Unused silicon is removed from the substrate, leaving isolated islands of silicon that form individual isolated tubs. The process tends to be expensive and has limited application.

MICROPROCESSOR TECHNOLOGY CONSIDERATIONS

The current microprocessor technology provides a wide spectrum of devices, which vary in speed and capabilities. These devices may be divided into groups according to their speed. In the slower group are chips that operate at instruction speeds of greater than .5 us or more such as the n-channel metal oxide semiconductor microprocessors; in the faster group are those that operate at .5 us or less, such as the bipolar data slices. This group of high speed chips can be used as building blocks in constructing microprocessor systems for use in realtime signal processing tasks such as speech or image processing. To understand the difficulties encountered in fabricating high speed circuits, three factors which tend to limit how much logic may be put on a single chip must be considered: (1) power dissipation, (2) pin count, and (3) cost.

Power Dissipation

In high speed semiconductor technologies, the maximum number of gates is a direct function of the chip's maximum allowed power dissipation. This, in turn, depends on the maximum allowable junction temperature of the die, the ambient temperature, and the ability of the package to dissipate heat. The effective thermal resistance between the package and the ambient environment may be reduced by mounting the package to a heat sink and cooling it by forced air. If operation is limited to the commercial temperature range with a maximum of $70\,^\circ$ C, the maximum power dissipation is then about 7 W for a package rated 1.25 W in free air. While this dissipation value may be feasible, ICs with this power output are undesirable since they may be surrounded by ICs that dissipate less than 1 W, thus generating a hot board area that may be difficult to cool.

The maximum number of gates that can be integrated into a chip is equal to the maximum allowed power dissipation divided by the power dissipation for each gate. A technology like low power Schottky TTL dissipates about 2mW/gate; an LSI chip of this type then may contain about 600 gates. For a particular technology, the power dissipation of each gate is proportional to its physical size, which is a function of the number of active elements and the geometry of the individual transistor. In the case of random access memories, the early advances from 1k to 16k bits were largely due to reducing the 3 transistor cell to a 1 transistor cell and cutting the line width used for the fabrication pattern.

Pin Count

The number of available pins in the IC package can limit arithmetic with longer word lengths as well as chip testing. Multiplying 16 bit operands may require a package with 66 pins. Even custom ICs do not exceed this by much.

Multiplexing the functions of the pins reduces the pin requirement, but also reduces performance. Pin count limitations have resulted in multiplexing as well as the bit slice approach, partitioning the system into identical parts that have common control and interconnections.

Cost

If power dissipation is not a limiting factor, as with metal oxide semiconductors and integrated injection logic, cost will tend to limit the die size. Since yield is a function of random chip defects, cost increases exponentially with die size. For example, if a given chip size yields 10% of a good die area, a chip twice as large will yield only 1%. The cost for twice the functions may then be 20 times as large. Other cost elements, such as testing and assembly, are usually fixed per chip. As the number of functions (N) per chip increases, the assembly cost decreases proportionally as $1/N$. The minimum cost per function tends to be at the crossover point between the silicon die cost and the assembly and test cost. The relationship of chip area to the degree of integration depends on several factors:

1. Layout strategy and efficiency.
2. Levels of interconnect.
3. Active chip area to total chip area ratio.
4. Type of circuit random logic or memory.

Variation is considerable in the gate density (or gate area) among the different IC processes. Also, most active chip area estimates assume the entire active circuit is made up of a gate of average size. This may not be accurate for the size of many random logic chips. Die cost is determined to a large extent by yield. For a fixed volume of product in the manufacturing line, die cost tends to be inversely proportional to yield.

Technology Trends

To evaluate the future microprocessor ICs, semiconductor technologies may be examined in terms of potential density. Among the major density improvements has been an increase of the wafer diameters. Some companies have increased the chip size while maintaining the same approximate costs, since the cost of processing a wafer is almost independent of its diameter.

A more revolutionary improvement is the new methods of drawing the patterns for the fabrication of integrated circuits. Resolution in optional techniques is limited by diffraction effects that occur between the mask and the wafer. Electron beam lithography provides up to 20 times the resolution of optical lithography methods. With these improvements, it is possible to achieve a larger chip size with more components. With this level of integration, a single chip, high speed microcomputer that performs multibyte addition, multiplication, and division, and contains enough memory for many applications is possible. Thus, one future possibility is the integration of multichip systems into a single chip while maintaining the same processing power. This is already occurring in mature NMOS microprocessors.

SYSTEM COST CONSIDERATIONS

The relationship of integrated circuit cost to total electronic system cost, and various parameters which affect integrated circuit cost will be discussed here. Emphasis will be on the development of an understanding of the cost of very large scale integrated circuits (VLSI). A cost analysis of an electronic system can be developed which assumes that the system under consideration is partitioned into subsystems consisting of integrated circuit chips. The analysis should be sufficiently general to include different aspects, such as choice of device type, process technology, packaging, and the effects of quantity level.

Consider an electronic system which is partitioned into LSI chips. The system is assumed to be composed of subassemblies which contain LSI in some degree with some SSI and MSI. The cost of the entire system then can be related to the subassembly cost. The number of subassemblies is a variable which depends on the level of circuit integration.

The impact of VLSI microprocessors on electronic systems can be understood through a consideration of the different cost factors at the system level. These cost factors are the sum total of the various costs that occur during the design and manufacture of an electronic system. Some of these factors follow:

1. *Component costs* include all electronic elements such as resistors, diodes, and LSI chips along with all mechanical components.
2. *Tooling costs* include the engineering labor costs and materials, and expenses such as printed circuit board layout and procurement, i.e., basically, all one time engineering charges.
3. *Assembly costs* include all labor and overhead incurred in the physical structuring of the system.
4. *System testing* includes the debugging of all system operations along with the establishment of system reliability.
5. *System repair and maintenance costs* extend over the life of the product.
6. *Inventory* includes requirement for stocking component spare parts for the system.

The relative importance of each of these cost factors depends upon the particular type of system to be manufactured, the facilities and manpower available, the total number of systems to be built, and the schedule.

The design and manufacture of an IC is a costly process containing hundreds of sequential steps. The difficulty and level of sophistication necessary in the manufacturing process is realized by limiting errors in any one of the sequential steps that can cause a circuit failure. Thus, in the manufacture of ICs, the manufacturer must strive to improve quality control and reduce errors by the use of such techniques as:

1. Automated equipment using computer or microprocessor control.
2. Redundancy of fault tolerance in the actual fabrication process.
3. Close supervision and monitoring at all levels of work in process.

4. Increased operator training and the use of higher skill levels.
5. Fault tolerant or process tolerant circuit designs.

Volume Considerations

Because of the major investment required in capital equipment and personnel necessary to manufactuure ICs, it is impractical to manufacture most LSIs in a low volume line. A major cost item is the equipment to manufacture LSIs. The growth in alignment equipment costs has risen sharply, which is typical of most IC equipment.

IC technology improvements have permitted higher levels of integration which result in reduced component costs. These cost reductions are possible with the use of high volume and advanced technology. The level of integration has increased and IC cost is a strong function of the technology used. This is true for large volume production. If one considers low volume, custom LSI, the cost curves show the effects of nonrecurring costs which become dominant for low volume LSI. With equipment costs so high, low volume LSIs are usually manufactured in a line which runs a large volume of related products. Thus the production line will be required to handle a number of product types. If the manufacturing line is handling a large volume of products, then some of the large, fixed, nonrecurring costs can be amortized. In this way, low volume manufacturing can approach volume fabrication costs. It can never be the same because of the paper work involved in tracing lots in the line, but this scheme works well for a typical high reliability product. To reduce manufacturing costs, the line should operate as a lot traceable line with a product mix and perhaps a few process modifications. To reduce design/layout costs, computer aided designs with standard gate arrays or logic cells which the computer will interconnect can be used. Generally this will result in lower parts costs but in larger (nonoptimum density) diex area.

For most VLSI design/layouts, a CAD system is used to simplify design time and lower costs. Following this CAD layout, a period of manual or semimanual die size optimization is undertaken because of the inadequacy of present day CAD software. If the computer generated layout yields a chip of area A' and the absolute theoretical minimum valve of die area is A'' (no wasted chip area), the manual labor to reduce the die area is an exponential relationship between time spent vs. area reduction. The area depends on the remaining IC operations being small and the number of pins on the chip package or the number of bond pads on the chip. The level of integration is generally related to the square of the number of pin-outs. The overall assembly cost of the IC is a function of the number of bonds required, the package cost, and the fixed labor costs, plus overhead. The testing cost of complex ICs is a function of:

1. Equipment costs.
2. Labor.
3. Circuit complexity and design which determines test requirements.

The number of gates affects the cost function in that more complex functions are more difficult to test, thus increasing test time. Test time is more dependent on the number of pins on the IC package than on the actual die area. For a volume testing operation, the test cost will be determined by test time.

MICROPROCESSOR CHIP EVOLUTION

As with other computer systems, microprocessors have gone through several generations. The first generation of microprocessors is typified by the Intel 4004 and 8008. These calculator type chips had a limited instruction set and were based on p-channel MOS technology. The second generation is typified by the 4040 and 8080. These chips contained a larger instruction set with up to 80 instructions, and used the faster n-channel MOS technology. The third generation is typified by the Intel 8086 or Zilog Z-8000 which used a more sophisticated instruction architecture and improved high density MOS technology. Other devices have also been introduced which may feature programmable logic array microprogramming, integrated injection logic (I^2L), memory, analog-to-digital conversion, and other functions on the chip.

The growth in both bipolar and MOS technology is demonstrated by the increase in the number of gates per chip, and the memory size per chip.

Microprocessors, as other bipolar or MOS components, have decreased in price. Important factors in determining microprocessor feasibility are technological obsolescence and price/performance changes. To analyze the problem of technological obsolescence, consider the factors which may affect the cost of microprocessors:

1. Size of the chip.
2. Number of gates on the chip.
3. Clock rate and other physical or environmental requirements.
4. Number of processing steps for the technology used.

Each of these factors can affect the cost of producing the microprocessor chip and ultimately can affect the price of the microprocessor to the user.

Technology obsolescence is caused by the outdating and outpricing of products due to innovation and technological changes. An example is shown here, which compares the cost of doing 100,000 multiplications over a 25 year period:

1955	$0.85
1958	0.26
1964	0.12
1970	0.05
1975	0.01
1980	0.005

Other improvements have occurred in speed, size, and reliability in digital technology. These low computing costs, along with wide availability, enable new applications and allow more efficient use of existing applications.

Consider the design of data terminals. In the past, manufacturers would purchase components and design a terminal according to the desired specifications with the available components. Many of the necessary functions are available now in LSI components, and the design task is reduced considerably. Now large scale users of these terminals use custom or standard microprocessors, which use in-house microprocessor development facilities.

In another trend, microprocessor manufacturers have begun directing their efforts to develop more mature software to accompany their microprocessor systems. Also, many manufacturers have used the microprocessor for selling more profitable memory and peripheral products for a complete system. This trend is expected to continue with more integrated products to make the microprocessor easier to use.

RANDOM ACCESS MEMORIES (RAM)

The memory section of microcomputer contains storage units, some of which consist of semiconductor cells. These storage units are binary, i.e., they use two states to represent the values of 0 or 1. The memory is organized into bytes, which are the shortest groupings of bits that the microcomputer can handle at one time, and words, which have the same bit length as the microcomputer's data registers, data buses, and arithmetic unit. A byte usually consists of 8 bits, and a word may be from 4 to 37 bits in length. The memory is arranged sequentially into words, each of which has its own unique address. The address of a word in memory should not be confused with the memory contents. A memory location might contain any value and represent a number, an instruction code, alphanumeric characters, or other binary coded information.

Memory Unit Characteristics

The communication between a memory unit and the microprocessor is achieved through control lines, address selection lines, and data input and output lines. The control signals specify the direction of transfer. A word to be stored in a memory register is a write operation, and a word previously stored to be transferred out of a memory register is a read operation. The address lines specify the particular word chosen of all the available words in that section. The input lines supply the information to be stored in the memory and the output lines supply the information coming from the memory. LSI components, and the design task is reduced considerably. Now large scale users of these terminals use custom or standard microprocessors, which use in-house microprocessor development facilities.

A memory unit is specified by the number of words (m) it contains and the number of bits in each word (n). The address selection lines select a particular word out of all the m words. Each word is assigned an identification address, starting from o and continuing up to m–1. The selection of a specific word inside the memory occurs by placing its address value on the selection lines or address bus. A decoder in the memory accepts this address and connects the paths required to select the word specified. Thus, k address bits can select one of $2^k = m$ words. Microprocessor memories might range from 1024 words, requiring an address of 10 bits, to $1,048,576 = 2^{20}$ words or more, requiring 20 address bits. One refers to the number of words in a memory with the unit K, where K refers to 1024 $= 2^{10}$ words; then 1K = 1024 words, and 64K = 65,536 words.

Buses are used to connect the microprocessor and memory. The address bus carries the address of the memory location while the Read and Write lines determine the direction of the transfer. Data buses carry the data between the units. A single bus may carry data in different directions or carry data and addresses at different times. When a bus is used for more than one purpose, it must be multiplexed.

Memory units which lose stored information with time or when the power is turned off are volatile. Most IC memories are volatile since their cells require power to maintain the stored information. Nonvolatile memory units, such as magnetic cores or disks, retain their stored information after power is removed. The stored information is determined by the direction of magnetization, which is retained when power is off. Microcomputers with volatile memories can use backup batteries or power supplies that continue to deliver power for some time after a power interruption.

Semiconductor RAMs

A semiconductor memory is a collection of storage registers, together with the circuits required to transfer information in and out of the registers. When the memory can be accessed for information as required, it is a random access memory, or RAM. Integrated circuit RAMs sometimes use a single line for the read/write control. One binary state specifies the read operation and the other state specifies the write operation. Enable lines are included in the IC to provide a means for expanding a number of packages into a memory with a larger capacity. In a random access memory an address enables a specific location on each data plane of the memory, allowing data to be written in or read out, depending on the state of the read and write enable lines.

RAM Construction

The construction of a semiconductor RAM of m words with n bits per word consists of an m x n binary cell matrix along with the logic circuits required to select a word for a write or read operation. The binary cell is the basic building block of a memory unit. A typical binary cell for one bit of information might use transistor flip-flop with multiple inputs. The cell is made as small as possible to pack many cells in the area available on the chip.

A select input is used to choose one cell out of all those available. With the select line true, a 1 at the read/write input allows a path from the ouput of the flip-flop to the read output. When the read/write terminal is 0, a bit at the input is transferred into the flip-flop. The input and output are disabled when the select line is false. The flip-flop usually operates without a clock.

A 4 by 3 RAM has four words of 3 bits each for a total of 12 storage cells. Each cell contains the circuit of a binary cell similar to the one discussed. The address lines can use a 2 by 4 decoder with a memory enable input. When the memory enable is false, all the outputs of the decoder will be false and none of the words is selected. When the memory enable input is true, one of the four words is selected, depending on the bit combination of the two address lines. When the read/write is true (read), the bits of the selected word go through OR gates to the output. The nonselected cells produce 0's at the inputs of the OR gate and have no affect on the outputs. When the read/write control is false, the data at the input lines are transferred into the flip-flops of the selected word. The nonselected cells in the other words are disabled by their address selection line so their previous values remain the same. When the memory enable is true, the read/write line initiates the required read and write operations. An inhibited state is obtained when the memory enable is 0. This condition leaves the contents of all words in memory as they were, regardless of the read/write control input.

RAMs sometimes use cells with outputs that can be tied to form a wired-OR or a wired-AND function. Other RAMs may provide tri-state outputs. These outputs are useful when a high impedance path is desired for isolation from the other integrated circuits in the system.

MOS RAMs

MOS memories use circuits of field effect transistors to store addressable sequences of binary information. A static MOS memory cell may be operated at any speed up to its rated maximum and it does not require a periodic refreshing signal to maintain the storage of bits as in the case of the dynamic MOS memory cell. The static MOS RAM memory cell normally uses a flip-flop circuit with two MOS transistors which act as loads. These transistors are used to enable the sense and select lines. The static memory cell is easier to drive and thus requires simpler external circuitry than the dynamic MOS cell, but it tends to be more expensive due to its additional components when compared to a dynamic MOS RAM circuit.

A static CMOS memory cell uses a flip-flop arrangement, but the transistors which function as the loads are combined in the complementary manner. A static CMOS memory allows lower power dissipation, but with less density than standard MOS. It is also compatible with the TTL logic for interfacing. This type of MOS memory is less expensive to produce but it requires the additional circuitry for the refreshing function (Figure 1-13).

RAM Technology Evolution

The leading edge of process and device technology has historically been applied to dynamic RAMs. The success of dynamic RAMs has led the progress in other semiconductor memories and microprocessors.

A potential problem in memories has been alpha radiation. Positively charged alpha particles have been found to be a source of "soft errors" in many

FIGURE 1-13   This dynamic MOS memory has the required refreshing circuits along with jumper selectable addressing for the starting address. (*Courtesy Intel Corporation.*)

early dynamic RAMs. The phenomenon has also been observed in some static RAMs. No memory may be completely immune from alpha radiation, which is a special case of "noise" which causes errors. As signal levels inside memory and logic devices continue to shrink, other noise sources, such as thermal noise, may be isolated as error generators.

Not all errors can always be prevented, thus new interest has been created in error correction and detection. Simple parity checks can spot single bit errors and more elaborate schemes, such as the Hamming code, can correct single bit errors and detect double bit errors.

The Hamming code is increasing in popularity as 16 bit microprocessor systems emerge since the percent of overhead needed to perform the correction declines as word size increases. Five extra bits are required for an 8 bit word, while only 6 bits are needed to correct a 16 bit word. Error correction systems are increasing, and the logic necessary to perform the corrections is available in more and more ICs.

An evolution in static RAMs has paralleled that of dynamic RAMs, except that the bit density is lower by a factor of four since a static RAM cell occupies more chip space than a single transistor dynamic RAM cell. A major portion of chip cost is the silicon area used, so a static RAM may cost about three to four times as much.

Since a static RAM requires no externally applied control signals to maintain data, it is easier to use than a dynamic RAM. In small memory systems the cost implementing the multiplexed addressing and refresh control circuitry may offset the per bit cost advantage of the dynamic RAM. To bring the cost effectiveness of dynamic RAM technology in to the smaller memory systems, the quasi-static RAM has been developed. Here, a dynamic RAM configured in a directly addressable byte-wide memory uses an on-chip refresh scheme that is almost transparent to the user.

Another technique not strictly limited to RAMs is the use of partially good devices. To lower the potential price per bit in a system, a memory device is used whose failure locations are known. Several methods can allow these devices to be usable: (1) identifying and bypassing the failures with a memory mapping scheme, (2) error correcting the system, or (3) using only the totally good quadrants. As memories and die sizes increase, so does the impact of random defects on device yields. Consequently, more partially good devices can be offered for sale using this scheme.

READ ONLY MEMORIES (ROM)

The read only memory (ROM) is a memory unit that performs the read operation only; it does not have a write capability (Figure 1-14). The binary information

FIGURE 1-14   This ROM memory has sockets for ROM or PROM chips, switches to enable or disable each memory block, buffered address data lines and jumper selectable addresses for each 8K block. (*Courtesy Intel Corporation.*)

stored in the ROM is permanent and cannot be altered by writing different words into it. A RAM is used as a general purpose device with contents that can be altered during the computational process, but a ROM is restricted to reading words that are permanently stored in the device. An m by n ROM is an array of cells organized into m words of n bits. A ROM also has k address lines to select one of $2^k = m$ words of memory and n output lines, one for each bit of the word. An IC ROM may also contain one or more enable lines for expanding a number of IC packages into a larger capacity memory. The ROM does not use a read control line since, at any given time, the output lines always provide the bits selected by the address. Since the outputs are a function of only the present inputs—the address lines—the ROM is classified as a combinational circuit. Thus, the ROM is constructed from decoders and sets of OR gates.

LSI circuits allow extremely low cost digital elements; this is particularly true when the chip is in the form of arrays, such as random access or read only memories. These arrays have a large number of functions per connection and are well suited to LSI implementation.

ROM Applications

ROMs have a wide range of applications in the design of microprocessor systems. ROMs are used for look-up tables for mathematical operations, for the display of characters, and in other applications requiring large numbers of inputs and outputs. They are also widely used in the design of control units for CPUs. Here, they are used to store the coded information that represents the sequence of control variables needed for enabling the various CPU operations.

Read only memories are not new, although in the past their use has been limited. The original form of ROM was the diode matrix, which has been in use since the early days of diode production. Gradually many technologies evolved for read only memories. The rapid development of semiconductor ROMs has produced a succession of faster, larger, and more flexible devices, for use in code translation, character generation, and microprocessor program storage.

Semiconductor ROMs

Between the input and output lines of the ROMs, the coupling elements may be diodes, or bipolar or MOS transistors. The presence or the absence of an element at each of the intersections of the matrix determines the memory logic. Almost any active or passive electrical component may serve as a coupling element, each of which is subject to a number of trade-offs with respect to cost, speed, reliability, and commercial availability.

Semiconductor read only memories are well suited to the LSI manufacturing process.

They are characterized by small volume, low cost, and nonvolatility. The circuitry is much less complex than read/write memories and is well suited to MOS or bipolar manufacturing processes. Data storage is indicated by the presence or absence of a diode at the intersection point of the row and column lines. A diode at the crosspoint will conduct, thus providing the readout. Both discrete or monolithe diode arrays have been used in ROMs.

To program MOS arrays, the manufacturer will use a custom mask that is computer generated to produce the required pattern of stored bits. The ROM manufacturers provide forms on which the users record their requirements. The data on the forms are translated into a computer program for automatic mask programming.

PROGRAMMABLE READ ONLY MEMORIES (PROM)

The extensive use of different programming techniques in the digital control and data processing fields created the need for versatile read only memories. As they are used in various applications, ROMs are usually programmed to suit individual applications. If the programming is a part of the manufacturing process, a special mask is required. This cost can only be justified for large production requirements.

ROMs may be divided into two basic categories: mask programmed ROMs and field programmed PROMs. For masked programmed devices, the truth tables, on tapes or punched cards, are submitted to the IC manufacturer who uses them to prepare the photomasks used for processing the IC wafers, and the test program for testing the wafer. The presence or absence of the connections is established during the metallization process. Using computer aided mask making techniques, design costs and turn-around times can be minimized, but the overall costs become large. Also, to completely specify the memory configuration in the early design phases of the microprocessor system is not always possible. Because of these limitations for mask programmable ROMs, many designers have turned to programmable ROMs (PROMs) in which the information pattern is recorded by the application of an electrical signal.

The electrically programmable ROM may be divided in two categories: (1) ROMs in which a permanent change in the memory interconnection pattern is induced by an electrical pulse, and (2) reprogrammable ROMs in which a reversible change in active memory device characteristics is produced electrically.

The PROM is preferred where a system design is tentative or when memory variability is a requirement. These field programmable ROMs allow the users to enter data into the devices as desired.

Fusible Link PROMs

Programming is accomplished by fusible links; field programmable ROMs were initially of the fusible type. The connections represented by these links are opened for the corresponding bits by pulsing current through the appropriate segments. The disadvantage of using fusible links is that they cannot be reprogrammed. Although fusible link programmable ROMs are more expensive, they do not require a special mask and are convenient in moderate quantity applications.

The electrically erasable PROMs eliminate the disadvantages of the fusible PROMs. The reprogrammable feature permits the later correction of errors made during the write operation.

These devices rely on storage in a dielectric that is part of the insulated gate field effect transistor structure. The charge is removed by the application of ultraviolet light.

PROMs function as an engineering development aid. While the information in a ROM is written in permanently during the process of fabrication, in a PROM, it is programmed after the chip is packaged using one of a number of techniques. Fused links and avalanche induced migration are used with bipolar PROMs while MOS reprogrammable PROMs or EAROMs may use one of a number of processes.

Bipolar devices are selected for speed, since these have shorter access times compared to MOS devices. The MOS devices consume less power and higher density parts are available with this technology. Programmable bipolar devices are limited to the nonerasable devices. Programmable bipolar devices are available predominately as TTL Schottky with a few devices offered in emitter coupled logic (ECL). Programmable MOS devices may be fabricated in PMOS, NMOS, CMOS, and MNOS. Most programmable MOS devices are made to be compatible with TTL logic levels.

The nonerasable PROMs use some form of fusing process in which a link is fused open or closed. Three types of fusing technologies are in use: (1) metal links, (2) silicon links, and (3) shorted junctions. All the fusible devices have a relatively short programming time.

The fusible link devices are completely nonvolatile and once programmed cannot be erased. Improvements in the technology have resulted in the various types of fuse material in use: nichrome, platinum silicide, polycrystalline silicon, and titanium tungsten.

All fusible link technologies are similar. The fuse material is deposited as a thin film linked to the column lines of the PROM. The memory cell is constructed as a transistor switch. The fuses are blown during programming by saturating the transistor through the selection of the row and column by a decoding circuit. A large current is switched through the transistor and through the fuse in the emitter leg. The emitter fuse link is open circuited by the current, resulting in the programming of the bit location.

A problem in the use of metal link PROMs has been the regrowth of opened fused links over a period of time. In the regrowth process cells go from a programmed or open state back to the unprogrammed or closed state. Manufacturers have refined the process of programming to achieve improved yields and reliability from regrowth.

Silicon link PROMs use notched strips of polycrystalline silicon material. Programming of these devices involves melting the links using a similar technique to that in metal fuses. A current of 20-30 mA is used to blow the fuse link. This current generates a heat of about 1400 °C. At this temperature the silicon oxidizes, forming insulating material around the open link. The use of silicon results in the absence of contact problems or other difficulties caused by the use of a dissimilar material.

The use of silicon eliminates conductive materials in the open gap between the formerly linked polysilicon fuse ends. Thus, grow-back is greatly reduced but not completely eliminated. One may achieve improved reliability and simpler programming with this technology.

Shorted Junction PROMs

Shorted junction devices may also be referred to as the AIM (Avalanche Induced Migration) technology. This technology uses two semiconductor junctions which appear as a high impedance circuit of back-to-back diodes. Generally an npn double diffused transistor structure is used and only the emitter and collector contacts are metallized. The base is left open, thus forming the back-to-back diodes. One diode is reverse biased and the flow of electrons in the reverse direction causes aluminum atoms from the emitter contact to migrate through the emitter to the base, causing an emitter-to-base short. This technique requires a higher voltage and current than the fuse method. The remaining junction is useable as a forward biased diode and represents a programmed data bit. The programming involved in this type of technology requires some precision. To program devices in which the second junction is damaged is possible, thus making the PROM unusable. The impedance of each link can be measured once it has been blown, which is not possible with open circuit devices. A potential advantage in reliability can exist if the impedance is measured after the devices are programmed.

The use of fusible link PROMs is not practical if the PROM program is expected to be changed several times before it is complete. Erasable PROMs, or EPROMs, are available for this purpose.

Erasable PROMs are implemented using an isolated gate structure in a NMOS field effect transistor. The isolated transistor gate has no electrical connections. It can store an electrical charge for indefinite periods under normal conditions. This stored charge becomes the memory mechanism. Under the erasing conditions the charge on the isolated gate is removed, clearing the contents of the memory cell. The two types of erasable MOS PROMs are ultraviolet (UV) light erasable and electrically erasable.

UV PROMs

The ultraviolet (UV) light erasable PROM uses a floating silicon gate which is erasable by ultraviolet light, and reprogrammable after each erasure. The floating gate is located in a silicon dioxide layer. It effectively isolates the source from the drain under normal conditions.

During programming a high negative voltage is applied which forces the junction of the desired cell into a breakdown condition, causing the injection of electrons in the floating gate area. After the voltage is removed, the gate retains this charge since it is electrically isolated by the silicon dioxide layer. The negative charge on the gate forms a conductive inversion region in the substrate, which acts as a channel between the source and drain. The presence or absence of this conductive channel determines if a 1 or 0 is stored in the memory cell.

A lid in the sealed package of the chip allows erasure by illuminating the chip surface with ultraviolet light. The lids can be opaque or transparent in appearance but they pass UV light for the erasing process.

The technique for erasing these UV EPROMs is to illuminate the window with an ultraviolet lamp which has a wave length of 2537 A.

The UV source is placed at a distance of 2–3 cm from the lid and the radiation is allowed to fall on the element for 10–45 minutes, depending on the type of device and source. The radiation raises the conductivity of the silicon dioxide and allows the floating gate charge to leak away. The erasing process is not selective; it results in resetting all cells in the device.

The UV source may age with time and its intensity may diminish, which means that an EPROM left to erase for the usual time may not be completely erased. Incomplete erasure can result in increased access time or unstable bits.

UV PROMs have seen an increase in popularity, much of which is due to their use in microprocessor systems. When properly erased and programmed they are reliable. Most UV EPROMs are second sourced and larger, faster, and easier-to-use devices are planned for the future. UV EPROM technology is a popular choice for microprocessor system designs needing an erasable memory.

Electrically Erasable PROMs

In some microprocessor applications, particularly in evaluating new system designs, to make changes in the memory content without the removal and replacement of the memory chips is desirable. In this type of application the electrically erasable PROM (EEPROM) is used. EEPROMs are also referred to as electrically alterable ROMs (EAROM) since they can be electrically erased and electrically written into. This feature allows in-circuit programming since the device can be erased, written, and read without being removed from the circuit. The two types of electrically erasable PROMs are floating gate MOS devices and metal nitride oxide silicon (MNOS) devices.

Floating gate EEPROMs are similar to UV EPROMs with the addition of a control gate. A lid is not needed to allow erasing, the control gate is used to control erasing and writing. Devices of this type use special voltages and complex voltage sequencing, making them somewhat impractical for in-circuit programming. These EEPROMs have access times and data retention similar to UV EPROMs. Like UV EPROMs they are usually removed from the circuit for erasing and reprogramming.

A newer floating gate structure employs Fowler-Nordheim tunneling to write and erase data. These devices erase and write by causing electrons to tunnel across a thin layer of silicon dioxide. The cells hold their charge in the same way as conventional EPROMs. They can hold data for years even at extended temperatures. The operation is fully static and refreshing is not required at any read frequency. A byte erase/write or chip erase requires the application of a 21 volt pulse for 10 milliseconds. Any 2 kilobytes of the device can be erased and rewritten in 20 milliseconds.

The MNOS memory uses a MOS transistor with alterable threshold voltage. The threshold voltage can be altered between two levels by applying large gate voltages of either polarity. Erase times run from 10 to 100 milliseconds and write times are from 1 to 10 milliseconds. Devices are available in word erasable, row erasable, or device erasable configurations. A PROM configuration with serial addressing and serial data output is also available.

MNOS EPROMs are not widely used. A variety of devices are available but limitations in data retention have reduced their use.

Metal nitride oxide silicon (MNOS) memories also require complicated manufacturing process.

MAGNETIC BUBBLE MEMORIES

Magnetic bubble memories (MBM) have the capabilities of a high bit density and large capacity per chip; they can also be less expensive. Based on solid state technology, they do not have the mechanical motion problems associated with the present disk and tape systems. Long term nonvolatility of the stored information is provided much more easily than with most semiconductor memories. The ability to store, transmit, and process information in the same medium also reduces the problems of interconnections and signal conversion.

The generation, propagation, and sensing of magnetic bubbles involve action of the bubble domain with its magnetic environment.

Bubble Memory Operation

Magnetic bubble memories are made by growing a thin layer of crystalline material which exhibits anisotropic (i.e., assuming different positions in response to external stimuli) properties on a nonmagnetic substrate. The magnetic domains in this anisotropic material tend to orient themselves in fields perpendicular to the surface of the material. When a magnetic field is applied perpendicular to the crystalline layer, the magnetic domains tend to position themselves in the same direction as the applied field. Domains of the opposite polarity of this field will form small cylindrical bubbles. The stronger the applied field, the smaller the bubble. At a certain field strength, these reverse domains vanish. A smaller applied field allows them to increase in diameter. The perpendicular field is produced by permanent magnets on top of and below the bubble chip.

The bubbles are mobile and travel across the memory by magnetic attraction. Strips of permalloy on the surface which are isolated from the bubble medium by a layer of silicon dioxide are used to control the bubble positioning. Applying the magnetic field in parallel with the device surface produces magnetic poles in the permalloy elements. Bubbles are attracted to the poles of the magnets. By rotating the direction of the horizontal field and with the proper element design, the bubbles will propagate from one element to another.

A rotating horizontal magnetic field varies the polarity of the permalloy element, with the bubble continuously moving to a position under the nearest positive pole. Each element is a cell, and each rotation of the magnetic field is a cycle.

Data are indicated by the presence or absence of a bubble in a cell and data are propagated through the memory like a shift register.

Reading data requires detecting the presence of bubbles under the permalloy elements. A resistor network compares the resistance of a detector element with that of a dummy element with no bubble.

Wrapping the memory with two orthogonal coils and driving an ac current into each that is 90° out of phase creates the rotating field. The field is used only for propagating bubbles. However, without a horizontal component, the bubbles will tend to drift. To prevent this, tilting the die in the package 4° to 5° creates the required horizontal component.

Loop Functions

The electrical interface to the memory is through current loop functions. At certain points within a bubble track, conductors are deposited under the permalloy element in the form of single loops. Passing a pulse of current through these loops alters the level of the field under the permalloy element. The direction and the extent of the localized field and the horizontal field as well as the design of the permalloy element determine the effect on the bubbles. The basic current loop functions are:

1. Generate—Inverts the bias field to allow the creation of a new bubble.
2. Annihilate—Intensifies the field which causes any bubble under the permalloy element to collapse.
3. Transfer—Moves bubbles on one track to an element in an adjoining track.

Architectures

Single loop memories are the simplest to operate, since they require the minimum number of current loop functions. They also have a higher data access time, but a single defective location in the loop makes the memory inoperative.

A memory with smaller minor loops is an improved architecture. The added tracks increase the number of current loops. This configuration reads or writes data in blocks or pages and a block of data equals the number of minor loops. The block addresses refer to the location of a bit of data in a minor loop. A normal transfer-in function accumulates data in the minor loops. Another type of transfer gate function uses a swap function to interchange bubbles from one track to another. Bubbles swapped into the input track move to the end where they normally collapse.

In a multiple loop memory a defective element in a minor loop affects only the operation of that loop. An increase in yield can be obtained by discarding some of the minor loops while still qualifying the device as an operational memory.

Each memory chip can have a unique loop error map to define the unusable loops. A programmed memory controller can bypass these loops when reading or writing data.

EXERCISES

1-1. List the registers in a basic microcomputer. State, in one sentence, the function of each register.

1-2. Describe how the expression below would be evaluated on a microcomputer having one accumulator and on a microcomputer having two accumulators:
A x (B + C) x (D − E)

1-3. Which status bit would be used for the following:
(a) To check if a counter has been decremented to zero.
(b) To check if a binary addition resulted in an answer that could be represented in a single word.
(c) To determine if the result of a subtraction is positive.
(d) To determine if two numbers are both positive or both negative.

1-4. A CPU is available without a program counter. Instructions contain three parts: an op code, the address of an operand, and the address of the next instruction. The operation code consists of 8 bits and the memory has 8,192 words.
(a) How many bits must be in a memory word if an instruction is stored in one word? Draw the instruction word format.
(b) What other register is required in the CPU besides an operation register?

1-5. A computer has 16 processor registers of 16 bits and the instructions use 4 bits. Write a possible instruction code format for an operation with the content of any register with any other register and the result placed in any third register.

1-6. Explain how index registers can be used as pointers and counters in a micro-processor.

1-7. A first-in-first-out (FIFO) is a memory unit that stores information such that the item first in is the item first out. Show how a FIFO memory can operate with three counters: (1) a write counter WC that holds the address for the memory write microcomputer; (2) a read counter RC that holds the address for the read operation; and (3) a storage counter SC that indicates the number of items stored. SC is incremented for every item stored and decremented for every item that is retrieved.

1-8. An 8 bit microprocessor communicates with up to 2^{16} words of memory. It has an 8 bit bidirectional data bus for the transfer of instructions, data, and addresses. A type of instruction format consists of three 8-bit words: the first word contains the operation field and the other two contain the address field. Write the type of information (instruction, data, or address), the direction of transfer (to or from the microprocessor), and the number of times that the data bus is used in order to fetch and execute the following:

(a) Load from memory to accumulator.

(b) Store content of accumulator in memory.

1-9. If the last operation performed with an 8 bit word was an addition in which the operands were 2 and 8, what are the values of the following flags:

(a) CARRY

(b) ZERO

(c) OVERFLOW

(d) SIGN

(e) EVEN PARITY

(f) HALF-CARRY

1-10. How many bits long should the stack pointer be in the following cases?

(a) The stack is in the first 256 words of memory.

(b) The stack can be anywhere in a 16K memory.

1-11. A memory has a capacity of 65,536 words of 24 bits each. The instruction code is divided into three parts: (1) operation code, (2) two bits that specify a processor register, and (3) an address part.

(a) What is the maximum number of operations that can be incorporated in the CPU if an instruction is stored in one memory word?

(b) Draw the instruction word format indicating the number of bits and the function of each part.

(c) How many processor registers are in the CPU and how many bits in each?

1-12. Discuss the fan-out, power dissipation, propagation delay, and noise margin for the following logic families: TTL, ECL, NMOS, and CMOS.

1-13. Discuss the type of system applications where you would use:

(a) Bipolar microprocessor.

(b) An I^2L microprocessor.

(c) A CMOS microprocessor.

(d) An NMOS microprocessor.

(e) A PMOS microprocessor.

(f) ECL microprocessor.

1-14. What are the basic characteristics of ECL technology?

1-15. Compare the utility of a RAM that is bit organized with one that is word organized.

1-16. Describe the technology and define the operational characteristics of a static MOS ROM. What are the advantages and disadvantages of this type of memory?

1-17. Describe the type of bipolar PROM implementation technique known as shorted junction. What are its disadvantages?

1-18. Describe the approach to a reversable PROM in which one alters the program using ultraviolet light.

1-19. Describe how ROMs could be programmed in the field using equipment designed for this purpose. What are some ways for the programmer to determine the memory locations?

1-20. What problems can alpha radiation cause in memories?

1-21. What are the advantages of the tunnel oxide EPROM over the UV EPROM?

1-22. Discuss the major factors that control the type of memory used in a system.

1-23. Discuss the operation of a bubble memory. What control functions might be required in a bubble memory controller?

1-24. What are the major cost factors for an electronic system? Discuss their relative importance in a small, high volume microprocessor system.

2

Microprocessor System Development

A number of stages occur in the development of microprocessor systems, and a number of methods may be applied during each stage. This book is concerned with problem definition, program design, coding, debugging, testing, documentation, maintenance, and redesign. A variety of methods and the trends should be understood by the system designer, who should be aware of the range of problems encountered in microprocessor development, and thus be able to select the methods most effective for the system application.

It may be useful to apply some principles of modern programming techniques. No single development method is widely used across the spectrum of applications. However, some of the development systems and equipment now available for microprocessors are becoming widely accepted.

The stages of system development follow.

Product Definition

This stage represents the initial definition of the task, including the specification of inputs and outputs, and the processing and system requirements such as execution times, word length, response times, and error checking requirements.

A basic flow chart showing the sequence of steps involved in the task can be formulated to illustrate the overall structure of the task and subtasks.

The initial phase of development is always an evaluation phase, when the solutions to the problems are proposed and evaluated. The detailed design involves the selection and assembly of the microprocessor system and the design of the software program. The essential consideration, at this point, is to define performance of the hardware and software for the system application.

Software/Hardware System Design

This stage includes the design of a software package which meets the requirements of the product definition phase. The techniques may include top down design, and structured or modular programming. Flowcharting may be used extensively to define the structure of the software package. Definition of the software design responsibilities and flow and the required development tools is also considered at this time.

The hardware design and the software design may be accomplished in parallel, which is a major difference in microprocessor system development compared to other electronic hardware development.

The hardware design can be independent of the software development. The hardware design is usually simple, especially when standard microcomputer hardware is used. It can be more complex when nonstandard interfaces or bit slice hardware are needed.

The most significant task is in the software design. A number of microprocessor development tools and systems are available which allow the software to be designed independently of the hardware.

The designer decides what functions to implement in the form of chips, and how many to implement in the form of software. This is called hardware/software partitioning. A major consideration is always the quantity of systems produced. If the quantity is large, the number of hardware components should be minimized for the lowest system cost. Then more can be accomplished through software design. If a small number of systems are to be produced, more hardware can be used to decrease the complexity of the program as well as development costs. The evaluation requires estimating the complexity of the software and the time and cost involved. When the decrease in software cost and time is significant, then chips and board area can be added.

Program Development

This stage involves coding the algorithms into a program. Coding involves the actual translation of the program design into computer instructions. These instructions then form the actual program or software product. Assembly language programming or alternative higher level languages can be used. In either case the techniques of software development and structure are emphasized.

Program development requires a precise formulation of the program logic, structure, and timing. Flowcharting, modular or structured programming, and top down techniques are techniques for formulating programs that may be easily coded, debugged, and tested. The proper coding simplifies the later stages of development. The emphasis should always be towards greater clarity and comprehensibility.

A simple program structure with the proper documentation can make the program easy to debug and test. The designer can make the software more efficient once the program is running.

Program Verification and Debugging

This stage allows the discovery and the correction of programming errors. Only very simple programs run correctly the first time, thus debugging is a critical and often time consuming task in development.

Editors, debugging software packages, simulators, emulators, logic analyzers, and other tools can be used during debugging. The types of errors that are most common in microprocessor programs, as well as techniques to reduce their occurrence, are discussed later in this chapter. System testing and program debugging are closely related. System testing is a later stage in which the program is validated for a number of test cases. Some of the test cases may be the same ones used in the debugging phase.

System Testing

This stage involves the validation of the hardware and the program. System testing proves that the program performs the required tasks in conjunction with the system hardware. The important consideration here includes the selection of the test criteria from the specification and the development of the testing methods.

Evaluation and Redesign

This stage can be the logical extension of the design for requirements beyond those described in the initial product definition. Thus the developed programs and techniques can be used for additional tasks and capabilities.

Any one task is not completely in isolation from other tasks that follow. Thus each stage of development affects the other stages. Problem definition can include the consideration of a test plan, documentation, and maintenance provisions with the possible extension to other tasks. Program design could involve similar extended provisions for debugging, testing, and documentation. Coding, debugging, testing, and documentation are very often concurrent activities.

PRODUCT DEFINITION

Many microprocessor system applications require a number of operations rather than a single task and thus require considerable definition effort. The use of a microprocessor to control an electrical or mechanical system such as a scale, terminal, or print wheel, requires a variety of calculations, and the device can generate a variety of outputs. The application may require the solution of a particular set of equations, the search for particular data records, or performance of other tasks. This initial stage of development must define the tasks to be performed and the requirements to be met.

A major consideration is the form of the inputs and outputs, which concern the devices to be attached to the microprocessor and the form that they will send and receive the data. Data rates, error checking procedures, and the control signals which the input/output devices will use to indicate the availability of the data or readiness of the system to accept data must also be considered. Word lengths, format requirements, and protocols must be decided. In many control tasks the input/output requirements are major factors in the definition phase.

Other major concerns involve the processing requirements. The sequence of operations on the input data and the order in which all other tasks will be performed must be decided. The order of operations can be critical, if the input/output signals must be sent or received in a particular time sequence. The system can have critical time requirements such as minimum data rates within the system itself. Some circuits and most interconnect wiring must be designed just for this purpose.

The use of the microprocessor in systems introduces a new element: the control program. The microprocessor and the control program replace many circuits and interconnects used in conventional designs, resulting in major reductions in hardware costs and increased system flexibility.

Design Tools

In conventional logic design, the design tools may be block diagrams, schematics, assembly drawings, and parts and wire lists. For microprocessor systems some old tools and some additional tools and disciplines must be used due to the abstract nature of the program design.

The new tools which are borrowed from the computer industry are flow charts, memory maps, and program listings. The disciplined use of these tools, coupled with the techniques of hardware/software partitioning, produces the design approach.

The block diagram can be an essential tool. In many microprocessor designs the block diagram can be the key link between the program and the hardware. It is at this level that partitioning of the problem can begin.

A block diagram can serve several purposes. One main purpose is to condense the overall concepts of the design which should provide maximum commu-

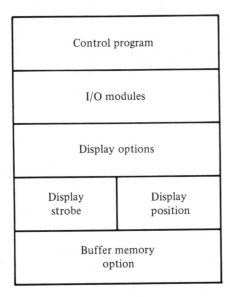

FIGURE 2-1 A Memory Map.

nication with as little detail as possible. In the initial design the block diagram can provide the foundation for collecting and communicating design ideas. In the production stage the block diagram can be used as a training tool for test and service personnel.

The designer uses the block diagram as a reference for the system's resources. It can illustrate all elements of the system and is useful for interfacing details such as port assignments and memory allocation.

Flow charts can provide a visual statement of the problem solution using sequentially interconnected symbols to illustrate the program sequence. Flow charts can be used to complement the block diagram. The block diagram illustrates the interconnection of the hardware while the flow chart shows the interconnection of the program flow. Flow charts can be a key tool for the program partitioning, reduction, and simplification.

To share system resources, the microprocessor accesses its resources by addressing. To avoid addressing conflicts, the programmer must map out the resources available and make the allocation assignments as required. The resources such as registers, memory, or I/O must be allocated. To keep track of the resource allocation, mapping is used. These are visual aids or maps to allow register, memory, or I/O space assignments as they are required (Figure 2-1).

The program listing is a step-by-step list of the program operation. The listing may contain the machine language bit patterns, their numerical representations, or verbal assembly statements.

Flow charts have some advantages over program listings in that they can:

1. Show the order of operations as well as the relationships between sections of the program much more easily.
2. Be independent of a particular microprocessor or language.
3. Utilize a standard set of symbols as shown in Figure 2-2.

Flow charts have most of the advantages of pictures over words. They can be a useful way for nonprogrammers and programmers to communicate.

Several levels of flow charts may be used—general charts which show the flow of the program and others which provide more detailed information. Figure 2-3 shows an example using a program for an editor. Too much detail can make a flow chart difficult to understand and a very detailed flow chart has little advantage over a program listing.

Flow charts can be helpful when they are general; however, they may have only a small connection to the actual structure of the data or hardware. These general flow charts may only describe the program flow. They will not show the relationships of the data structures or the hardware elements.

To design good flow charts, think first of the task the microcomputer must perform. After the task is established then consider how the microcomputer will carry it out.

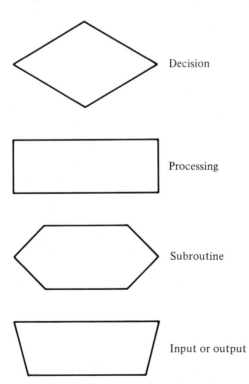

Decision

Processing

Subroutine

Input or output

FIGURE 2-2   Flow Chart Symbols.

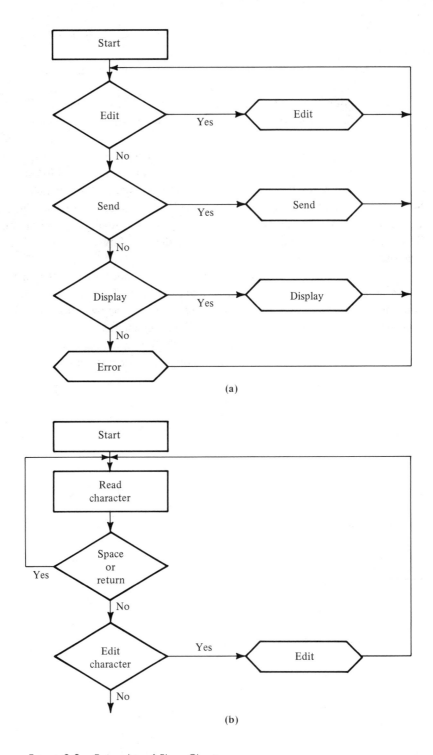

FIGURE 2-3 Examples of Flow Charts.

The flow chart can be used to calculate what is needed in the way of mechanical or electrical delays, holding and settling times, or enable and disable rates. Latches and timing circuits may be needed to satisfy these requirements.

Paper checking is a technique used to check these designs on paper. It can be used to test the logical design and the program itself.

In paper checking, the program is executed by hand and entries are filled out in a table corresponding to the values of critical registers or outputs. This requires no development hardware but it can be a long and tedious process.

Paper checking is more often done at the flow chart level to verify the overall design. It may not result in a reasonable evaluation of actual performance.

To evaluate performance normally, development tools must be used. In many cases, paper checking is used to evaluate different microprocessors or different control schemes using a common benchmark or set of benchmarks.

The benchmark program is a program or routine which is used to test the efficiency of a given microprocessor in a system. Some benchmark programs, for example, accomplish a block transfer or similar function. Unless they are carefully optimized the programs may not be valid benchmarks.

A valid benchmark is one that is written for the application. A benchmark should also be representative of the programmer's skill, if it is to be a true measure of the efficiency of the final program.

A major difficulty is that many applications cannot be characterized by simple programs; therefore, they require some mix of instructions to be executed. Since a typical mix may not exist for the application, it is sometimes difficult to decide if a particular program has true benchmark value during the evaluation phase.

The advantage of benchmarks occurs when the application can be clearly defined. It is then possible to use a sample program with the execution times of different microprocessors. The designer can then decide which microprocessor would be best for the application.

Design engineering involves the creation, implementation, and documentation of products. A documentation discipline allows these product ideas to be produced and maintained. Manufacturers may make control systems, instrumentation, or test equipment with electronic or electromechanical components. The electronics inside sense the outside world through its inputs and cause something to happen by driving the outputs. The input signals might come from switch contact, transducers, or other devices. The output loads can include relays, motors, and displays.

The way the system functions when power is on is determined by the components used and the interconnects between the components. The internal circuits may include input and output drivers, timing, logic, arithmetic, and data storage. These circuits may be in the form of standard modules used in systems of this type and are fully specified, documented, and tested apart from the subject system, or the circuits may be implemented from chips used to build up boards or modules. In either case a definition of the task is begun and an analysis of a system which will meets the requirements continues. The next step is an implementation which accomplishes the result.

Design Application

Let's consider a particular application in some detail. Along with the design concept we'll consider the application characteristics which provide the design foundation.

The use of microprocessors for traffic control requires the regulation of traffic lights at intersections which had been accomplished using electromechanical control in simple situations and electronic control in more complex situations.

The algorithms for controlling the flow of traffic in urban areas have grown in complexity, requiring complex logic for the proper sequencing. Some of the functions that might be performed by a traffic controller are:

1. Sequencing of traffic lights.
2. Timing cycles which may include cycles such as normal day sequencing, night cycle, rush hour cycle, or others, depending on the time of the day or the traffic flow.
3. Initialization sequence from the time power is applied.
4. Preemption facilities by police or other emergency vehicles.
5. Pedestrians or vehicle actuation.
6. Computing traffic as density or volumes.

Some major considerations involving the design are:

1. Cost.
2. Complexity and number of the algorithms to be implemented.
3. Adaption schemes for different intersections and different algorithms.
4. Reliability.

An important consideration is that each traffic controller must be custom developed for each intersection. The critical parameters are the geometry of the intersection, the number of phases and lights, as well as the combinations of algorithms which must be implemented. By customizing at the software level, one can produce standard hardware for traffic control. The adaptation and programming are done on a development system at the software level. Because of the production of identical hardware units, the hardware costs of the controller decrease.

Software features such as status monitoring can allow greater reliability with the detection of faulty relays or bulbs. The program might also allow the implementation of a variety of algorithms such as those for green traffic streams. This allows vehicles to proceed through all or most of the signals on a street without stopping. It is accomplished by transmitting information from one traffic control to the next one in the network and coordinating the operation of the network from a central CPU. The modes of operation might be:

1. Start-up begins from the restart mode; the microprocessor assumes that no information is available. A power up sequence is used until the system parameters become available. These input parameters might include the time of day and traffic measurements.
2. Operational programs are used during the various segments of the day such as the morning and afternoon rush hours.
3. Parameter actuated modes are used for operation in between these times. Parameters such as the speed of vehicles, density, and the distance between vehicles are used to calculate and optimize the traffic flow.

The definition phase must consider the human interface, including control switches for timing as well as the display of information.

Control might be provided to specify the timing intervals and the selection for the various modes of operation. The status of the system must also be provided.

Manual actuation or preemption facilities for emergency operations must also be considered. These facilities are used when an accident occurs. The intersections are then switched into emergency modes or sequenced manually.

All of these functions can affect cost. Because of the cost of the various functional hardware and software modules in such a system, the benefits of each capability must be evaluated.

Fail soft facilities may be provided in case of hardware or software malfunctions. A conflict monitor can be used to monitor the status continually for all lights at an intersection. When two conflicting lights are on simultaneously, the conflict monitor detects this condition and disconnects the microprocessor while turning on the caution flasher unit. This produces alternate red and amber signals in the directions involved.

When a software or hardware malfunction occurs which can result in a conflict, it can be detected and the system can take action. The complete system does not have to be disabled, only those safety related or critical disruptive functions should be disabled because of the malfunction. Techniques can also be used in the microprocessor itself to diagnose and correct error conditions.

The lines used for loop detector information and status may be transmitted to a central processor in a network of traffic control processors. Information displayed at this control center could include the status of each intersection. The density of traffic might then be evaluated at the control center.

If a malfunction occurs, another microprocessor could take over the functions if the system had the required software and signal lines. In a dynamic self-optimizing system, the microprocessor can start in a time-of-day mode, then switch to a parameter actuated mode as the parameters become available, and then switch into a self-optimizing mode.

These considerations for traffic control are similar to those encountered in many other microprocessor control applications.

SOFTWARE/HARDWARE SYSTEM DESIGN

After the overall system performance task is completed, the next step is to begin development of the software/hardware system. Some designers use extensive flowcharting at this stage.

Actually, flowcharting techniques are helpful when describing program structures and in explaining the program to others. They are also very useful for documentation. However, few programs ever result from a detailed flow chart, because drawing the detailed flow chart can be as difficult as writing a program and is less useful since the program must then be derived from it.

Design techniques that are more useful in writing programs include:

1. *Modular programming*—This is a technique in which programs are divided into smaller programs or modules which are then designed, coded, and debugged separately and linked together later.
2. *Top down design*—Here, the overall task is first defined by generalized subtasks which are broken down into fully defined tasks. This process continues until all the subtasks are defined in a form suitable for programming. The opposite of this is bottom up design in which all subtasks are first coded and then integrated into an overall program design.
3. *Structured design*—In this method programs are written according to specific defined forms. Only certain types of program logic may be allowed, but some routines can be nested within one another to handle complex situations. Structured programming is often written in sections with a single entry and a single exit.

These software design techniques will be considered in more detail in a later section of this chapter, but first some basic techniques for system development at the hardware level, the software level, and at the systems level will be discussed. Some major choices at these levels include:

1. Selecting the microprocessor or microprocessor board.
2. Performing the hardware/software partitioning.
3. Selecting a programming language.
4. Selecting the necessary development tools.

Choosing the microprocessor involves the selection of a product which can perform sufficiently for the application. Other considerations may include the availability of system supply components, development support equipment, and availability of trained personnel.

The hardware/software partitioning task should be based on the evaluation of the cost and performance aspects of the techniques or components to be selected. The techniques used to assemble the microprocessor system, as well as the software techniques used for the system implementation, facilitate this choice. Hardware/software partitioning is one of the more delicate tasks to be accomplished during the system design. The allocation should be reevaluated throughout the design and the trade-offs must be carefully considered. The partitioning has a major impact on the software design.

Throughout the design, a trade-off evaluation should be pursued as shown in Figure 2-4, and the partitions that were made should be reconsidered.

It may become necessary to go from polling to interrupts, or to add hardware in the form of encoders or other circuitry. Buffers, shift registers, latches, counters, level shifters, and interrupt control units are among the hardware that may be required.

In the past designers have tried to minimize memory usage and external hardware, since hardware and memory were more expensive than software. But today the cost of memory and external hardware is less than the cost of developed software.

The tasks that once had to be performed by hardware may now be performed by software with the reduced costs of processors and memory. Recent additional hardware which can substantially reduce the complexity of programming is also available at relatively low cost.

Software costs remain the same regardless of the number of units produced, but hardware costs are proportional to the number of units. These trade-offs between hardware and software differ today from the past when hardware was more expensive.

The time and cost required to develop the software are always important factors. Processors, memories, and other hardware are less expensive but the cost of programming has risen. This change in the relative costs of hardware and software is a major reason for the attention on techniques such as structured programming to increase programmer productivity.

The proper design, debugging, testing, and documentation methods can do

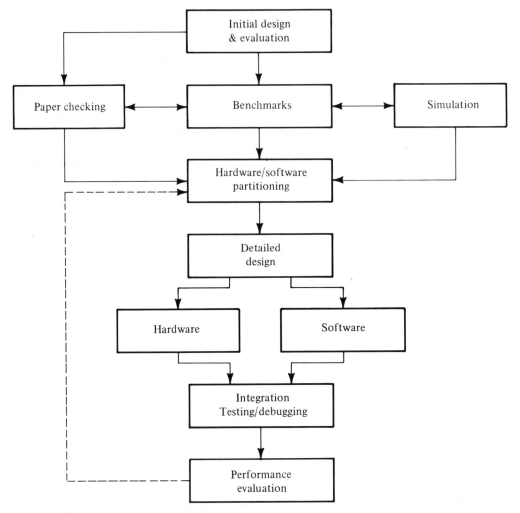

FIGURE 2-4   Development Tasks.

much to reduce the overall program development costs. Writing programs in high level languages can increase programmer productivity. It can also simplify parts of the development. But there is also a trade-off between the software and hardware costs involved. The programmer can write and debug the program faster in a high level language, but the final program will require more memory than one written in assembly language. One solution is to first code in a high level language and then edit the repetitive loops to reduce the memory requirements as shown in Figure 2-5.

The product specification is used to define the problem, then block diagrams are used to break the problem into sections. The hardware and software design

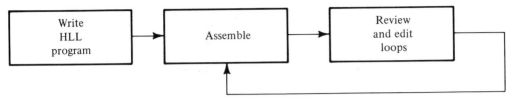

FIGURE 2-5   Using Assembly Language to Minimize Memory Requirements.

and debugging are first done at the module level, then the subsystem level, and finally at the system level. Field trials can later be used to verify the design solution in the application environment.

An important characteristic of the design is that it be modular. The basic level of hardware modules are the circuit modules. The basic procedure is first a design on paper, then a prototype can be built from these documents for testing. When this testing and debugging are complete, the original documents are updated for the manufacturing of the modules.

Selecting the Hardware

A number of hardware choices are available when designing a microprocessor system. Single chip microcomputers may contain most of the elements required in one package. A fully assembled board containing a microprocessor and support circuitry can be used. Performance and cost are key considerations along with the support software and hardware available.

The microcomputer's performance is often a direct function of the speed of its microprocessor. However, speed is not always the most important factor in some applications. It should not always be given a high priority. Other more important considerations may include the number of input and output lines, the memory capacity, the types of interfaces to other circuitry and systems, and the expansion capabilities which may be needed as the system grows.

In applications where the processing demands a fast microcomputer, multiple microprocessors can share these processing requirements.

System expandability and support hardware can allow features to be added and design problems to be corrected without having to redesign the system or change to another microprocessor.

To consider the relative trade-offs between single chip and single board designs, compare a single chip microcomputer like the 8048 and a single board microprocessor system. The 8048 provides a CPU, ROM, RAM, and I/O in one DIP package. Other ICs can be added as required. In the single board approach, much of the design has already been done. The 8048's pin compatible EPROM version, the 8748, may be used for development and low quantity production.

After the choice of microprocessor and system development technique, the other components of the system must be evaluated: memory, input/output, and peripheral devices. Intelligent peripheral devices like floppy disc controllers or CRT display controllers may help reduce the support circuitry. These devices can also simplify the programming needed for the peripheral devices.

The denser ROMs and RAMs can reduce the number of parts in the memory system as well as the system power consumption. EPROMs are always useful during the development phase. For small systems, integrating memory, I/O, and other functions in a single chip can reduce the system complexity. Chips like the 8155, for example, have 256 bytes of RAM, 22 I/O lines, and a 14 bit timer.

When preassembled boards are used the system can expand easily. Boards which are compatible with bus systems such as the Intel Multibus allow memory and I/O expansion or multiple microcomputers to be used for added throughput. The Multibus system is supported by many different products from a number of manufacturers. The functions include disc and tape memory controllers (Figure 2-6), IEEE-488 bus interfaces, and wire wrapped boards for custom interfaces. Some single board computers have no built-in expansion bus, but they may be hardware and software compatible with larger boards using the same microprocessor which can be expanded.

System expandability is important when choosing a single board computer or selecting components for the design. If the design can be upgraded for im-

FIGURE 2-6   A floppy disc memory controller circuit board like this one allows a microprocessor system to be expanded easily. (*Courtesy Intel Corporation.*)

proved performance, then a major portion of development costs can be saved.

The cost of the system depends not only on the price of the parts; the design costs for nonrecurring engineering (NRE) and the production costs for each unit must be included. These NRE costs include preliminary design, breadboarding, checkout, and other engineering costs for the first production unit. The initial design includes the microprocessor clock, buffers, memory system, and the most critical part of the design—the software.

After the initial design of the hardware and software, these systems must be combined and checked out together. In this system integration phase, problems will occur from the new breadboard hardware and the new, untested software. If an error occurs in the software, it may seem that the hardware is at fault. Development systems, incircuit emulators, and high level programming languages can reduce these problems during system integration.

Once the system is complete, the documentation must be updated. This includes the schematics, timing diagrams, program listings, flow charts, and maintenance procedures.

The NRE must be amortized over the production phase and should be recaptured within a certain time for the required return of the initial investment. When one includes the development expenses in the total system cost, the NRE has little impact on the unit cost for large production runs (of thousands). But the impact is much greater on a small number of systems in the lower hundreds.

A common technique to minimize the design cost is to use a preassembled microcomputer system in the final design (Figure 2-7). Even though the design would have approximately the same system capabilities, it's usually more expensive to completely duplicate the purchased single board computer.

The proper support can also reduce development costs. This support comes in various forms and should be convenient at all stages of implementation.

Initially, support is in the form of documentation and training. The more documentation, the easier it is to evaluate the system parameters which affect the application. Users' manuals, programming manuals, and application notes can help to evaluate the capabilities of the microcomputer and also aid in the design.

Programming development time sometimes may be reduced by extensions of commonly used software routines. User programs may be available. Some of these routines can usually be incorporated into the application program.

Seminars, workshops, and courses may also be available. Discussions with other engineers or consultants can always help to evaluate some of the cost tradeoffs as well as the actual availability of components. This additional advice can be used during all phases of design to discuss the advantages and disadvantages of prospective microcomputers and associated components in light of the application. Outside support can also be used to periodically review the design to optimize the end product.

For example, if well hidden techniques can be used to upgrade the system,

FIGURE 2-7   Microcomputer systems such as this contain the necessary controls, terminations and power to allow one to assemble the configuration desired. (*Courtesy Intel Corporation.*)

then the designer does not have to change to another microcomputer for improved performance in future product versions.

Designing the Program

Program design should be a modular process. The design starts by dividing the problem into smaller modules using flow charts which act as a tool for describing the sequence of events. Only two symbols are usually necessary: a box to define a process and a diamond to define a decision. The number of levels in the flow chart depends on the complexity of the program. After the flow chart defines a basic module in enough detail, the program coding can be started for the module. When this program module or subroutine is designed and documented, it can be integrated with the other modules into the overall program. The integration step is similar in some respects to the process of connecting circuits or modules together into systems.

The three basic choices the designer has for a programming language are:

1. Direct binary or hexadecimal coding.
2. Assembly language.
3. High level language.

Writing in binary or in hexadecimal form does not require much support in terms of hardware or software. This technique is used on simple systems that do not require much coding and where the coding must be very compact. The user communicates with the system through a hexadecimal keyboard and LED display. The instructions and data are transferred into the system using the keys in a hexadecimal format.

The direct coding technique is cost efficient in terms of the hardware, but it is slow and tedious from the programmer's viewpoint. Short programs may be designed and entered into the system by this method, but for longer programs much more efficient alternatives are available. This method is useful to change a few instructions in the memory, as the user is not required to go back to the editor to enter the changes, then reassemble and load the memory.

The trend in software development is towards more powerful aids with increasing sophistication. These should be considered unless the cost is prohibitive for the application.

Assembly level language uses a mnemonic or symbolic representation of the binary code. In terms of efficiency for the user program, it is efficient but it requires the manipulating of the registers along with a solid understanding of the hardware structure of the system.

The assembler converts the symbolic programs which are coded by the user into an executable binary format. Many assemblers will also detect the gross syntax errors and flag them for the user. A major disadvantage for large programs is the tediousness of the programming required in assembly language and the programmer time involved.

Assemblers may produce either absolute or relocatable code. Absolute code must be loaded into a fixed location in memory, but relocatable code can be loaded at any point. Relocatable code is an advantage since several program modules may then be developed independently and linked together by a linkage-editor program into an executable module.

High level languages such as FORTRAN, PL/M, or PASCAL allow the programmer to use more powerful instructions to specify the algorithm. The high level programming language is closer to the algebraic conventions used in writing algorithms. Using it allows coding of the algorithms in a shorter time.

Programming in a high level language may be considerably faster than programming in assembly level language. This is true especially for long programs where getting the program to run initially can be time consuming. A disadvantage is that the use of most compilers which will produce the object code will be inefficient compared to assembly language coding. As it compiles a high level instruction into the machine level binary instructions, it will not optimize the use of registers, thus causing many unnecessary register transfers.

A compiler might generate two to five times more instructions than would be generated at the assembly level of programming. This results in additional memory requirements and in execution speeds two to five times slower. This may not be objectionable in some applications. If the program required is very long and complex, it may not ever be implemented correctly if it is coded in assembly level language. By coding this type of program in a high level language, a working program may be written and debugged much faster. Later, an assembly language version can be produced. High level languages can allow a faster implementation of working programs by more users with less programming experience.

Some interpreters can analyze the source code as a program is being executed. An interactive interpreter allows the user to execute instructions one at a time and verify immediately if the syntax is correct in the instruction. This type of interpreter translates each instruction into binary code at once, thus each part of a program can be executed as it is keyed in. This feature provides a valuable diagnostic capability and can result in faster system debugging.

Many of the compilers available let the user first develop a program in high level language and then substitute assembly language modules within the high level program. When the system computing performance is insufficient or the memory space too large, the user can then recode the suspect sections of the high level modules using assembly language to optimize them. These new modules are then substituted into the program.

Thus, the programmer codes in a high level language to get the program up and running correctly. A software package is then available to prove that the system will indeed perform the required system tasks. A recoding of selected modules then takes place in assembly language. The efficiency of the program is improved and the memory usage is reduced. This technique produces a product early. The program then can be optimized before entering volume production.

To decide if a high level or an assembly level language should be used, the programmer must evaluate the system complexity and the resources available based on the number of units to be produced. Software cost is the dominant factor for low volume quantities. Software costs can be reduced in many cases by an order of magnitude using a high level language. This will result in additional memory, but the increased cost may still be small compared to the savings in programming costs. In large quantity applications, this additional memory may be much more significant while the programming costs are distributed over a large number of units. The user should thus calculate the programming cost and divide it by the number of units to obtain the software cost per unit. Then compare this to the hardware savings per unit from using assembly language as shown in Figure 2-5.

SOFTWARE DESIGN METHODS

The increase in software design methods is partly due to evolution but is mostly due to the increasing complexity of the problems involved. Use of these design methods requires the understanding to fit the proper technique to the problem. The designer is forced to think intuitively and procedurally at the same time during the design effort. As the effort progresses, the emphasis tends to shift. In the initial stages, the ideas which formulate the design are conceived. The designer may have intuitive feelings for a solution but suspects that there may be a better answer. So the ideas are examined and conclusions are drawn. The process has been characterized as divergence, transformation, and convergency.

Modular Design

When the bigger problem is divided into smaller problems, the programmer is using *modular design.* In a modular hardware design, the modules tend to isolate and separate the functions, thus the designing, troubleshooting, and debugging tasks are easier.

Several advantages accrue in using program modules. The programming efforts can be distributed among several individuals. The modules can be run and tested separately before they are linked into the main program. The modules can be designed for open end locations to allow changes without affecting the other parts of the program.

Modular programming is based on techniques in which the programs modules are written, tested, and debugged in smaller units that are then combined. Top down design may use modular programming, but modular programming is more common and has been in use longer. It is often used independently of any other techniques.

The modules are often divided along functional lines. In microprocessor systems, this division is most useful, since the modules can be used to form a library of programs for use in other designs.

By partitioning the program into small modules and interconnecting these modules into larger programs, larger complex programs can be created easily.

The modular programming technique limits the size of programs that must be debugged and tested at any one time. It also provides basic programs which may be reused and allows a division of functional tasks.

The disadvantages include the additional program interfacing that is required and the extra memory needed to transfer control to and from the modules.

The need for separate testing of modules can also be more time consuming without duplicate facilities. Modular programming can be difficult to apply if the structure of the data is critical and cannot be modified.

Modular programming can be used for developing microprocessor software in conjunction with other design techniques, some of which are:

1. A program that signals an A/D converter to begin conversion, waits for the conversion to be completed, and then places the results in memory.
2. A program to read a keyboard for key closures.
3. A program that divides two signed decimal numbers.

The modular program design technique coupled with the proper documentation discipline tools is a simple but effective microprocessor design technique.

The designer usually uses a combination of techniques. Basic techniques like top down design, structured programming, and modular programming are not mutually exclusive. The idea is to produce a working program, not to follow the restrictions of any particular design method.

Top Down Design

Top down design is based on the philosophy that the problem can be divided into smaller tasks which can become modules. In top down design the testing and integration can occur along the way rather than at the end. Therefore, problems may be discovered early during the development stage. Testing can be performed within the actual system environment instead of using driver programs. The top down design technique tends to combine the design, coding, debugging, and testing stages.

A disadvantage is that top down design sometimes requires the hardware to perform tasks that it does not do well.

The technique can be difficult to apply when the same task occurs in different places. The routine that performs the task must then interface properly at each place. The proper interface stub is not always easily written, especially if the program does not have the simple tree structures that tend to mesh easily with the top down approach.

The sharing of data by several routines may also present some problems. Also, errors at the top level may have major effects on the entire project if not discovered early.

Top down design has improved program development considerably in some cases. However, it should not be used if it interferes with the development of workable programs which allow the efficient use of a particular microprocessor hardware system.

An example of top down design for A/D conversion follows:

1. The flow chart is written as shown in Figure 2-8. The program initially calls the A/D input routine which is a program stub, and then calls the other routines or program stubs if the input datum is not 0, and returns to reading the A/D input.
2. The routine which reads the A/D input is expanded. The input from the converter consists of three BCD digits that the CPU fetches one at a time. The expansion results in the following tasks:
   (a) Send a START CONVERSION signal to the A/D converter.
   (b) Check the CONVERSION COMPLETE line. Wait if the conversion is not complete.
   (c) Fetch a digit.
   (d) Check the digit for 0.
   (e) Repeat steps (c) and (d) three times.
   (f) If all the digits are 0, repeat starting with step (a).
   (g) Check if the converter has reached the final value by waiting and then repeating steps (a) through (f).

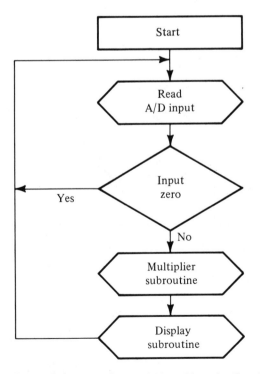

FIGURE 2-8   A/D General Flow Chart for Top Down Design.

     (h) If the inputs are not equal, repeat step (g) until equal within the converter accuracy requirement.

     (i) Save the final input value.

A flow chart of the expanded routine except for steps (g) through (i) is shown in Figure 2-9. The procedure continues expanding each block until the detail is sufficient for coding.

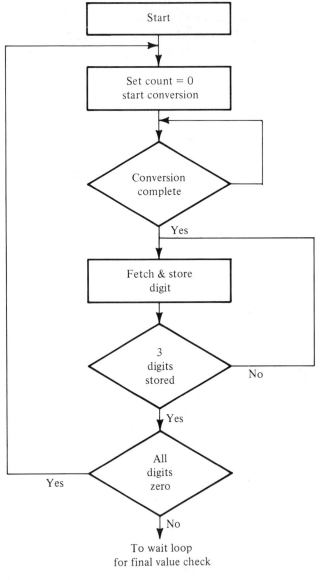

FIGURE 2-9   Expanded Flow Chart for A/D Read Subroutine.

Structured Design

Structured design uses measures, analysis techniques, and guidelines to control the flow of data through the system, which in turn is used to formulate the program design. The data flow is traced by charting each data transformation, transforming process, and order of occurrence. A system specification is then used to produce a data flow diagram and the diagram is used to develop a structure chart. The structure chart is then used to develop the data structure and these results are used to reinterpret the system specification. The process is iterative, but the order of iteration is not rigid.

Structured programming tends to use only simple logic structures. Some of which are discussed here.

In sequential structures, the instructions or routines are executed in the order written.

Conditional structures of the IF-THEN-ELSE type operate as follows: IF A THEN R_1 ELSE R_2, where A is a logical expression and R_1 and R_2 are routines consisting of the permitted structures. If A is true, the processor executes R_1; if A is false, the computer executes program R_2. R_2 is not necessary if the computer is to do nothing if A is false.

An example of the structure for THEN AND ELSE follows:

$$\text{IF } X \neq 0, \text{ THEN } Y = 1/X, \text{ ELSE } Y = 0$$

This structure ensures that the processor will never try to divide by zero and it defines Y in the case where X is zero.

An example of the structure for THEN only follows:

$$\text{IF CENTS 50 THEN DOLLARS} = \text{DOLLARS} + 1$$

This structure rounds DOLLARS to the nearest dollar. No action occurs if CENTS 50.

Loop structures of the DO-WHILE TYPE work as follows: DO WHILE X, where X is a logical expression and R is a routine with the permitted structures. The processor checks A and executes R if A is true and then returns to check A again. The processor executes R for as long as X is true. An example of the structure for DO-WHILE is:

```
INDEX = 1
DO WHILE INDEX MAX
    BLKA (INDEX) = BLKB (INDEX)
    INDEX = INDEX + 1
    END
```

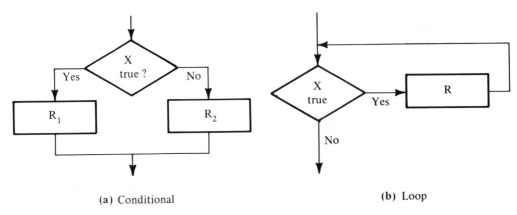

(a) Conditional

(b) Loop

FIGURE 2-10   Logic Structures Used in Structured Programming.

This structure moves the number of elements specified by MAX from memory locations in one array (BLKA) to the memory locations in another array (BLKB).

A structured program for the application is written using only this set or some other set of structures. Flow charts of the conditional and loop structures are shown in Figure 2-10. Each structure has a single entry point and a single exit.

When an error occurs during the execution of R_2 in Figure 2-10, the user can trace how the processor reached this point. If the program was written as shown in Figure 2-11, the error could be in any one of the sequences leading up to R_2. Also, a correction here can affect the other sequences.

The use of structured programming in a simple editor program allows the user to space or backspace along a line on a CRT, delete or replace characters,

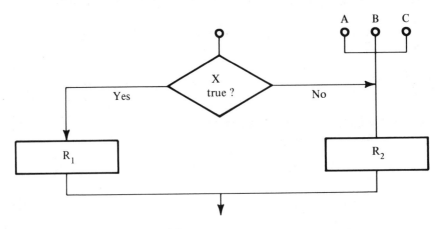

FIGURE 2-11   An Unstructured Program.

and end the edit by pressing the carriage return key. A structured program for this task might take the following form:

```
TASK - SET POINTER TO FIRST CHARACTER IN LINE AND
    READ FIRST KEYBOARD INPUT
STRUCTURE - CHARACTER POINTER = 1
    READ INPUT
EXAMINE CHARACTERS UNTIL CARRIAGE RETURN FOUND
    DO WHILE INPUT ≠ CARRIAGE RETURN
MOVE POINTER IF INPUT IS SPACE AND NOT ALREADY AT
    END OF LINE
IF INPUT = SPACE THEN →
    IF CHARACTER POINTER = 80 THEN
    CHARACTER POINTER = CHARACTER POINTER + 1
MOVE POINTER IF INPUT IS BACKSPACE AND NOT ALREADY
    AT START OF LINE
    ELSE IF INPUT = BACKSPACE THEN
    IF CHARACTER POINTER ≠ 1 THEN
    CHARACTER POINTER = CHARACTER POINTER + 1
DELETE CHARACTER BY REPLACING WITH SPACE
    ELSE IF INPUT = DELETE TURN
    CHARACTER (CHARACTER POINTER) = SPACE
    REPLACE WITH INPUT CHARACTER
    ELSE
    CHARACTER (CHARACTER POINTER) = INPUT
    IF CHARACTER POINTER ≠ 80 THEN
    CHARACTER POINTER = CHARACTER POINTER + 1
READ NEXT INPUT
    READ INPUT
    END
```

This structured editor has no unconditional jump or GO TO statements. The main loop is based on a DO-WHILE structure which examines the characters until a carriage return is found. The loop is not executed if the first character is a carriage return. The loop uses nested IF-THEN-ELSE statements. The indentation indicates the nesting levels. If some large programming projects improvements of 50 to 100% have been reported in programmer productivity, but how applicable are these techniques to microprocessor programming?

Structured programming has primarily been used in large programming projects involving teams of programmers with tens of thousands of instructions. The structured programming techniques are most easily applied to programs written in high level languages. In the past most microprocessor programs have been written in assembly or even machine language, but microprocessor programs are becoming longer and techniques will be used to make debugging and testing simpler.

Most microprocessor programs are written as application programs to be used as an integral part of the system. They are to be tested, documented, maintained, and extended just as the hardware in the system hardware. Structured programming can help in defining the system software.

The structured process seems simple, but difficulties can be encountered. Consistently identifying the transformations of data is one problem area. It is easy to be too detailed in some parts of the data flow and much less detailed in other parts.

Identifying the incoming and outgoing flow boundaries is critical to the definition of the modules and the relationships between them. If the boundaries of the modules are moved, this leads to different system structures.

The structured design method has been an aid in the rapid definition and refinement of data flows. This has been true in military command and control applications which use some structured design techniques.

The method can also reveal previously unknown properties of some systems, such as the generation of information already furnished elsewhere in the system.

Languages that contain the actual structures of structured programming include the high level languages based on PL/I such as Intel's PL/M and Motorola's MPL. Structured programming could be used in the design stage and then the structured program translated to assembly language.

Structured programs are usually slower and require more memory than unstructured programs. However, in large programs execution time and memory usage are not as critical as the time required for program development.

The use of structured programming can result in substantial savings in program development time. The method is best suited to designs where a well defined data flow can be derived from the specifications. Some characteristics that allow the data flow to be well defined are: (1) the inputs and outputs are easily distinguished from each other, and (2) transformations of data occur in incremental steps so the character of the data is not greatly modified. Many microprocessor systems fit into this category.

The same structured design may be used when the program is rewritten in a different language or even for a different processor. Tracing errors is a problem with unstructured programs. The programmer must first find all the statements that could cause a branch to the statement being questioned either by examining the entire program or by using a cross-reference table. Then which sequence of statements caused the error must be determined. Any corrections must not affect other sequences that use the revised statements.

Structured programs can make the debugging, testing, and maintenance stages of software development much simpler. Two major difficulties with FOR-TRAN are the GO TO and IF statements, which can complicate the program structure since the programmer may not know how the program reached a particular point. Many users of structured programming do not use GO TO statements, which cause unconditional transfer of program control. The simpler flow of control produces clearer, more reliable, and more easily traced programs. Use of structured programming in various applications has been rewarding, especially in large programming projects.

Software design methods are collections of techniques based upon a common concept. Some other well known design methods include:

1. The Jackson Method.
2. The Logical Construction of Programs (LCP).
3. The Meta Stepwise Refinement (MSR).
4. Higher Order Software (HOS).

Each of these methods has a set of techniques which are intended to make a software development more successful. The concept of either the Jackson Method or the Logical Construction of Programs (also known as the Warnier Methodology) is that the identification of the data structure is critical and that this structure may be used to derive the structure as well as certain details of the program.

The Jackson Method

In the Jackson Method, which was popularized in England, the program is viewed as the means by which the input data are transformed into output data as shown in Figure 2-12. Paralleling the structure of the input data and output report will

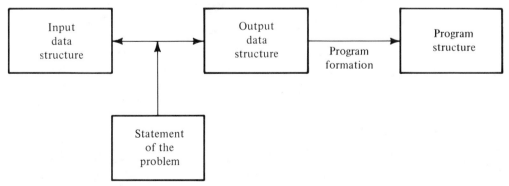

FIGURE 2-12   The Jackson Methodology of Software Design.

ensure a well designed system. Some other assumptions are that the resulting data structure will be compatible with a rational program structure, that only serial files will be used, and that the user of the method knows how to structure the data.

Some features of the Jackson Method are:

1. It does not depend on either the designer's experience or creativity.
2. Principles are used which allow each design step to be verified.
3. It is not difficult to learn and use correctly, so any two designers working independently should arrive at nearly the same program design.

The process appears to be simple; however, some required structures for practical implementation are lacking. For example, error processing must be fitted in, since erroneous data do not exist in the structure.

File accessing and manipulation are restricted. Data structuring is usually dictated by the data base management system used. In this case, the basic method may not be valid.

The Logical Construction of Programs Method

The Logical Construction of Programs (LCP) Method is shown in Figure 2-13. As with the Jackson Method, LCP assumes the data structure is the key to software design. However, the LCP Method is more procedure oriented. This approach involves the following steps:

1. Identify and organize the input data in a hierarchical manner.
2. Define and note the number of times each element of the input file occurs and use variable names to note the ratio of occurrences, such as N records.
3. Repeat steps 1 and 2 for the desired output.
4. Obtain the program details by identifying the types of instructions required for the design using a specific order: (a) read instructions, (b) branches, (c) calculations, (d) outputs, and (e) subroutine calls.

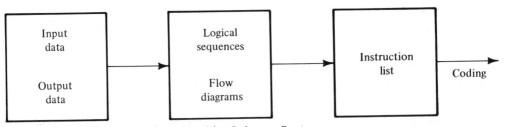

FIGURE 2-13   The LCP Method for Software Design.

5. Using flow chart techniques graph the logical sequence of instructions using Begin Process, End Process, Branching, and Nesting labels.
6. Number the elements of the logical sequence and expand each one using step 4. Additional guidelines exist for data structure conflicts.

Many of the difficulties in the use of this method are similar to those of the Jackson Method. The LCP Method forces the programmer to contrive a hierarchical data structure which was previously apparent. Run environment or file access methods are not considered.

The hierarchical data structure can result in a pseudo-code statement of the program early in its development, but the program may not be near optimum. The method is actually well suited to problems involving a few modules where the data are tree structured. It is susceptible to the same type of problems as the Jackson Method.

Meta Stepwise Refinement Method

Meta Stepwise Refinement (MSR) uses the philosophy that the more times one does something, the better the final result is. The designer assumes a simple solution to the problem and then gradually builds in detail until a final solution is derived. Several refinements at the same level of detail are used by the designer each time the additional detail is required as shown in Figure 2-14. The best of these refinements is used for building more detailed versions.

The specific features of the MSR Method include:

1. Use of an exact, fixed problem definition.
2. Independence of language in the early stages.
3. Design is developed in levels.
4. Details are postponed to lower levels.
5. Design is successively refined in a stepwise manner.

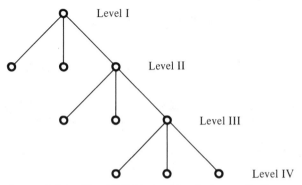

FIGURE 2-14   The MSR Method for Software Design.

MSR is a combination and refinement of top-down design, the stepwise refinement concept, and level structuring. It results in a tree structured program using level structuring concepts. Using the proper program organization, the designer separates functionally independent levels or layers into programs as shown in Figure 2-14. The higher levels remain close to the problem statement while the lower ones contain the implementation detail.

A problem appears in applications that require evaluation and modification. Since the solution at any one level depends on prior higher levels, any change in the problem affects prior levels. If changes are delayed until the design is completed, then the solution and the requirements do not agree.

The handling of multiple solutions are also a problem; selecting the best solution may be difficult.

Due to the number of times the problem will be solved, this approach works best on smaller problems. It is useful where the problem specification is fixed and an elegant solution is required, such as developing an executive for an operating system.

Higher Order Software Technique

Higher order software (HOS) was developed on NASA projects as a means of defining reliable, large scale, multiprogrammed multiprocessor systems. Its basic features include:

1. A set of formal laws.
2. A specification language.
3. An automated analysis of the system interfaces using an analyzer.
4. Layers of system architecture which are produced from the analyzer output.
5. Transparent hardware.

This method is based on rules which define a hierarchy of software control, in which the control is the specified effect of one software object on another, as follows:

1. A module controls the operation of functions on only its immediate, lower level.
2. A module is only responsible for elements of its own output space.
3. A module controls the access rights to a set of variables which define the output space for each immediate, lower level function.
4. A module can reject invalid elements of only its own, input set.
5. A module controls the ordering of the tree for only the immediate, lower levels.

The technique has been used with an automated analyzer program which checks the solution as expressed in a metal language. The analyzer is not required to implement the method since it can also use pseudo-code. HOS is useful when the accuracy and the auditing of algorithms are concerns, such as in large scientific and detailed financial computations. HOS attempts to ensure a reliability consistency by the use of interface definitions and attention to detail.

No single design method is best for every design. A design problem may be well suited to a particular technique but it also might be unique. Various software design methods can only assist in solving some aspects of a given problem. A particular method can reveal design problems and allow one more time to address them.

Planning, scheduling, and control must also be integrated into the design methodology. A balance between methods and environments is also critical. Merging these concerns may be the next evolutionary stage in software design.

WRITING PROGRAMS FOR MICROPROCESSORS

Programming costs are among the most important factors in software development for microprocessors. A primary objective is to develop a program that works in a reliable and cost effective manner. Improvements can always be made later.

Since the major cost in most projects is program development time, methods which minimize that time are desirable.

Now that we have larger low cost memories and low cost processors, the memory and hardware constraints are not as critical for microprocessor software development as they were in the development of software for earlier generations of computers.

A major emphasis should be on writing and documenting a reliable program in a reasonable time. We'll discuss techniques for writing shorter and faster executing programs for specific microprocessors.

Criteria for evaluating programs are necessary. These criteria will help determine the aims, methods, and relative importance of the various stages of software development.

One important criterion to judge a program is whether it works reliably. The structure, the efficient use of time and memory, the design time, and documentation are useless if the program does not run.

Design and documentation are also important factors in determining whether a program is easy to use. Because human factors are important when the program requires human interactions, many microprocessor systems are designed to simplify tasks for human operators.

A program that tolerates errors will be easier to maintain than one that does not. Thus, the program may be designed to react to errors whose occurrence cannot be foreseen.

Error tolerance may be critical in applications where human operators will change the data or equipment. The program can make the operator or system element aware of erroneous inputs or malfunctions without shutting down the system.

A program that can be extended to tasks other than the one for which it is designed is superior to one that can be used for only one task. Design and documentation are particularly important in meeting this objective. Modular programming can be helpful, although structured programming and other design methods can also be useful for extending the program's capabilities later.

Programming Style

Some specific guidelines, such as ones concerning programming style, will be useful regardless of the particular language or processor involved.

For example, keep the modules short by dividing long modules into sections. A shorter module is easier to debug and correct and it is more likely to be used again since its function is more likely to recur. One or two pages of code with 50 to 80 lines is a maximum size for a module.

Modules may also be implemented as macros or copied into the program to avoid a large number of calls and returns. In many microprocessor systems the calls not only require extra time and memory but they can also overflow the limited stack use for storing the return addresses.

Try to make the modules general. Modules that are too specific, such as one that sorts only a limited number of elements or searches for the particular characteristic, may find little repeated use. Greater generality can often be achieved with little extra code.

Program modules should be tested to see that they work under conditions that reflect the actual operating conditions. The selection and execution of a test plan are not always simple tasks but the problem definition and program design stages can be used to produce a test plan for the program so that it can be easily and thoroughly tested.

Speed can sometimes determine if the program works at all, since critical timing requirements may exist. If the speed of the system depends on external factors, such as operator response times, the input or output data rates of sensors, or displays and converters, the program speed may not be as important.

Each chip needed for the program memory can add to the system cost. The additional memory requires additional interconnections, board space, decoding circuitry, and increased power.

As larger semiconductor memories have become available, the importance of memory size has decreased. But memory size must still be considered, particularly in the small applications in which the cost of a single memory chip is significant.

A program that is easy to work with is more valuable than one that is relatively hard to use. Complicated data formats and unclear error messages can make a program difficult and expensive to debug, use, and maintain. General routines which may require large amounts of extra code, such as a code conversion routine for both ASCII and EBCDIC characters, should be avoided. Some simple ways to achieve generality include the use of names instead of specific addresses or data, the collection of definitions at the start of a routine, and the use of names which suggest the purpose or identities of the item. The consistent use of these techniques will tend to produce programs that can be used frequently and be easily modified.

The Use of Documentation Techniques

Documentation is a part of development that is often underestimated. It is very useful in the debugging and testing stages, and essential for maintenance and redesign. The properly documented program can be used again when necessary. The undocumented program can require so much extra work to use that one might just as well start over from the beginning.

The techniques most used for documentation are flow charts, program listings with comments, memory maps, and parameter and definition lists. Structured programming and some of the other design techniques have developed their own documentation forms.

Flow charts act as a visual aid for program documentation. A general flow chart may serve as a pictorial description of a program while a more detailed flow chart can be invaluable to the user who must use or maintain the program.

Comments are an important part of program documentation. A program with a clear structure and well chosen names can be almost self-documenting.

The comments should explain the purpose of the instructions; they should not merely repeat the meaning of the instructions, for then they add nothing to the documentation.

The following general rules should be used for comments:

1. The comments should explain the purpose of instructions or instruction sequences, not define the operational codes.
2. The comments should be clear and brief; shorthand and obscure abbreviations should be avoided. Complete sentences are not necessary.
3. Comments should be limited to the important points of program flow. Too many comments can make the program difficult to follow. Standard sequences are not necessary to explain unless they are used in some unusual manner.
4. The comments should be placed close to the statements to which they apply.

5. The comments should be up-to-date. Comments referring to previous versions of a program should be deleted.

To use comments properly, ask what explanation is necessary to understand the program. When such comments are provided systematically, they are helpful in all stages of the software development.

Memory maps list the memory assignments made for the program. These maps prevent different routines from interfering with each other and help in determining the amount of memory needed as well as locating subroutines and tables.

Memory maps are particularly important in microprocessor systems because of the use of separate program and data memory (ROM and RAM). The addresses are assigned as part of the hardware design and conserving memory, particularly for RAM, may be necessary. Memory mapping is an aid in knowing the precise locations of parameters that may have to be changed. Mapping is also used for register allocation and I/O assignments. These are all important reasons for proper memory mapping.

Parameter and definition lists explain the function of each parameter and its meaning. The parameters can also be explained in the program.

Program forms describe the subroutines. The programmer should provide the purpose of the program, the form of the input and output data, the requirements for memory of the program, and a description of the parameters.

Proper software documentation should combine all or most of the methods discussed. The total documentation for a program can include:

1. General flow charts.
2. Detailed programmer's flow charts.
3. A description of the test plan.
4. A written description of the program.
5. A listing for each program module.
6. A list of the parameters and definitions.
7. Memory and I/O maps.

Documentation is best developed during the design, coding, debugging, and testing stages of software development. Good design and coding techniques make the program easier to document, and good documentation, in turn, simplifies any maintenance and any redesign required.

Program redesign may involve adding new features or meeting new requirements. The redesign should follow the same paths as the previous design stages of software development.

The redesign process may involve making a program meet critical time or memory requirements. When increases of 25% or less in speed or reductions of the same order in memory are desired, the program can often be reorganized, but the program structure may be sacrificed. Such a task can require large amounts of time. The proper use of names can do much to reduce the confusion between addresses and data. Meaningful names and labels can be helpful in documentation and maintenance as well as in debugging and integrating programs.

A block search routine with and without meaningful names and labels is shown next:

Block Search Routine without Meaningful Names			Block Search Routine with Meaningful Names		
	LXI	H, X		LXI	H, BLOCK
Z:	MOV	A, M	NEWMX:	MOV	A, M
W:	INX	H	NEXTE:	INX	H
	DCR	B		DCR	B
	JZ	Y		JZ	DONE
	CMP	M		CMP	M
	JC	Z		JC	NEWMX
	JMP	W		JMP	NEXTE
Y:	STA	V	DONE:	STA	MAX
	HLT			HLT	

The names should be simple and straightforward, for example, *MAX* for a maximum value, *START* for the beginning of a program. Use of well thought-out names can save time in all stages of software development.

Definitions should be grouped at the start of the program, where they can be easily located, checked, and changed. It is also useful to describe each definition with a comment.

General guidelines for programming microprocessors follow:

1. Use names or labels instead of specific memory addresses, constants, or numerical factors.
2. Use names that suggest the actual purpose or meaning of the particular address or data.
3. Do repetitive operations outside loops.
4. Use short forms of addressing when possible. The use of these short forms may require that the data be organized.
5. Try to reduce the number of jump statements. They can require too much time and memory.
6. Take advantage of the addresses that are 8 bit quantities. This includes the addresses such as the even multiples of 100 in hexadecimal.

7. Use stack addressing instead of direct addressing to move data between the memory and registers.

Controlling the Execution Time and Memory Size

Many of these techniques minimize the execution time and memory size since longer programs require more memory accesses and more execution time. Subroutines can allow memory savings at the cost of the execution time required for the call and return. Loops require a similar trade-off. If minimum execution time is required, subroutines and loops can be replaced by repeated copies of the same instructions. A minimum memory system requires the opposite approach. The gains from program optimization using these techniques can result in a speed increase or memory reduction of about 25%. If larger gains are required, the following methods may be used:

1. The development of new algorithms can provide larger increases in speed or decreases in memory use.
2. The use of microprogramming with a microprogrammable processor might enable execution of the algorithms at significantly higher speeds.
3. The clock speed provided might be increased so that the processor and memory can be adapted to the higher speeds.
4. Additional hardware might be used, such as multipliers which can increase the throughput by removing some of the processing burden from the microprocessor.
5. Parallel or multiple processing by two or more processors may be able to perform the tasks at higher speeds without greatly affecting the system cost.
6. Distributed processing with two or more processors dividing the tasks may also increase speed.

If the microprocessor is at the upper limits of its performance range, the techniques described may be more helpful than attempts to obtain large increases in performance by program optimization.

General rules for increasing execution speed using programming techniques follow:

1. Find the loops that are executed frequently and reduce the number of instructions in these loops.
2. Use register operations when possible since these operations are faster than others; however, remember that they may require extra initialization.

3. Emphasize simplicity and comprehensibility. The main objective is to write a program that works. Saving a few microseconds or a few memory locations is seldom critical initially since the coding may be optimized later.

4. The initial program should always be obvious rather than clever. Some practices to avoid are performing operations out of order, using multiple word instructions for unrelated items, using leftover results for calculations or to initialize variables, and using parameters as fixed data.

An example of a problem which occurs for unrelated items using the 8080 instructions are:

LXI   D, 280   ; INITIALIZE D TO 4, B TO 8

This statement should be avoided when A and B are not related. A better format is:

MVI   A, 4   ; INITIALIZE A TO 4
MVI   B, 8   ; INITIALIZE B TO 8

Distinctive names and labels can also reduce confusion. The numbers 0, 1, and 2 and the letters O, I, and Z are often confused as in MINI and MIN1. The chance of error in this case is great. Such names which may confuse the programmer should be avoided.

Obscure constructions, such as the use of many offsets from the program counter in microprocessors that use multiple word instructions to build complex conditional assemblies, should be minimized.

Some microprocessors use a subroutine call instruction for restarts (RST). This instruction is the same as CALL ; where i = restart number, O − N. The difference between a restart call and its equivalent subroutine call is that the subroutine call can use 3 bytes of memory while the restart call (RST) may use only 1 byte.

An efficient way to use restarts involves converting the most frequently used subroutine calls to restart calls. Each conversion could save 2 memory bytes and six clock cycles.

To reduce memory usage, 1 byte call may also be used to produce a large number of 2 byte calls. This is 1 byte longer than a restart but 1 byte shorter than the standard call. This is done as follows: create a table with the subroutine address.

```
TAB      DW      ROUT1
         DW      ROUT2
         •       •
         •       •
         DW      ROUTn
```

To call any of these routines, use:

```
RST      x       ;restart call
DB       n*2;n = routine number
```

TAB is positioned in the memory such that addition carry corrections are not required and the restart routine transfers control to routine n in the table.

Arithmetic computations may be coded in many ways. Although the following example applies to the 8080, the same techniques may be applied to other microprocessors.

Consider the task of clearing a single register or a number of selected registers. One method for doing this uses the following three sequences:

```
SEQUENCE 1    MVI A,O ; load accumulator with zero, A = O
SEQUENCE 2    LXI II, L ; load register pair with zero, II, L = O
              MVI E,O ; load register with zero, E = O
SEQUENCE 3    LOOP •
                     •
              DCR B    ; decrement register
              JNZ LOOP
                     •
                     •
              MVI B,O ; load register B with zero, B = O
```

These sequences all have inefficiencies.

SEQUENCE 1   The accumulator is loaded with zero, but a faster method is to compute zero.

SEQUENCE 2   The register is loaded with zero, a better method is to move zero into the register, since H and L are already zero.

SEQUENCE 3   The loop is not required since the register already contains zero.

One method of computing zero is $A = A - A$, but an even faster way is $A = A$ XOR A. Many other arithmetic computations can be improved using the same type of critical analysis.

Some microprocessors, such as the 6800, have a software interrupt instruction (SWI) that can be used to decrease memory requirements. This instruction uses the subroutine address which is stored in memory locations FFFC and FFFD. It is quite similar to restart except that its address is not fixed since it comes from the memory.

SWI can be used to simulate hardware interrupts to allow debugging without special hardware.

If a table for subroutine addresses is stored in memory which can be modified, like a PROM, then the PROM can be used to correct errors. Once an error is detected, the table in the PROM is modified so that control for the detected subroutine goes to the PROM. Then the corrected subroutine is placed in the PROM and that subroutine then returns control to the main program.

We have tried to emphasize defensive programming, in which changes can be made easily and in which misinterpretations and other errors are minimized.

Defensive programming can take time and the programmer can never anticipate all the problems that might occur. But careful programming can result in fewer errors as well as programs that are easier to use and maintain. All the suggestions presented here do not always have to be followed, but the use of these techniques can make software development for microprocessors much easier.

TEST CONSIDERATIONS

A microcomputer system is most easily tested using the system in which it was developed—either a development system or a larger computer. The software testing should identify and correct many of the program bugs before the software is introduced into the hardware environment. If the software is to be placed on a ROM, then it must be thoroughly tested before the ROMs are programmed by the factory.

The program is then loaded into the system and a system checkout of both hardware and software is made. The timing and synchronization between the processor and the other elements of the system tend to be critical factors.

The peripherals are checked next. These may be sensors, storage facilities, or actuators that provide the data or transmit the control response. Timing and synchronization between fast and slow peripherals and the processor are key factors to be resolved in this phase.

Diagnostic software can be provided on ROMs which are then plugged into the system in place of the ROM containing the program. This software can check the operation of the program and provide an error message when a failure is detected. A monitor program that can place TRAP instructions at specified addresses can also be used for diagnostics.

Checklists are a useful tool that can be used with flow charts. With them, the programmer can check that each variable has been initialized, each flow chart element has been coded, the definitions are correct, and all paths are connected properly.

A good checklist can save time, but long or complicated programs should not be hand checked since the programmer may be more likely to make additional mistakes when checking the program.

Some loops and sections of programs can be hand checked to see that the flow of control is correct. In the case of loops, the programmer can always check to see if the loop performs the first and last iterations correctly. These are the sources of many loop errors. The program can also be hand checked for the trivial cases, such as tables with no elements.

Program checking and debugging should always be done using a systematic method. Do not assume that the first error found is the only one in the program.

Error Sources

Some common sources to check for errors are:

1. Failure to initialize all variables, particularly in counters and pointers. Registers, flags, and memory locations should not be assumed to contain all zeros at the start of the program.
2. Incrementing of counters and pointers before it is required or not incrementing them at all.
3. Problems of the program with trivial cases, such as tables with no elements.
4. Inversion of conditions, such as jump on zero instead of not on zero.
5. Reversing the order of the operands, such as move A to B instead of B to A.
6. Jumping on conditions that can change after they are set. An example is using flags for jump conditions when they can be changed by intermediate instructions.
7. Lack of follow-through conditions, such as a data item that is never found in a table or a condition that is never met. This usually causes an endless loop.
8. Failure to save the contents of the accumulator or other register before using the register again.
9. Inversion of addresses and data, such as immediate addresses in which the data are part of the instruction and direct addresses in which the address of the data is part of the instruction.
10. Exchange of registers or memory locations without using intermediate storage, for example:

H = L
L = H

sets both H and L to the previous contents of L, since the first statement destroys the previous contents of H. The use of the following sequence will exchange the registers:

A = H
H = L
L = A

11. Confusing numbers and characters, for example, ASCII zero or EBC-DIC zero is not the number zero.
12. Confusing numerical codes, for example, BCD 61 and binary 61.
13. Counting the length of a data block incorrectly, for example, locations 30 through 38 have nine and not eight words.
14. Ignoring the direction of noncommutative operations, for example, SUB C which subtracts the contents of register C from the contents of the accumulator.
15. Confusing two's complement and sign magnitude notations.
16. Ignoring the overflow in signed arithmetic.
17. Ignoring the effects of subroutines which can change flags, registers, and memory locations.

Other errors may exist, but this list can be used as a guide for where to search.

Test problems occur in microprocessor systems because of the inability to observe the register contents directly, the close interactions between hardware and software, the dependence of programs on timing, and the difficulty of obtaining adequate data in real-time applications.

Diagnostic Tools

Some of the tools that can be used to help test and debug programs are:

1. Simulators.
2. Logic analyzers.
3. Breakpoints.
4. Trace routines.
5. Memory dumps.
6. Software interrupts.

The simulator is a program which allows the user to simulate the execution of programs on another computer. It tends to act as a programmer would to trace the effects of instructions.

The simulator usually runs on a larger computer for the smaller microcomputer which may lack all the facilities needed for testing. Most simulators are large programs. These programs allow the programmer to change data, examine registers, and use other debugging facilities.

Many simulators cannot fully model the input/output paths or provide much help with timing problems. Sometimes the simulator is used to test the concept of the real computer before it is built.

Logic analyzers are test instruments which act as a digital bus-oriented version of an oscilloscope (Figure 2-15). The logic analyzer can detect the states of the digital signals during an instruction cycle and store them in its memory. It then displays the information on a CRT. Several events may be monitored and displayed at once and the desired events can be defined and triggered at selected thresholds.

FIGURE 2-15 A logic analyzer such as this can be triggered to store machine states after a desired program event occurs. (*Courtesy Gould Biomation.*)

Logic analyzers provide a convenient display for changing parallel digital signals. Most analyzers have the ability to trigger on a particular instruction or sequence of instructions and recall previous data; some can capture short noise spikes. Logic analyzers can be a complement to software simulators, since the analyzers can be used in solving timing problems.

A breakpoint is a place in the program at which the execution is halted to examine the current contents of registers, memory locations, or I/O ports. Most microcomputer development systems and many simulator programs have facilities for setting breakpoints (Figure 2-16).

Breakpoints are often created with a TRAP instruction or a conditional jump instruction dependent on some external input which is controlled by the programmer. For example, the instruction JUMP ON TEST causes a jump to itself until the TEST input is off. This can allow the contents of registers or memory locations to be examined if stored.

Some microprocessors produce special status information on the buses while halted. This information may include the current contents of the accumulator, program counter, or other registers.

The trace is a program which prints information concerning the status of the

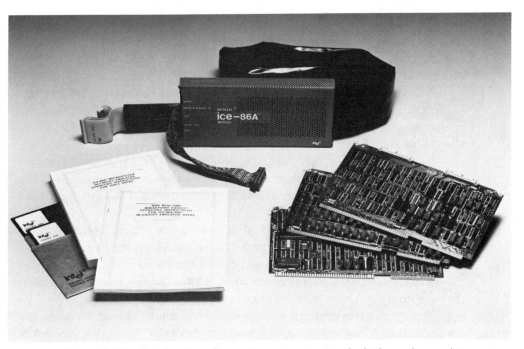

FIGURE 2-16   Breakpoint software is an important tool which can be used in conjunction with an emulator probe to examine the operation of microprocessor boards. (*Courtesy Intel Corporation.*)

processor at selected times. Many simulator programs and development systems have trace facilities.

Sometimes the trace will print the complete status of registers and flags after the execution of each instruction. Other systems have tracing of particular registers or memory locations only when the contents change. Traces can result in large listings unless the programmer selects variables and formats carefully.

A memory dump is a listing of the current contents of a section of memory. Simulator programs, development systems, and monitors may all produce memory dumps.

A complete memory dump can be long and difficult to interpret. The memory dump is not usually an effective technique for debugging, but sometimes it may be the only tool available. A complete memory dump is normally used when all other methods fail.

SOFTWARE INTERRUPT and TRAP instructions can both be used for debugging purposes. The instruction usually saves the current value of the program counter and then branches to a specified memory location. This memory location may be used as the starting point for a debugging program that then lists or displays selected status information. These breakpoints can be inserted with TRAP instructions. When the 6800 executes a SOFTWARE INTERRUPT, it automatically saves the contents of all registers in the stack. The programmer can then observe the contents directly. An 8080 interrupt or RESTART instruction only saves the program counter in the stack. The programmer enters a TRAP instruction into the program and provides the debugging routine.

Along with the software design and the program coding, the program debugging phase will use a major part of the software development costs. Generating the documentation necessary for maintaining and updating the software can also contribute to these costs.

For debugging purposes the listing produced by the assembler or compiler should be easy to follow and should provide line numbers, addresses, and source and object code in the same line. The error messages should be easy to read and understand.

A good statement should be clear and self-documenting. This will reduce coding errors produced by obscure mnemonics and eliminate comments which would be required to explain mnemonics. Self-documentation and readability can also reduce the program maintenance costs. It frees the programmer generating the code from explaining how the program operates.

During program debugging, the syntax errors can be flagged during assembly and the error can be indicated by an error message. A linked list of errors, called a syntax error list, can also be provided so the program can be scanned to locate the errors.

Program and data memory allocation can be shown on a load map produced by a linking loader. The program modules are listed in the sequence in which they are loaded, along with the absolute starting and ending memory addresses with the memory address limits defined within each module as shown next:

PROGRAM NAME	PROGRAM LIMITS		DATA LIMITS	
MAIN	0000	016A	2400	2464
ARITH	016B	02CF	2468	2478

A symbol cross-reference table of all identifiers or entry points used in the program modules may also be provided by the linking loader. The identifiers can be listed in alphabetical order, followed by the name of the module in which they are defined as well as the modules referencing them.

Combined with the assembler cross-reference listings, this table can provide good traceability of the identifiers and their references as shown below:

Identif.	Addr.	Defined	Referenced
ADD	016B	ARITH	MAIN
DIVIDE	01A6	ARITH	MAIN
MULTIPLY	028B	ARITH	MAIN
SUBTRACT	0270	ARITH	MAIN

The value of this traceability becomes obvious as an error is detected within a subroutine. Before the correction is made to the subroutine, the effect of the correction can be traced to each program module that calls the subroutine. If the calling sequence to the subroutine is modified, the program modules affected by this change and the locations within each module can be determined from the symbol cross-reference tables, thus preventing the same error from being debugged twice.

While a program is being debugged under the control of the debugger, the diagnostics are normally being printed by a printer or displayed on a CRT. Some corrections may then be possible. Otherwise, the user must go back to the beginning of the process, type the corrections, and reassemble.

An essential facility of the debugger is to provide breakpoints, which are the addresses specified by the user where the program will stop automatically. The user can then examine the value of variables in the memory or the contents of registers. Program execution is suspended at these points.

Program testing is more than a matter of exercising a program a few times.

The testing of all cases is usually impractical. A routine that uses 16 bits of data to produce a 16-bit result would require 4 billion possible combinations of inputs and outputs. Most formal methods use applications of simple programs.

Most program testing will require a choice of test cases. Also, many micro-computer systems may depend on real-time inputs which can be hard to simulate. Some tools which can be used to help with this task include:

1. Input/output simulations which allow a number of devices to be simulated from a single input and a single output device. These simulations may also provide inputs for external timing and other controls. Most development systems have some facilities or I/O simulation and many software simulators may provide I/O simulation but not in real-time.
2. In-circuit emulators which allow the microprocessor prototype system to be attached to the hardware development system and tested.
3. Simulators which provide read/write memory for programs. The timing characteristics of the ROM can be the same as used in the final design.
4. Real-time operating systems or monitors which control real-time events, provide interrupts, and allow real-time traces and breakpoints.
5. Emulations which execute the instruction set at close to real-time speeds.
6. Special interfaces to allow another computer or programmable controller to test the program by externally controlling the inputs and outputs.
7. Testing programs or exercisers that check branches in the program to find the logical errors. When the program is not executed on the real microprocessor, an emulator or simulator is used. A simulator is a program which runs on another machine like a 370 and executes 8086 code to simulate an 8086. The simulator cannot operate in real-time because of the software interpretation involved. An emulator is a simulator which runs nearly in real-time. Emulation implies that the behavior is identical and not similar to the target.

Testing the Software

Even with the use of some of these tools, the testing of microprocessor software can be a difficult and lengthy task. Some testing may require special equipment to be built.

Some general rules that can aid in program testing are:

1. Make the test plan part of the program design. Testing should be a consideration in the early definition and design stages.
2. Always check for the trivial and special cases. These can include zero inputs with no data, warning and alarm inputs, and other special situations, which can lead to problems.

3. Select the test data on a random basis, which will eliminate any bias. Random number tables are available and many computers have random number generator programs.
4. Plan and document the software testing through all development stages.
5. Use maximum and minimum values for variables as test data. Extreme values are the source of many special errors.
6. Use statistical methods for complex tests. Optimization techniques may be used to set the system parameters.

When the program is debugged and its execution is assumed to be correct, the object code must be placed in the actual system memory, usually ROM or PROM. During debugging, it was residing in RAM.

A PROM programmer is then connected to the development system and the binary contents of the program transferred onto the PROM chips. The PROM chips are then inserted into the system. Stand-alone PROM programmers are also available (Figure 2-17).

For efficient program development most software facilities should be avail-

FIGURE 2-17   A self-contained PROM programmer can be used to transfer programs which are developed on a modular microcomputer with the aid of a floppy disc memory. (*Courtesy Pro-Log Corporation.*)

able for any program involving more than a few hundred instructions. They are a necessity for any program involving several hundred or thousands of instructions.

Using the support programs requires additional hardware facilities. The programs must be readily accessible so they are stored on a convenient medium which is easily accessible to the processor. This implies a general file system.

The file system can be tape cassettes or floppy discs. Tape cassettes are low in cost, but they are slow and it can take several minutes to access information. The access to information on the cassette is sequential and if data are accessed at several points of a tape, winding and rewinding time is always involved.

One alternative is to use floppy or hard discs, which are better storage media, although the costs are higher. The floppy disc allows access to any location of the disk within milliseconds and the hard disc within 50–100 microseconds.

Microprocessor Board Test Problems

Even after the test and debug phases, manufacturing problems may remain. These problems usually fall into three categories:

1. Circuit design.
2. Board assembly.
3. Device specifications.

Circuit design problems are presented by many boards. In the long run, testing cannot improve a poor design. For instance, a circuit may synchronize the computer master clock to a slave subsystem during direct memory access. It could contain several asynchronously driven flip-flops and extensive feedback. Microprocessor designs often incorporate such circuits, thereby increasing the difficulty of testing the boards. Most of these are time dependent; the board relies on the presence of signals which arrive concurrently with other signals. Frequently, signals must occur within a particular interval in the basic clock cycle.

The bus-oriented architecture of microprocessor systems also makes testing and fault isolation strategies a problem. Conventional techniques may not always be successful in diagnosing the system or in isolating faults. The use of LSI chips reduces the number of test points available to the test systems.

To deal with the microprocessor system test problem, the test strategies must be evaluated to select the best techniques. The strategy selected will depend upon a number of considerations and trade-offs. Guidelines for incorporating self-test capabilities should be considered as the foundation of any strategy.

Today's microprocessor circuit boards compare in complexity with the minicomputers of just a few years ago. This rapid increase in board complexity has surpassed the ability of many available test systems to effectively test these boards.

The design interval of microprocessor boards has decreased with the use of LSI devices, but test development time and cost have increased. The tests for microprocessor boards can often be ineffective if the microprocessor board test lags behind the board design.

When a test strategy for microprocessor boards is selected, some limitations of standard logic techniques can be avoided by relying on system self-test methods. The thoroughness of the self-test simplifies the task of the test engineer. Also, the sophistication of modern microprocessors can be applied to provide improved self-test capability.

A typical board may contain a CPU, peripheral controllers, RAMs, and ROMs. These LSI chips are typically tied together by common buses. The input and output circuitry may contain analog elements such as line drivers/receivers and D/A or A/D converters. Evaluating these circuits with existing functional testers may prove to be difficult.

The major troubleshooting problems in a typical system design result from:

1. The replacement of hardware with firmware or software.
2. Jump and similar operations difficult to trace in software programs.
3. Multiplexed operation.
4. Bidirectional bus systems.
5. Limited number of test points.

Troubleshooting the failures in these assemblies poses problems, since one cannot diagnose a failure to a single part with a single function. Microprocessor products require more sophisticated methods for evaluating these buried signals.

No existing troubleshooting methodology is ideal, but some strategies combined with evolving techniques may provide solutions.

Troubleshooting

The logic analyzer can be used to evaluate the unit under test by monitoring eight or more bit streams at once (Figure 2-18), allowing the examining of timing and states. Status information on the display identifies what a microprocessor was doing before, during, and after a data acquisition.

Logic analysis is useful with prototype boards. Using the analyzer requires a thorough understanding of the unit being developed; more test points are generally available on the prototype boards.

Board and part swapping is a basic isolation method, which offers a quick solution under certain conditions. Even semi-skilled personnel can use it. A disadvantage is that often good as well as bad parts get replaced, tying up inventory.

Signature analysis uses a test mode or pattern to exercise the system. Four

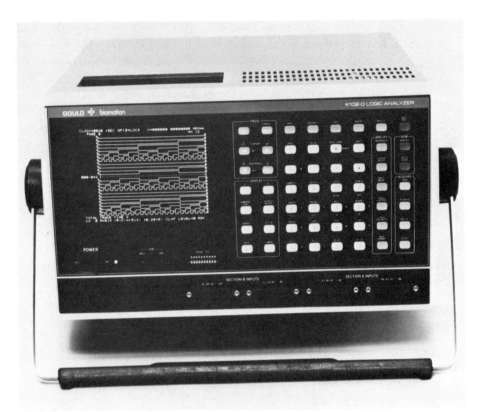

FIGURE 2-18   This logic analyzer allows the examination of a number of bit streams to compare logic states and timing. (*Courtesy Gould Biomation.*)

hex digits can be used to represent a pattern for 16 bits. A test sequence will use these signatures within a certain window of time at a desired clock rate to detect timing-related faults.

Several basic methods can be used for isolating faults using digital signatures. The signal can be traced from the point of application to the output. As the test point moves, the signature will change at a particular point. The faulty component is then connected to the point at which the signature change appears.

The half-splitting technique starts at a point where a fault is equally likely to exist ahead of or behind this point in the circuit path. The faulty half is found and then split in half again until the fault is isolated. The tracing from input to output can be time-consuming and costly. Half-splitting is faster and can be used for automatic board testers. Several commercial automatic board testers use signature analysis with add-on modules or as a standard operation.

In systems where a signature can be unstable and hard to trace, time domain measurements can be used. Here frequency measurements on clocks and timed circuits are taken and compared over a period of time.

The in-circuit emulation technique replaces the microprocessor in the board under test with an emulator probe which is inserted into the microprocessor socket (Figure 2-19). The probe is connected to the emulator, which simulates the microprocessor.

The emulator controls the board, while test circuitry monitors the microprocessor signals. The test program resides in the tester memory rather than in the microprocessor memory. The use of the socket as a test connector allows the emulator access to the other modules in the system. This allows tests of the modules in succession until the defective module is found. The operator does not have to move test probes and test lines from point to point and the elimination of connector interfaces reduces test costs. But, many users question the reliability of the socketed microprocessor, especially those in military hardware where shock and vibration exist.

In-circuit emulation can identify a bad module, but it is not as useful for finding the problem within the module.

Self-test techniques use the microprocessor in the system to test the system functions. Self-test uses time-outs, start-up routines, or other programs stored in a self-test ROM.

Many simple products can easily be tested upon each reset or power-up. The self-test program can usually detect major faults in the system or its boards. The use of the system bus solves the problem of insufficient test points. The bus can also be used to connect other devices which can generate test stimuli as well as provide the verification for proper operation.

A number of trade-off considerations can be made to deal with microproc-

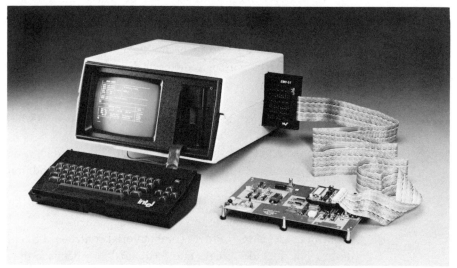

FIGURE 2-19 This in-circuit emulation system is shown connected to a microcomputer evaluation board. (*Courtesy Intel Corporation.*)

essor system testing. The strategy should depend upon the following considerations:

1. In-house capabilities, including facilities and personnel.
2. The application and nature of the product.
3. The desired operator skill level and product volume.
4. Required self-test capability of the system.
5. The hardware and software cost trade-offs.

This evaluation and study phase allows definition of the most effective technique or techniques for testing the microprocessor-based system.

Testability

The key to success lies in a product design with testability. More and more attention is being devoted to test techniques which provide solutions to microprocessor system test problems.

The advantages of self-testing are well known. By verifying that a system operates correctly, customers have more confidence in the product. Enhanced software techniques can also aid in diagnosis. By incorporating product testability during the design stage, the evaluation of microprocessor systems can be much easier.

As board complexity increases at a rate similar to the increases in chip complexities, the design for testability will become an integral part of circuit design.

In the future even more emphasis will be placed on testability. Today's boards are already hard to test because of difficulties with automatic test program generation (ATG) systems and the limitations in test and probe units. Even major automatic test equipment advances will not make testing much easier unless design for testability efforts are used. Some basic design for testability recommendations include:

1. Initialization—Provides simple and short reset sequences which the tester may control.
2. Electric isolation—Designs the circuits to enable independent tri-state ICs and maintains this condition regardless of test conditions for other circuits on the board.
3. Buses—Provide access to the main buses, as well as a capability for getting all ICs off the bus to test the bus independently.
4. Control of LSI chips—Proves combinatorial control of LSI chips with the tester. This includes breaks and feedback loops between the control circuitry and the LSI chips. Do not hardwire the control lines of the LSI

chips, but allow the test system to operate the chip modes which the actual system may not use.

5. Dynamic circuit states—Isolate the dynamic circuitry from the static circuitry to set up each section for independent testing and allow for a static mode of operation for dynamic circuits.

Increasing demands may force designers to incorporate more testability requirements. The major burden of testing the board may be designed into the system using built-in test techniques. Examples are more error checking circuits and ROMs which exercise test sequences that the CPU can execute to test the equipment.

These built-in tests can reduce the time required to troubleshoot and repair most malfunctions, resulting in major savings in time and cost with the proper internal tests. But internal tests add costs to the selling price of the product, which must reflect the costs of self-test design and components. By the careful implementation of internal tests, the design and hardware costs can be minimized so that diagnostics enhance the product's value.

A test design philosophy must be developed early by treating self-tests not as an afterthought, but as a part of product design. The following guidelines may be helpful:

1. Early in the design phase, decide which functions to test and how to test them.
2. Determine which functions the product's microprocessors can perform and which functions require external assistance.
3. If feasible, design the software and documentation for malfunction detection.

The economics of low cost, powerful chips demand consideration as a part of any self-test strategy. Also, multiple processor configurations can distribute product capability, provide greater system reliability, and allow easier fault detection and maintenance.

In the past some mainframe machines have used a dual-processor approach in high reliability applications. Here, two processors operate in a step-by-step match mode. The two processor outputs are matched and tested for both the operation and machine state. The overall processor speed is not affected. When a mismatch occurs, a fault isolation routine is used to detect the failed processor and switch it out of the operational system. Self-test programs are then used in the test system to isolate the fault to a group of replaceable circuits.

All test techniques require some test stimulus. Self-test techniques may use the in-circuit microprocessor and a test ROM. This same source can also provide signature analysis stimuli. The same is true for emulation which produces the test stimuli in a manner similar to self-test, but does not use a resident test memory. A combination of techniques can be used. A well constructed self-test can check that the system is working properly. If an operation is incorrect, proceed to an evaluation by more elaborate SA or emulation test techniques.

The test and diagnostic requirements of many products require that failures are isolated to at least an individual module. Even if tests are developed to exercise each module individually and to evaluate the bus, this problem will not necessarily be resolved.

The bus architecture found on microprocessor boards makes it difficult to apply many of the diagnostic techniques used in the past for logic boards. Some techniques such as the stored input/output method can be implemented by adding test logic to the system. Other techniques such as simulation can require a high degree of knowledge concerning the functional aspect of the board to be successful.

The advance in integrated circuits will continue to outstrip the advances in test equipment. In the past, the cost of developing test programs used to be controlled for digital logic boards by the use of simulators and automatic test pattern generation. The effectiveness of these techniques has diminished because of their inadequacy to handle the newer LSI circuits in a cost-effective manner. In many cases the programmer time becomes excessive.

These problems can be overcome with internal error detection that may use hardware, firmware, and software methods. Fault diagnostic techniques can then be used to localize or pinpoint the failure to a replaceable module or unit. The type of checking techniques used will depend on the logical structure of the machine, the types of data and control signals available, and the facilities and personnel available.

Fault Detection Techniques

Error detection codes have been effectively used in many bus-oriented systems. The information transfers are conducted with one or more added parity bits to provide error detectability through the use of parity checking circuits which are placed strategically in the microprocessor system. The parity circuits are used to count the number of ones or zeros in the data word.

The residue of a number is calculated by dividing it by the modules, resulting in an integer and a remainder. The remainder is the residue. Residue codes can be used to detect or even correct errors from an arithmetic unit or faults produced by data transfers or memory operations.

For control functions it may be necessary to use fixed-weight or m-out-of-n codes. Here the weight of the code word is the number of nonzero components in the code word. A 2-out-of-5 code is illustrated in Table 2-1.

TABLE 2-1 2-OUT-OF-5 CODE REPRESENTATION

Decimal Digit	2-out-of-5 Code
0	00011
1	00110
2	01100
3	11000
4	10001
5	00101

The control circuits in a microprocessor can vary greatly in their logical implementations. This irregularity of implementation makes the control section of a microprocessor the most difficult to test and diagnose. The control section usually requires the most test hardware.

In a microprogrammed control section, the more regular structure of the control store and the sequencing logic allows the fault detection circuits to be integrated into the system. The errors may be detected here at the microsteps level, where one or more microsteps equals a microinstruction.

In the conventional control mechanization the sequencing and timing are embedded in the control logic and error checking must be done at the macroinstruction level. One can thus verify that data have been gated from one register to another or operated on properly at the register level. In either case, the most complete fault detection scheme is to duplicate each logical unit and match the outputs of the two units after each step is completed. This technique is becoming more feasible as the cost of hardware decreases. For example, this duplication and match philosophy has been used for telephone switching systems.

Diagnostic Techniques

A basic diagnostic technique is the stored input/output method in which a truth table with the input/output responses is stored in memory and executes the truth table. Most automatic functional test systems use some variation of this method. The principal differences are in the mechanization of the hardware or software. A major disadvantage in many systems is cost since thousands of digital states must be generated to provide the stimulus.

In some microprocessor systems this technique is used at the functional block level by adding additional interface circuits. However it is usually easier to consider built-in diagnostics for the microprocessor system and specify these early in the design stage.

A watchdog or safeguard timer can also be incorporated into the system for fault detection. Here a hardware timer is run continuously. It is periodically reset if nothing delays the main program within its appointed window of time. If an error causes the main program to be delayed, then the timer is not reset and an error is detected.

Another technique is to test some portion or kernal of the processor which is then called the hardcore section. This kernal is validated by an external processor or other hardware. The hardcore section is then used for the validation of another small portion of the system. The process is expanded until each successively larger level is validated.

To minimize the hardware in the hardcore section, a portion of the diagnostic program is kept in ROM to avoid bootstrapping all the diagnostics from I/O devices. The hardcore section then includes only that hardware required to load the diagnostic.

The use of a maintenance processor to remotely check the microprocessor can allow the checking of memory or register contents. The maintenance processor can also be used to single-step the clock of the microprocessor to search out faulty register contents. Other circuitry could be checked in a similar manner.

Simulators can be used for systems that can be evaluated in a relatively slow static mode. This can be time-consuming and many microprocessor systems must be tested in a dynamic mode.

Functional or hierarchical simulation can be used where the microprocessor system acts as a number of functional building blocks. Tests for these blocks can be in the form of digital signatures. The signature and the timing window should be as close as possible to the environment that the microprocessor system will experience in its actual application.

The functional simulation approach is a macro technique and may not exercise each and every device unless it is carefully applied. In this technique we consider the microprocessor system as a number of functional building blocks. Tests are then provided for these blocks on the functional level. A well designed hierarchical simulation can combine the advantages of both gate level and functional simulation with several levels of detail. For example, gate level models can be used at functional interfaces or for critical internal faults. In some systems it is useful to simulate only critical sections that are known to cause most of the faults. This can be achieved with some of the more advanced microprocessor development systems in a semi-automatic mode.

Since a number of diagnostic techniques can be used, how does the best technique for a particular microprocessor system be selected? Two of the more important considerations are the nature of the product and the application, i.e., how does it lend itself to self-test techniques? Could additional hardware be added without much impact? How can the application software be modified to include the self-test features? In addition, the in-house capabilities should be considered with the possibility of expansion in the areas of facilities and personnel.

The operator skill level should be evaluated with the possibility of improvement for more complex products. The impact of the product volume on the need for more automation should be considered. Many microprocessor products are designed with test diagnostics as an afterthought. With an early evaluation and trade-off study, the most effective techniques for test diagnostics can be defined. If this is done, the diagnostics will be simpler and easier to apply and the test costs as a whole can also be lower.

TABLE 2-2 DIAGNOSTIC TECHNIQUES FOR MICROPROCESSOR SYSTEMS

	Application	*Characteristics*
Parity Checking	Bus Transfers	Requires additional hardware/software
Residue Codes	Bus Transfers	” ” ”
Fixed-Weight or m-out-of-n Codes	Arithmetic Control Operations	” ” ”
Duplicate Hardware and Match Step	Critical Functions or Complete System	Requires much additional hardware, allows most complete detection
Stored Input/Output	Logic or Memory Test	Becomes complex for modeling microprocessor systems
Hardcore Validation	Complete System	Allows a small kernal of the system to be checked and expanded
Remote Maintenance Processor	Complete System	Can require complex software
Functional or Hierarchical Simulation	Critical Functions or Complete System	Can require complex software and special hardware

The incorporation of systems with complex LSI chips has greatly reduced the usefulness of digital test systems. A number of more effective techniques for testing and diagnosing microprocessor systems has evolved. The real key to product evaluation lies in a design for testability at the earliest possible stage.

MICROPROCESSOR DEVELOPMENT SYSTEMS AND AIDS

A variety of development systems and aids are available which allow programs to be developed with the aid of special software and peripherals and later to be transferred to the actual microprocessor on which they will run. Some of the techniques which have been used are the microcomputer development systems, time-sharing development services, and the development systems which are based on other computers.

Microprocessor manufacturers have realized that the development of a microcomputer system starting with a microprocessor and associated parts can be a formidable task.

To simplify the problem of hardware and software development, many of the manufacturers package their products in ways which make the microprocessor more accessible and usable. These packages have the following forms: parts families, ready-to-assemble kits, prewired printed circuit (PC) cards, assembled kits, and development systems.

A parts family is a set of closely related and compatible IC components, such as a clock driver, microprocessor, RAM, ROM, and I/O chips. These parts are available as a kit and sold in single quantities by the manufacturer.

A ready-to-assemble kit may have one or more PC cards and a number of ICs and discrete components for assembly on the cards. Once assembled, it is an operative microcomputer. Kits are also available in the form of factory assembled systems. In some kits, the program and data entry uses keyboard switches and the data output on an LED display.

Development Techniques

Of all the program development techniques, only one does not require any significant tools—hand programming. For a program developed directly in binary or hexadecimal, a hexadecimal keyboard and LED displays are required for the output.

This type of kit or assembled board is useful for gaining familiarity with simple programming. Most are not equipped with an assembler, so the programming must be done in hexadecimal. This limits the length of programs that can be developed on them to a few hundred instructions. The usual mass memory for such a board is a cassette recorder interfaced to the board. Although such boards are a valuable educational tool, they cannot be used alone as a true development facility for most applications.

Many modular systems exist which allow the user to plug in additional boards and to build up progressively to a true development system. These frequently used alternatives are more powerful and more efficient. Prewired printed circuit cards, including a microprocessor and related interface ICs, are available from a number of microprocessor manufacturers for those who already have some microcomputer development equipment.

Another technique for developing microprocessor software involves the use of other computers. Assembler and simulator programs may be purchased to use on an in-house computer. This technique is less expensive than using time-sharing services over the long term, but it requires the purchase of software and possible modifications to meet the requirements of the in-house computer. Any computer may be used to execute these cross programs. When an in-house computer is available, it may be possible to run the cross assembler, cross-compiler, and other programs on it. The facilities of the local computer can be less extensive than those of a time-sharing service. Also, working in the batch mode without an editor can be considerably more difficult than using an interactive system.

If the computer is used in the batch mode, the program may be submitted on cards or some other medium and the results returned hours or days later. This slow response could be a major obstacle to program development. Developing a program for a microprocessor may involve many program changes and it is time-consuming to require hours or even days to perform these changes.

Some development software runs on minicomputers or microcomputers. These computers usually have a printer, editor, interactive display terminal, and a large disc system. Such a system can be used as effectively as time-sharing. In addition the computer I/O bus can be interfaced to allow the object code which the computer transfers directly into the read/write program memory of the microcomputer being developed. Figure 2-20 shows a block diagram for this type of system.

If an in-house system is used in a time-sharing mode, its advantages are similar to using a mini or microcomputer except that a time-sharing system will not allow hardware checking for system integration. The only tool that allows both software and hardware debugging is the dedicated development system.

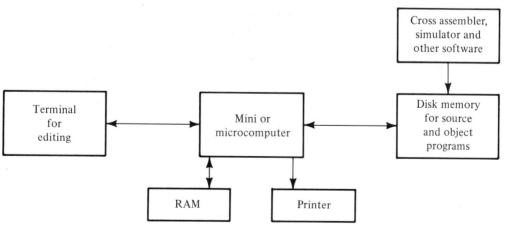

FIGURE 2-20   Mini or Microcomputer Development System.

Time-Sharing Systems

A time-sharing system is a general purpose computer used in the time-sharing mode. *Time-sharing* means that a number of individuals are using the computer at the same time in an interactive mode where the system can respond almost immediately, on-line. The dialogue with the machine is fast since there is little or no waiting time.

Most time-sharing systems have large peripherals and extensive software packages, including powerful file systems and editors. The microprocessor program can be effectively developed and debugged on such a system. Most systems can be provided with the cross programs necessary to generate the code required for many of the existing microprocessors.

To execute the programs, using a simulator program may be necessary. When user programs are executed by a simulator, it is possible to perform debugging of the program and to measure the program timing by counting the simulated duration of the instructions. The actual architecture of the I/O operations as they will be performed cannot be simulated. Also, many input-output chips are now processor equipped and to simulate them it would be necessary to have the complete results of the program execution.

In many projects it is desirable to have several persons working simultaneously on the same program. The time-sharing technique allows several users to work simultaneously. Many large development systems also offer this facility for multiple users. The time-sharing technique permits fast development of the program itself. Most time-sharing services have simulators and cross-assemblers for the popular microprocessors. They offer larger amounts of storage, interactive facilities, and high speed peripherals. Figure 2-21 illustrates the procedures for developing microprocessor software using time-sharing services. The advantages are the software facilities, low initial cost, independence of a particular processor, fast mass storage, and access to high speed peripherals. The disadvantages include high continuing costs, an inability to test programs in the real computer environment, and the need to use facilities that are located at some distance and that may involve extra turnaround time. Time-sharing is often a convenient way to get started on microprocessor software development before other equipment is available.

Packaged Development Systems

The microprocessor manufacturers provide a packaged microprocessor development system. This type of system includes the basic hardware and software necessary for the development of a prototype microcomputer system and has simplified the task for the system designer.

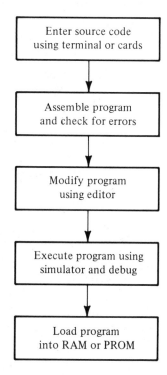

FIGURE 2-21   Time-sharing Development.

One of the key advantages of the microprocessor is its capability to replace cost-recurring random logic hardware with non-cost-recurring software. This advantage is less significant when a hardware system must be built for the software development.

To simplify the software development, the microprocessor development system includes most of the hardware needed with the software development aids. The system designer can then concentrate on the software capabilities of the microprocessor. After the software has been developed and optimized, the hardware design required to implement the software can be developed.

Microprocessor development systems are available from the chip manufacturers as well as from independent test hardware vendors. Most microcomputer development systems use an actual microcomputer or an emulation of one with additional hardware and software.

These development systems usually have the following facilities:

1. A display to allow the observation of the contents of registers and other memory locations.
2. Some capability for changing the contents of memory locations.

3. A reset control which can start the processor from a known state.
4. A single step capability to allow the program to be executed one step at a time for debugging.
5. A run control capability which allows a program to be executed from a specified location.
6. RAM or PROM memory that can be used to alter the program as required.
7. Interfaces for input/output devices, such as keyboards, displays, printers, and floppy disk systems.
8. A loader used to enter the programs into the microcomputer memory.
9. Utility programs that aid in developing the user programs.

Some development systems are constructed on a modular basis, with a number of boards for memory I/O and other functions. A block diagram of a typical development processor module is illustrated in Figure 2-22.

The various peripheral devices required can be connected to the I/O ports provided by the system. An I/O interface is also provided for program development to the system. This interface would not be used in the final system since it is only used to develop software for the system. The usual interface is to a keyboard and CRT terminal which allows the user or users to do the on-line programming and editing.

The modular development system is popular since the user may insert or delete modules from the system at will.

Some development systems may contain up to 15 or 20 modules. Thus the RAM or PROM memory can be expanded or special input-output interfaces can be added. The peripherals used in such a system might include:

1. An input device such as a keyboard.
2. An output device such as a CRT display.
3. A printing device for hard-copy records.
4. A mass memory for storing and merging files.
5. A PROM programmer for placing the program in the PROM memory chips.

Two basic methods are used for microprocessor development in the design of a system. One can develop the software and then, once the software is developed, it is implemented on the prototype hardware. A more common method is to develop both the hardware and software at the same time while using the development system for control and diagnostics. The first method is suitable for small systems where a large amount of interaction does not exist between the hardware and software. It can also be used for systems which are architecturally similar to

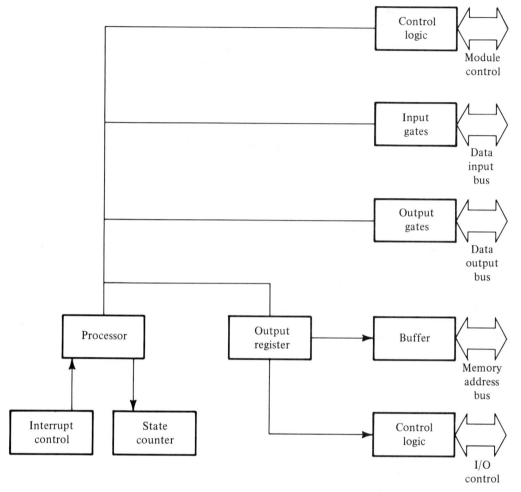

Figure 2-22 Processor Module Structure.

the development system itself. The second method is suitable for more complex systems in which there are more interactions between the hardware and software.

In the modular development system the designer has the option of selecting the particular modules most suited for the application. A typical system could include the following hardware modules in addition to the processor:

1. A read/write memory.
2. A programmable read only memory.
3. External event detection.
4. A monitor/control panel.
5. Power supply and reset circuits.

Physically, the development system looks like a traditional minicomputer. It is equipped with a front panel to facilitate the debugging function. When the development system is marketed by a microprocessor manufacturer, it uses the manufacturer's microprocessor. Development systems are also available which produce code for several microprocessors. These development systems offer the facilities necessary to effectively develop the application system. They should provide a file system, which can be connected to a variety of peripherals. The software should include programs, such as an editor, assembler, and other supporting programs such as utility routines for debugging. It is also desirable to have a high level compiler or interpreter to develop programs of any complexity.

In programs in which there are considerable interactions between the hardware and the software, the development system can be linked by means of an umbilical to the hardware system being developed.

When the basic hardware elements of the microprocessor system are provided in the development system, the designer can devote the major attention to the software. Since a key advantage of a microprocessor system is its ability to replace hardware components by software, the best system design optimizes the software and coding.

Software Development Aids

The task of microprocessor program development is improved due to the number of software development aids available, such as:

1. Assemblers.
2. Cross-assemblers.
3. High level languages.
4. Editors.
5. Loaders.
6. Debuggers.
7. Simulators.
8. PROM programmers.
9. In-circuit emulators.

The assembler program converts symbolic code written by the programmer into the machine language instructions which can then be executed by the processor. The assembler converts the symbolic instructions such as ADD, SUB, or MULT into machine bit patterns. It also converts the labelled machine addresses designated by the programmer into the real machine memory locations.

After a program has been typed into the development system, the programmer should print it to verify that it is complete. The source program is then printed as a listing.

The program is written in symbolic form and must be translated into a machine executable format.

The assembler translates the source code into a binary object program, which can then be directly executed by the machine. It substitutes the actual address in place of the symbolic ones, and it substitutes the actual binary encoding of the data instead of the symbolic names along with the binary code of instructions in place of the mnemonics. This creates the object program which is a sequence of binary words that can be directly executed by the microprocessor. The actual processor is not necessary to develop the object program. The functions that are necessary, such as the editing and the assembly facilities, can come from any program. The use of a larger processor can improve system capabilities for more powerful peripherals and software facilities. In this case the assembler acts as a cross-assembler.

The cross-assembler is a program for one machine which resides on another machine. A cross-assembler for a 8086 is an assembler which will produce 8086 code, but it executes on a larger minicomputer or other machine. The cross-assembler converts the symbolic code written by the programmer into machine language instructions executable not by the microprocessor, but another computer. The cross-assembler allows the programmer to design, develop, and refine the program on a larger and in some cases a more familiar computer. Cross-assemblers are available for minicomputers and large-scale computers.

Some assemblers have a macro capability, which allows an identifier to be associated with a block of text which then is substituted every time the macro is invoked. Parametric macros will allow a different parameter value to be used each time the macro is invoked, so the text can be varied.

There is an important trend in the increase of high level languages for use with development systems. The use of a high level language can greatly simplify the task of the designer. The efficiency of coding in a high level language is less than what can be achieved using assembly language and, with the use of high level languages, the development of microprocessor systems places less requirements on the software design.

In assembly language the programmer manipulates the registers to allocate memory space. The assembler mnemonics are sometimes hard to remember since each computer manufacturer uses a different set of mnemonics. To make the program understandable, a comment may be needed on nearly ever line.

High level languages allow the programmer to forget about allocating registers or memory space since the compiler or interpreter takes care of these details. The high level programs can be machine independent, but they tend to be slower than assembled programs because the interpreter's code is not as efficient. High level programs use fewer lines of source code and require less detailed comments than assembler programs.

Since the programmer may concentrate on the application more, the program takes less time to code, debug, and maintain. Languages such as PASCAL and PL/I use control structures with syntaxes that make it easier to follow the flow of a program. Such constructs as IF - THEN - ELSE, BEGIN - END and DO - WHILE encourage structured programming by eliminating GO - TO loops. Languages such as PASCAL are now available on some microcomputers. FORTRAN, which was the first high level language, lacks these transfer-of-control structures, but it still is widely used. Generating the software is much easier when it is written in an English-like, high level language such as FORTRAN or PL/M.

Compared with assembly language, the high level language reduces the time to write and check a program. With a compiler or assembler to translate the source language into machine code and a link program to allow software to be written in small, easily manageable modules, the design may even be distributed among several programmers working in different languages.

To modify the program for entry into the system, it is desirable to have an editor, which is a program that allows the convenient manipulation of text, i.e., it allows textual changes in the program without reloading or rewriting the complete text. The editor performs the functions of adding or deleting a line or a character automatically. Without the editor, a simple typing error in a program would have to be corrected by retyping the entire program. With the editor, it is possible to issue commands like "go to line 18 and insert the following word" or "look for B2 in the text and replace it by B8." A powerful editor can enhance the speed at which a program can be typed and also be modified once errors are located.

The file system can be modified using editor commands such as add, delete, copy, find, list, and replace as shown below:

ADD   Add lines entered from the CRT into the text file.

DELETE   Delete lines from the text file.

LIST   List lines on the CRT printer.

FIND   Find a character string or line.

REPLACE   Replace a character string with a new character string.

MODIFY   Modify character strings with subcommands.

COPY   Copy lines or a block of lines.

TEXT   Read a block of text from memory.

KEEP   Store a block of text on memory.

The editor assigns line numbers to the stored text so the individual lines can be identified and accessed. Some editors renumber the lines each time new ones are added or deleted. Others use incremental line numbers. For example, a line added between lines 102 and 103 would become 102.1.

A debugger is a diagnostic aid which allows the user to analyze the program. The debugger allows the programmer to insert breakpoints and obtain register and memory dumps at desired points of the program execution. When the program detects an error, some debugger programs allow the user to make a modification and continue running the program. This feature of a debugger—allowing examination of the contents of registers and changing them—is accomplished either executing display instructions on the microprocessor, or by executing them under the control of a simulator or emulator, which then stores a copy of the value of the registers in memory.

A loader is a program which initializes the processor to allow the application program to begin execution. The loader is often used with a hardware facility for a reset. When the processor is halted, the user restarts the program by pressing a reset button. The reset circuit then interrupts the processor so that control is taken away from the program formally being executed and the processing of the loader program begins. Some types of loaders permit separate groups of machine language code to be linked together and executed by the processor. These loaders thus act as linkage editors.

The simulator is a specialized program for the analysis of user programs. A simulator can be used to model or simulate the timing characteristics of hardware, such as peripherals, which may not be available for testing at the time the software is ready.

The PROM programmer is used with many development systems for the programming of ROMs. A PROM programmer simplifies the software development since the test programs can be placed on PROMs and run in the system without requiring the loading of the programs into the RAM for each test. The PROM programmer can include such functions as program listing, manual keyboard, duplication, and verification. The main role in most applications of the PROM programmer is to program the EPROMs or the PROMs on which the programs will reside. Many are equipped with a hexadecimal keyboard to allow the manual input of data. Some provide additional interfaces such as an RS232 connector, so that the device can be connected to a microcomputer system.

An in-circuit emulator can assist in debugging when the software is first integrated with the hardware in the early stages of debugging. Emulator devices with symbolic debugging capability can reduce the hardware/software checkout time. They can allow monitoring of internal functions that may be otherwise inaccessible.

The in-circuit emulator offers the capability of testing and debugging the actual system connected to the real input-output devices in real-time. This is one of the more powerful facilities to use in the debugging of the complete hardware/software system. It is a required device for effective debugging on any real-time system. In a typical microprocessor development system, the emulator executes about 10% slower than the actual microprocessor.

The in-circuit emulator performs a trace which is a recording capability to automatically record events during the previous machine cycles. This is analogous to a film of events within the previous cycles of the system. When an error is diagnosed at a breakpoint, it may be too late, since the error may have been caused by a previous instruction. The solution is to identify the instruction which caused a wrong value at the point where it was first diagnosed. In many programs, a number of branching points exist and determining which branch was being executed prior to the detection of the error can be difficult. It is therefore useful to record which path in the program was followed, up to the breakpoint. Then either the erroneous instruction can be identified or an earlier breakpoint can be set. The other machine cycles can then be recorded until the error is traced to the erroneous instruction.

The microprocessor manufacturer has versions of an in-circuit emulator but they are also available from independent vendors.

A number of development systems with equivalent capabilities are available from independent vendors. Since they are not microprocessor manufacturers, these systems are microprocessor independent. For example, the same system might be used for an 8086, 68000, and others. They are general purpose development systems and can be particularly helpful for a user who is reluctant to commit to a chip family. However, many of the development systems supplied by the manufacturers provide better software capabilities for a particular microprocessor, along with the prospect of continued improvements. Many of these development systems have cross programs for other families of microprocessors.

Diagnostic tools are available from a variety of sources. New instruments, such as the microprocessor analyzers, have evolved from digital analyzers. These can be extremely useful in the fine debugging required for complex hardware interfaces.

The Use of Operating Systems

The use of operating systems is also important in software development. An operating system is a set of prewritten programs that reside in the memory along with the user-written application programs. The operating system allows more efficient memory space by providing common services and more efficient time use by permitting parallel activities to occur, and more capability by providing services beyond the scope of the typical user program such as memory pool management. The operating system does not depend on the application; the system might be used for accounting or controlling a complex process.

The microcomputer without an operating system can be used as a controller, but an operating system can allow it to perform much better. Some operating systems may be small enough to fit in ROM, but for the flexibility needed for program development or multitasking, a floppy disc is required.

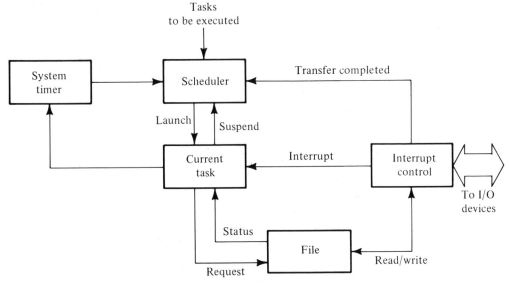

FIGURE 2-23   Basic Tasks of an Operating System.

An operating system acts as the manager of the system's resources: memory, terminals, communications links, and other peripherals. It contains a program or routine for each task and links these task programs together using other routines to perform the housekeeping operations as shown in Figure 2-23.

The two basic types of systems are the disk operating systems (DOS) and the multitasking operating systems (MTOS).

Some disc operating systems are designed to support users in preparing, debugging, testing, and running programs on another computer. This computer may be part of a standard development system or it may be a larger separate system.

When the DOS supports several peripherals such as two discs, printers, and several terminals and required memory, it can have a fairly large program in memory at all times. For many services additional programs are loaded into overlaid or common areas. The services provided are accessed either from terminal hardware or from the user program.

The terminal services which aid the program preparation are the text editing, assembling, compiling, linking, and loading functions. File manipulation functions include the copying, merging, deleting, and reformatting operations. Debugging aids include inspecting and changing memory, setting breakpoints, and single-step facilities.

In program initialization the variables such as pointers and flags are set to the appropriate starting values. The initialization usually involves interactions between the user and the operating memory program. After the initialization, the operating system enters a loop to read, analyze, and write the data. To read new information from a disk file, the user enters a command and receives a response. Some typical commands are:

LOGON   Set the default disc and user name for referencing files.
FILE   List the user's directory file.
RENAME   Change the name of a file.
PURGE   Delete a file.
RUN   Load a program file from disc and execute it.
EDIT   Run the Text Editor.
LINK   Run the Relocating/Linking Loader.
ABORT   Terminate the program.
RESUME   Resume execution of the program.
SUBMIT   Change input from the terminal to another device.
END   Terminate the session.

The time between the command entry and a reply is proportional to the difficulty of the task and the load on the system at the time of request.

A multitasking operating system is desirable in applications which have a significant number of random, asynchronous inputs. Such an application is illustrated in Figure 2-24. Most of these applications have sections that are functionally similar and an overall organization that is structurally similar.

A DOS provides a means to open, read, write, and close files. It also handles the reading and writing of input messages along with the output of results to the printer and supplies utilities to load the program from the disc and to start the execution.

A multitasking OS must also provide the switching from one task to another. To do this it assigns the various tasks different priorities. A line printer I/O driver would normally receive high priority because a line printer is relatively slow and in high demand. This high priority keeps the line printer busy as much as possible.

A number of techniques are used to decide which task is to run next. The slice procedure runs each task for a fraction of CPU time. Event-driven systems switch over when a significant event such as an interrupt occurs. Some event-driven systems are interactive. Data may be sent to the program from the terminal during execution.

In some systems data which are presented from the peripherals data must be captured or they will be lost. Some current activities must then be suspended and resumed when time is available. If high priority data arrive while a low priority task is taking place, the low priority task must yield. The OS gives the CPU the highest priority task that is ready to be executed. When the task reaches a pause such as waiting for a disc access, the next highest priority task is given a chance to

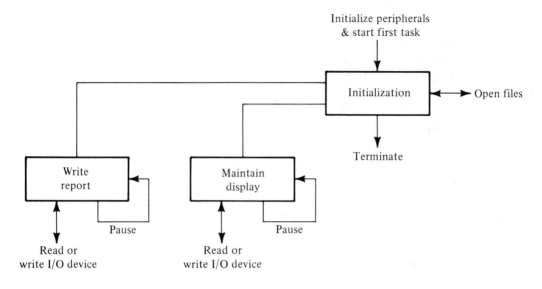

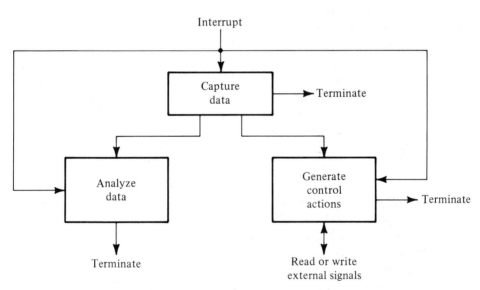

FIGURE 2-24   Multitasking Operating System.

continue for as long as it can, during the pause. In the switching from task to task, the OS swaps the program counter and other registers. This keeps the CPU busy while the tasks move along in parallel. This juggling is transparent to the individual task programs.

This dynamic management of the CPU distinguishes the multitasking OS from a DOS. The multitasking OS functions in a real-time, on-line environment.

The services include the starting and coordinating of tasks, and controlling the pauses and centralized I/O handling. The applications that can benefit are normally real-time, as opposed to applications in which the physical time is unimportant. Thus, multitasking is also a real-time operating system or RTOS.

Real-time applications may also be developed with general hardware using a DOS with the multitasking as a part of the application program. The application program can then be transferred to the dedicated hardware and run under the multitasking OS alone.

Operating systems and application programs are not always distinct levels of software. Programmers do not agree on which functions should be included in the OS and which in the user programs.

In a real-time application which requires some multitasking and the application is small, one can consider writing a special multitasking which is compact and does not waste overhead on unnecessary features. However using an available multitasking OS may be better since it can take many hours for the design, coding, and debugging. Even a limited size OS requires working out all the subtle timing bugs and priority conflicts. Also, if a real-time application is successful, it can expand as new requirements arise, and then the small multitasking OS can become a patchworked program.

To manage information storage and retrieval, an OS uses a file system. Some systems treat all the peripheral devices, including the line printers, CRTs, and disc drives, as files. Here the programs treat all I/O alike with no concern for the particular characteristics of a device. For example, to print on a line printer, the system routine for an open command is called to open the line printer file. This command allocates the line printer to the user if no other program is controlling it.

Another type of operating system—the executive—is a real-time system with some sections that may be reconfigured for each application. This is done with relocatable program modules. After the software has been developed on a development system, it is linked with the desired system modules and stored in ROM, PROM, or RAM. A disc driver module gives the system file management capability.

Most microcomputer development systems use a variety of general-purpose application programs that may be associated with an OS. These can include editors, assemblers, language compilers, linkers, loaders, debuggers, and utilities. Some are written as subroutines that can be linked with the application program. Others are used as separate programs, handling the user programs as input.

Some of these may be less useful with microcomputer designs because of the nature of the system activity taking place.

In selecting a microprocessor development system, a few basic questions must be considered:

1. The complexity of the final system, in both hardware and software.
2. The configuration of the final system.
3. The amount of testing necessary.

Don't neglect the follow-on support or field service. While a development system could be used for field diagnosis, it's difficult to transport all the necessary components to remote field locations. A portable test instrument that can duplicate some system capabilities is a good alternative. Some equipment allows systems to be checked in the field using an in-circuit emulator and a control panel.

The microprocessor development system is intended to simplify the most common hardware and software design problems by providing a suitable test vehicle.

The essential task during development is usually the software phase. The most efficient tool for both hardware and software development has been shown to be the microprocessor development system. Alternatives are time-sharing systems, an in-house computer, and kits. A useful facility for debugging complex hardware systems with real-time applications is in-circuit emulation.

The costs required for a development system can be repaid rapidly by reduced programming time and faster completion of the project. The investment in a development system that expands with the product can pay for itself on the first product introduction. The development system can lessen the design time and put the product on the market much sooner.

EXERCISES

2-1. A digital thermometer system is to use diode temperature sensors connected to a microprocessor. It is to record and display temperatures to the nearest degree over a $0°$ to $160°C$ range. What type of converter could be used and how often should the temperature be sampled? Define for the input-output processing, accuracy, and memory.

2-2. Draw a general flow chart for the following task: The microprocessor examines an input port. If the value at the port is not zero, the microprocessor waits 1 ms and samples the port again. If the value is the same, the processor turns on an alarm. Otherwise, the processor discards the first result and repeats the delay and sampling process. When the value at the port reaches zero again, the light is turned off.

2-3. Draw a general flow chart for the following task: The microprocessor reads a character from an input port. If the character is not X, the microprocessor waits 1 ms and reads another character; if the character is X, the microprocessor performs a data handling routine and starts the read process again.

2-4. Discuss the basic sequence of steps necessary for developing a microprocessor system. Consider the problems involved in each of these steps; present some typical solutions along with some devices developed to facilitate the implementation of these solutions at the various steps.

2-5. Consider the design of railroad crossing monitors with a microprocessor. The system input is a switch that is closed by the weight of the train. The outputs activate warning lights, sound alarms, and lower the crossing gates. Describe the input, output, processing, and accuracy requirements; memory; and error handling. The crossing gates remain closed until the train has passed the crossing. The system must allow cars and people to leave the crossing before the gates are lowered. Describe the time considerations if the switch is .8 miles from the crossing and the train travels at 60 miles per hour.

2-6. Consider the problems involved in designing a digital scoreboard with a microprocessor. The system uses inputs from a keyboard and counts downtime. Describe the input, output, processing, and accuracy requirements; memory; and error handling. The timer displays minutes and seconds and is activated by start, stop, and reset buttons.

2-7. Define the requirements in designing a digital temperature control system using a microprocessor. The system inputs are an A/D converter connected to a resistive temperature sensor. The output is a relay which controls a heating element. Describe what is needed for the input-output processing, accuracy requirements, memory, and error handling.

2-8. Documentation is important in a microprocessor system to those who must use and maintain, understand, and extend it for further applications. Discuss how flow charts, comments, and memory maps are used as documentation techniques.

2-9. The maintenance stage allows the updating and correcting of the program to account for changing conditions or field experience. Discuss how the proper diagnostic testing can reduce the frequency and extent of maintenance required for a microprocessor system.

2-10. An object program is to be placed in the memory of the system in which it will execute. Discuss the loading phase when the loading is accomplished by a loader program.

2-11. In the case of complex programs, the complexity of coding in assembly language may be unreasonable and inefficient. The reasonable solution may be the use of a high level language. If the program is that complex, the programming cost will be high and it may not be reasonable to consider a microprocessor. Discuss this argument.

2-12. Some errors may be found in the late stages of program or field testing. Equality cases within loops and conditional jumps may be sources of these errors. The problem here is usually which value of a flag to use as the jump condition and what action to take when a variable is equal to a threshold rather than above or below it. Discuss some techniques for reducing these problems.

2-13. If microprocessor programming is to be accomplished quickly, three basic tools are available. Discuss the respective merits of using a time-sharing system, an in-house computer, and a development system.

2-14. Assume that a 2K program is being developed and that no disc file system is available. How can the user program as well as the support programs be stored?

2-15. A large scale computer is available with a cross-assembler. Discuss the execution testing and debugging of a 10K program using this compared to a microprocessor development system.

2-16. Discuss how the microprocessor system itself can have the facilities to analyze each program step for the detection of critical errors?

2-17. Some microprocessor development systems suffer from the low speeds, limited software, and peripherals that are usually available and their limitation to a single processor. External interfaces can be used, but they can complicate system development because they depend on facilities present in the development system but not in the final product. Discuss how more advanced and flexible microcomputer development systems might solve these problems.

2-18. Advocates of the META Stepwise Refinement (MSR) state that if the problem is solved several times, each solution is more detailed and complete than its precedessor. What are some drawbacks in this philosophy?

2-19. Supporters for Higher Order Software (HOS) provide a set of rules to use. Discuss some potential problems in applying these rules.

2-20. Discuss applications where simple fault detection techniques can provide enough insight into program operation compared to more sophisticated debugging.

2-21. Debugging real-time systems can be difficult and require special equipment. Discuss how an operating system can help.

2-22. If the microprocessor must interact with a very precise control system, how can data be generated and controlled to reduce the build-up of errors? Use a general flow diagram to demonstrate.

3

System Considerations

To accommodate the input or sensor voltage in the microprocessor system, some form of scaling and offsetting might have to be performed by an amplifier. Also, analog information must be converted and, if it is from more than one source, additional converters or a multiplexer will be necessary. To increase the speed at which the information may be accurately converted, a sample hold can be used and to compress analog signal information, a logarithmic amplifier might be required.

The system design should begin with the choice of sensor. If the systems engineer can help select the transducer, the design task will be eased.

In the monitoring or controlling of motor shafts, the designer may have the choice of signals from three different position sensing approaches: shaft encoders, synchros, or potentiometers. Temperature measurements might be accomplished by thermocouples or thermistors, while force can be measured by strain gauges, or obtained by integrating the output from accelerometers.

If the transducer signals must be scaled from millivolt levels to an A/D converter's typical ± 10-volt full scale input, an operational amplifier can be the best choice. When the system involves a number of sources, each transducer can be provided with a local amplifier so that the low level signals are amplified before being transferred. If the analog data are to be transmitted over any distance, the differences in ground potentials between the signal source and the final location can add additional errors to the system. Low level signals can be obscured by noise, rfi, ground loops, power line pickup, and transients coupled into signal lines from machinery. Separating the signals from these effects can be critical for the designer.

Systems can be separated into two basic categories: those suited to favorable environments such as laboratories, and those required in more hostile environments such as factories, vehicles, and military installations. This latter group includes industrial process control systems where temperature information may be developed by sensors on tanks, boilers, vats, or pipelines that may be spread over miles of facilities. The data can be sent to a central processor to provide real-time process control. The digital control of steel and chemical production, and machine tool manufacturing are characterized by this environment. The vulnerability of the data signals here leads to the requirements for isolation and other data retention techniques. Systems in hostile environments might require components for wide temperatures, shielding, common mode noise reduction, data conversion at an early stage, and redundant circuits for critical measurements.

In laboratory environment applications using systems such as gas chromatographs, mass spectrometers, and other sophisticated instruments, the design problems are more concerned with the performing of sensitive measurements under favorable conditions rather than protecting the integrity of collected data.

In the more conventional data acquisition system, the designer is typically concerned with the collection and processing of analog sensor data into digital form for any of the following purposes:

1. Storage for later use.
2. Transmission to other locations.
3. Processing to obtain additional information.
4. Display for analysis or recording.

The data could be stored in raw or processed form; they might be retained for short or long periods or transmitted over long or short distances. The display could be on a digital panel meter or a cathode ray tube screen.

The data processing might range from simple value comparisons to complex manipulations. The designer might be interested in collecting information, converting data to a more useful form, using the data for controlling a process, performing calculations, separating signals from noise, or generating information for displays. The designer should be familiar with the data acquisition configurations that have been used as well as the considerations involved in the choice of configuration, components, and other elements of the system.

Hardware Configurations

The choice of configuration and components in data acquisition design depends on a number of factors:

1. Resolution and accuracy required in the final format.
2. Number of analog sensors to be monitored.
3. Sampling rate desired.
4. Signal conditioning requirement due to environment and accuracy. Lab-

oratory systems can have narrower temperature ranges and less ambient noise, but the higher accuracies require sensitive devices, and a major effort in signal conditioning can be necessary for the required signal/noise ratios.

5. Cost trade-offs.

Some of the choices available for the basic data acquisition configuration include:

1. Single channel systems:
   a. Direct conversion.
   b. Preamplification and direct conversion.
   c. Sample hold and conversion.
   d. Preamplification, sample hold, and conversion.
   e. Preamplification, signal conditioning, and direct conversion.
   f. Preamplification, signal conditioning, sample hold, and conversion.
2. Multichannel systems techniques:
   a. Multiplexing the outputs of single channel converters.
   b. Multiplexing the outputs of sample holds.
   c. Multiplexing the inputs of sample holds.
   d. Multiplexing low level data.
   e. More than one tier of multiplexers.

The signal conditioning techniques to be considered include:

1. Ratiometric conversion techniques.
2. Wide dynamic range techniques:
   a. High resolution conversion.
   b. Range biasing.
   c. Automatic gain switching.
   d. Logarithmic compression.
3. Noise reduction techniques:
   a. Analog filtering.
   b. Integrating converters.
   c. Digital data processing.

These techniques will be discussed later but first some of the components used in these system configurations will be reviewed.

System Components

To separate the common mode interference from the signal to be recorded or processed, instrumentation amplifiers can be used. The instrumentation amplifier is characterized by common mode rejection capability, high input impedance, low drift, adjustable gain, and usually a greater cost than operational amplifiers. They range from monolithic ICs to potted modules, and larger rack mounted modules with manual scaling and null adjustments.

When a very high common mode voltage is present or the need for an extremely low common mode leakage current exists—such as in medical applications—an isolation amplifier is required. Isolation amplifiers may use optical or transformer isolation.

Analog function circuits are used to perform a variety of signal conditioning operations on signals in analog form. When their accuracy is adequate, they can relieve the microprocessor of time-consuming computations as well as reduce software. Among the operations that can be performed are multiplication, division, powers, roots, nonlinear functions which can be used for linearizing transducers, rms measurements, computing vector sums, integration and differentiation, and current-to-voltage or voltage-to-current conversion. Many of these operations can be found in available devices such as multiplier/dividers and log/antilog amplifiers.

When data from a number of independent analog signal sources must be processed by the same microcomputer, a multiplexer is used to channel the input signals into the A/D converter. Multiplexers can also be used in reverse. When a converter must distribute analog information to many different channels, the multiplexer can be fed by a D/A converter which will continually refresh the output channels with new information.

In many systems, the analog signal can change during the time that the converter takes to digitize an input signal. The change in this signal during the conversion process can result in errors since the conversion period can be completed after the conversion command, and the final value will not represent the data at the time when the conversion command was transmitted. Sample hold circuits can be used for an acquisition of the varying analog signals, and the holding of this signal for the duration of the conversion process. Sample hold circuits are commonly used in multi-channel systems where they allow each channel to receive and hold a signal until required.

To get the data in digital form, the designer can use an analog/digital converter which might be a shaft encoder, a small module with digital outputs or a high resolution, high speed panel instrument. These devices convert the analog input data, usually in a voltage form, into an equivalent digital format. The characteristics of A/D converters include the absolute and relative accuracy, linearity, monotonicity, resolution, conversion speed, and stability. A number of input ranges, output codes, and other features are available.

The successive approximation technique has been used in a number of applications and popular alternatives are the counter-comparator ramp approaches. The dual ramp is widely used in digital voltmeters.

D/A converters change digital signals into the analog representation. The basic converter circuit uses weighted resistance values which are controlled by a particular level or weight of the digital input data. This develops the output voltage or current in accordance with the digital input code.

A class of D/A converters (DAC) exists which is capable of handling variable reference sources. These multiplying DACs produce an output value which is the product of the number represented by the digital input code and the analog reference voltage.

A typical data acquisition system appears in Figure 3-1.

In the past data acquisition hardware has changed radically, due to the advances in semiconductors. However, what has not changed are the fundamental system problems confronting the designer.

INSTRUMENTATION AND ISOLATION AMPLIFIERS

In some industrial systems retrieving millivolts of analog data from volts of common mode interference may be necessary. It may also be necessary to galvanically isolate the amplifier's input from its output and the power source to protect the amplifier from high voltage or to protect the object or subject being measured. The components required in this case are the instrumentation amplifiers, which include the subclass of isolation amplifiers.

These amplifiers may contain operational amplifiers, but they are commit-

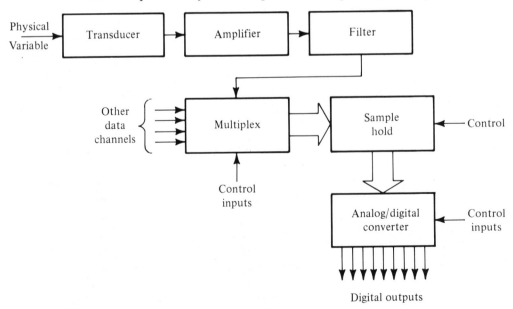

FIGURE 3-1   A Typical Data Acquisition System.

ted devices with a definite set of input-output relationships in a fixed configuration. They are designed with a high common mode rejection ratio (CMRR), low noise and drift, moderate bandwidth, and a limited gain range of about 1 to 1000, which is usually programmed by a fixed resistor.

The common mode rejection ratio (CMRR) is the ratio of the common mode voltage to the common mode error referred to the input. It is generally expressed in dBs.

CMRR is an important parameter in differential amplifiers. An ideal differential amplifier would respond to voltage differences between its input terminals without regard to the voltage level common to both inputs. Actually, a variation occurs in the balance of the differential amplifier due to the common mode voltage which results in an output even when the differential input is zero.

The usual configuration of the instrumentation amplifier does not use gain setting resistors or other components connected to the input terminals since this can degrade the input impedance. High impedance inputs will maintain a high common mode rejection even with low to moderate source impedances. Instrumentation amplifiers generally use high precision feedback networks.

The drift, linearity, and noise rejection capability of these amplifiers make them useful for extracting and amplifying low level signals in the presence of high common mode noise. They are commonly used as transducer amplifiers for thermocouples, strain gage bridges, and biological probes. In preamplifier applications, they may be used for extracting small differential signals which may be superimposed on large common mode voltages.

Some simple amplifiers which use one operational amplifier have the problem of poor source unbalance characteristics since CMRR depends on resistance matching. When very large values of resistance are used, noise and bandwidth problems can occur.

To reduce these problems, use additional op amps. Bipolar op amps can be used, but FET devices with inputs having a high source impedance are preferred. Matched input followers can provide a low drift and keep CMRR high, when the main amplifier's drift is low and the resistances are well matched.

The differential current feedback amplifier uses high impedance sense and reference input terminals. Any resistance in series with either of those terminals, unless matched, can cause common mode errors. The ability to match the transistors and current sources, due to the close spacing on an IC chip, makes this approach feasible in low cost ICs. These commonly used circuit design approaches use only one resistor which is adjusted to control the gain. Most commercially available types have feedback sense and reference terminals for lead compensation, current output sensing, and adjustable offset reference voltage.

In most systems, instrumentation amplifiers will be used for preamplification and for adapting the input signal range to the range of the A/D converter.

Since they respond only to the difference between two voltages, they can be used in both balanced and unbalanced systems. In balanced systems the output of the signal source appears on two lines, both having equal source resistances and output voltages in relation to ground or the common mode level. An unbalanced system does not use this symmetry. A major application of instrumentation amplifiers is in eliminating the effects of ground potential differences in single ended systems.

Since instrumentation amplifiers can measure voltage differences at any level within their range, they are useful in current measurements. Typically, they measure and amplify the voltage appearing across a low resistance shunt.

When the reference terminal is available, it can be used to bias out dc voltages, such as contact potentials, or it might be used to bias relay or comparator trip points. The reference terminal could be driven by an operational amplifier with either constant or variable voltage, or if the amplifier has high input impedance at the reference terminal, it can be driven by a voltage divider or potentiometer. The sense and reference points are usually connected to the specific points in the circuit at which the output is to be maintained.

Isolation Amplifiers

Isolation amplifiers are designed for applications that require the actual galvanic isolation of the amplifier's input circuit from the output and the power supply. Some typical applications include:

1. High common mode voltages between the input and the output.
2. Medical electronics equipment which makes physical contact for a measurement.
3. Two-wire inputs with no ground return for bias currents.

The most common techniques for obtaining the isolation required are transformer and optical coupling. Optical coupling uses a portion of the electromagnetic spectrum that completely eliminates any voltage, current, and magnetic flux which may transmit energy. A typical isolation amplifier with transformer coupling may have a capacitance of 10pF between input and output ground circuits, a CMR of 115dB at 60Hz, and a common mode voltage rating of 5KV. These devices have been used for medical applications which measure ECG waveforms to isolate the patient from ground fault currents.

As with the instrumentation amplifiers, these amplifiers use committed gain circuits with internal feedback networks and they can operate from dc to 2kHz. Normally, they are designed with two parts: an isolated amplifier section and an output section. The amplifier section includes a fixed gain op amp, a modulator,

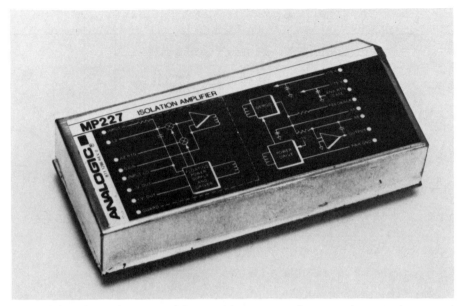

FIGURE 3-2   This isolation amplifier module uses transformer isolation with a guard-shield for the input section. (*Courtesy Analogic Corporation.*)

and a dc regulator enclosed in a floating guard shield. The output section contains the demodulator, filter, and power supply oscillator circuit operating from a single supply. Operating power is transformer coupled into the shielded input circuits and capacitively or magnetically coupled to the output demodulator circuit. A modular isolation amplifier is shown in Figure 3-2.

Filtering

Following the amplifier in the system, it may be necessary to use a low pass filter for one of two reasons: (1) to limit the bandwidth of the signal to less than half the sampling frequence to eliminate distortion frequency due to folding, and (2) to reduce electrical noise in the system. Most man-made noise has regular characteristics such as periodicity and a regular shape. It can be eliminated by specific techniques such as a notch filter. Thermal, or Johnson, noise is a random noise with a noise power proportional to the bandwidth. It may be minimized by restricting the bandwidth to the minimum required to pass the necessary signals. No filter is perfect for eliminating all noise or all undesirable frequency components. The choice of a filter is always a compromise.

Ideal filters have flat response, infinite cut-off attenuation, and a linear phase response. In practice one has the choice of a cut-off frequency, and an at-

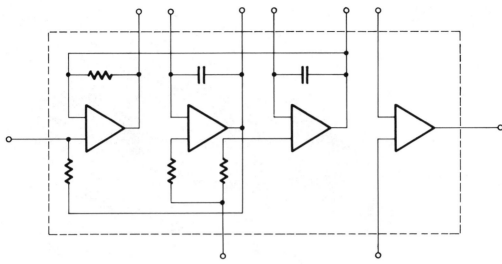

FIGURE 3-3   A Thin-Film Active Filter.

tenuation rate and phase response based on the number of poles and filter characteristics. The effect of overshoot and nonuniform phase delays must always be considered.

Active filters can have several advantages over passive filters. They eliminate inductors with their associated size, saturation, and temperature problems. The response of an active filter is set by a number of temperature-stable capacitors and resistors. Insertion loss and loading effects are minimized by the use of operational amplifiers.

A typical active filter may be manufactured with thick or thin film hybrid technology. It might use the state variable principle to implement a second order transfer function as does the device in Figure 3-3. Here three operational amplifiers are used for the second order function while a fourth uncommitted op amp could be used as a gain stage, summing amplifier, buffer amplifier, or to add another real pole. Two-pole lowpass, bandpass, and highpass output functions are available simultaneously from three different outputs. Notch and allpass functions are available by combining these outputs in the uncommitted op amp. To obtain higher order filters, several devices can be cascaded. Q ranges from 0.1 to 1,000 and the resonant frequency range is 0.001 Hz to 200 kHz. The frequency stability is 0.1%/C. Frequency tuning is done with two external resistors and Q tuning by a third external resistor. For resonant frequencies below 50 Hz, two external capacitors are added. By proper selection of the external components, any of the popular filter types such as Butterworth, Bessel, Chebyshev, or Elliptic may be realized.

To aid in the selection of an instrumentation or isolation amplifier, the following information may be considered:

1. The character of the application—differential or single ended, follower inverter, linear, or nonlinear.
2. Accurate description of the input signals—voltage, current, range of amplitude source impedance, and time and frequency characteristics.
3. Environmental conditions, including the maximum ranges of temperature, time, and supply voltage.
4. Accuracy required as a function of handwidth, static and dynamic parameters, and loading.

A good initial analysis of the problem should be made, using conservative design rules such as the best available data, reasonable tolerances on resolution, accuracy, and timing and the proper connection scheme. Where appropriate, breadboarding should be used to verify questionable points.

The design should always include features that ease testing and troubleshooting. Be sure that common mode, normal mode, and induced noise problems have been considered adequately. This includes the use of differential amplifiers, filtering, and lead locations. Grounding should be proper with no ground loops (by allowing ground current only one path). The digital and analog grounds should be separated along with high power and low signal grounds. One point where all grounds meet is advisable. Heavy ground conductors can be used to avoid voltage drops in signal return leads.

MULTIPLEXERS

When more than one channel in the system requires analog-to-digital signal conversion, or some other processing that may be costly to duplicate for each channel, it may be necessary to time-division multiplex the analog inputs to a single converter or processing circuit. The designer also may wish to provide a converter for each input and then combine the converter outputs by digital multiplexing into the processor.

Analog multiplexer circuits are normally used for the time-sharing of analog-to-digital converters between a number of analog information channels from the various sensors or system inputs. An analog multiplexer provides a group of switches with the inputs connected to the individual analog channels and a common output as shown in Figure 3-4a. The switches are usually controlled by a digital input.

MOS-FET switches are generally used and these are connected directly to the output load if it has a high enough impedance, or an output buffer amplifier

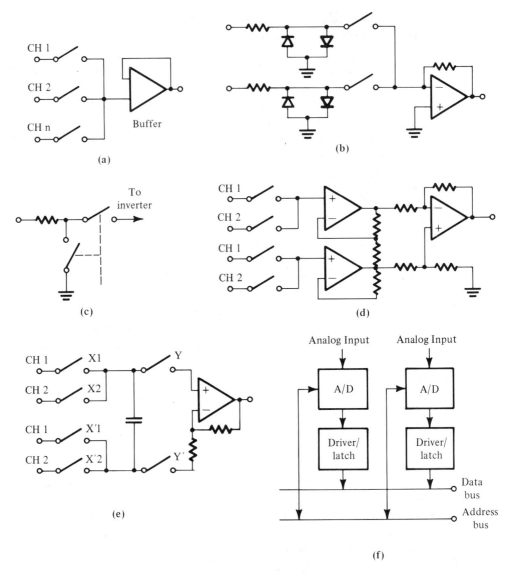

FIGURE 3-4   Multiplexer Configurations. (a) Analog Multiplexing; (b) Current Switch Multiplexing; (c) Constant Impedance Multiplexing; (d) Differential Multiplexing; (e) Flying Capacitor Multiplexing; (f) Digital Multiplexing onto the Bus System.

can be used to provide a higher impedance to the switches. A bipolar transistor follower buffer can furnish an input impedance of 10^9 ohms, resulting in a negligible transfer error across a switch resistance of 2K ohms.

MOS-FET multiplexers use reversed biased diodes to protect the input channels from being damaged by overvoltage signals. The input channels are usually protected for 20 V beyond the supply voltage. This can be increased by adding resistors in series for each channel. The input resistors limit the current flowing through the protection diodes.

Many different types of analog switches have been used in multiplexers. Electromechanical switches include relays, stepper switches, cross bar and mercury-wetted and dry reed relay switches. The fastest switching speed is provided by reed relays which switch in approximately 1 ms.

The mechanical switches are characterized by high dc isolation resistance, low contact resistance, voltage ratings to 1kV, and they can be inexpensive. Multiplexers using mechanical switches are limited to low speed applications as well as some with high resolution requirements.

The mechanical switches interface well with the slower A/D converters, such as the integrating dual slope type. They have a finite lifetime, which is usually expressed in the number of operations. A reed relay might have a life of 10^9 operations, which would allow a three year life time at ten operations/second.

The solid-state switches can operate at speeds of 30 ns, and they have a lifetime which exceeds most equipment requirements. Field effect transistors (FETs) are now used in most multiplexers. They have superseded bipolar transistors which can introduce large voltage offsets. The circuit in the multiplexer is usually arranged such that the gate voltage of a conducting junction-FET tracks the drain-source voltage to maintain $V_{GS} = O$. This ensures that the resistance of the switch is constant and not a function of the signal level being multiplexed. This is not the case of MOS-FET multiplexers where the insulated gate is driven by a fixed potential in the on condition. Here, V_{GS} and the on resistance can vary with the level of the applied signal. A device might have an R_{ON} of 200 ohms at +10V and an R_{ON} of 1000 ohms at −1OV. Enhancement-mode MOS FETs have an advantage that the switch turns off when power is removed from the MUX. Junction-FET multiplexers always turn on with the power off. Multiplexers that use CMOS (complementary MOS) switches have the advantage of being able to multiplex voltages up to and including the supply voltages. A ± 10V signal can then be handled with a ± 10V supply.

FET devices have a leakage from drain to source in the off state and a leakage from gate or substrate to drain and source in both the on and the off states. The gate leakage in MOS devices is small compared to other sources of leakage. When the device has a Zener-diode-protected gate, an additional leakage path exists between the gate and source.

Analog multiplexing has been a popular technique for low cost systems. The decreasing cost of A/D converters and the availability of low cost, digital integrated circuits specifically designed for multiplexing now provides an alternative with advantages for some applications. Since analog multiplexing may be used with sensors or other sources which may provide little overload capacity or poor settling after overloads, the switches should have a break-before-make action to prevent the possibility of shorting channels together. Some systems may be required to have all channels off with power down. In addition to the channel addressing lines, which are normally binary coded, it can also be desirable to have inhibit or enable lines to turn all switches off regardless of any of the inputs. This simplifies the external logic necessary to cascade multiplexers and is useful for certain schemes of channel addressing.

Multiplexing Configurations

Both analog and digital multiplexers must be able to absorb the overload transient energy and recover without damage. High level multiplexers are designed to operate with input signals greater than 1V. Most types use a bank of switches connected to a common output bus, as shown in Figure 3-4a. The bus output may be buffered as shown. The configuration is simple, and with an output amplifier, it offers a high input impedance. Depending on the switching device used, this type of multiplexer can operate over a wide variation of input voltage. With solid-state switches, the input voltage excursion is limited to about ± 20V. Most multiplexers are designed for the standard analog range of ± 10V.

For applications which require the switching of up to several hundred volts with solid-state speed, the inverting current switch multiplexer can be used (Figure 3-4b). Here the switching takes place at a summing junction, with protection diodes connected to ground. The actual switches are not subjected to high voltages and the circuit has a high immunity to transient voltages, a constant low input resistance while conducting, and it assumes a safe state when power is removed. This multiplexer is rugged and has been used for interfacing ± 100V analog computers and other high voltage inputs in the past.

If the diodes are replaced by FETs which are driven in a complementary mode, the input resistor is terminated in a real or virtual ground. Then the input resistance will be almost constant regardless of the channel selected (Figure 3-4c). This avoids any settling problems at the transducer during switching due to changes in the loading.

Another modification removes the input resistor to allow the multiplexing of current output transducers. When current output switching is used, the transfer accuracy is relatively unaffected by variations in line and connection resistance.

The multiplexing of voltages in the millivolt range, up to 1V, requires more critical circuits. Low level interference and thermal effects can be large here, so the lines are run in pairs and differential techniques are used to remove interference present as a common mode signal. When the common mode voltages are high, guarding techniques can be used along with 3 wire multiplexing of shielded input pairs.

A 2 wire differential multiplexer is constructed with pairs of switches, as shown in Figure 3-4d. The output amplifier is normally an instrument amplifier with a high common mode rejection. This rejection is only effective if the input lines are identical so twisted pairs of cabling and matching of the parameters for both the channels and switches are required. The use of integrated circuits and dual FET switches can allow the matching required. Switch leakages and thermal EMFs may also introduce errors in the low level inputs; drift is of greater concern. To reduce the effects of an unbalance in the input cables, shielded pairs may be used with the shield driven by the common mode voltage. This is done by either the source or a mid-position tap in the amplifier or between the input terminals. This type of shielding, or guarding, is common in high resolution systems.

Another low level multiplexer circuit that is effective for common mode interference is the flying capacitor multiplexer, which is a 2 wire sample hold circuit as shown in Figure 3-4e. Switches X and X' are on with Y and Y' to acquire the input signal. When the capacitor is completely charged, all switches are momentarily turned off. Then Y and Y' are turned on to transfer the signal to the output amplifier. No common mode voltage is transferred across the switches and the output amplifier may be single ended and noninverting.

This circuit is effective in eliminating common mode voltages, but if normal mode interference is present as well, a better rejection ratio of both normal mode and common mode is obtained by using straight multiplexer and a floating input integrating converter. An integrating converter used with the flying capacitor multiplexer integrates a sample of the input rather than the input, and the sample taken includes the variations from normal mode interference which will be integrated out.

Multiplexing can introduce both static (dc) and dynamic errors into the system. If these errors are large in relation to the resolution of the measurement, they will degrade the measurements.

Static voltage errors are caused by switch leakage and offsets in amplifiers. Gain errors may be caused by switch ON resistances, source resistances, amplifier input resistances, and amplifier gain nonlinearity.

Dynamic errors may come from charge injection of the switch control voltages, settling times of the bus and input sources, circuit time constants, crosstalk between channels, and output amplifier settling. Dynamic errors are also affected by the multiplexer circuit and system wiring layout. Other errors may be due to the characteristics of the multiplexer components, for example, reed relays may generate a thermal EMF. The errors found in high level multiplexers can also occur in low level multiplexers, but the effects are more serious because of the lower signal levels involved.

Digital Multiplexing

Digital multiplexers provide a solution to most of these problems, but the decision to use them should consider the trade-offs based on the following factors:

1. Resolution—The cost of an A/D converter rises steeply as the resolution increases, due to the cost of the precision elements required. At the 8 bit level, the per-channel cost of an analog multiplexer can be a large part of the cost of a converter. At resolutions of approximately 12 bits, the reverse tends to be true and analog multiplexing may be more economical.

2. Number of channels—This sets the size of the multiplexer required and the amount of wiring and interconnections. Digital multiplexing onto a common data bus reduces wiring to a minimum in many cases. Analog multiplexing is suited for 8 to 256 channels. Beyond this, the technique becomes unwieldy and the analog errors become very difficult to minimize. Analog and digital multiplexing can often be combined in large systems.

3. Speed of measurement or throughput—High speed A/D converters can have a considerable cost impact on the system. When analog multiplexing requires a high speed converter to achieve the desired sampling rate, a slower converter for each channel with digital multiplexing may be cost effective.

4. Signal level and conditioning—Wide dynamic ranges between various channels can cause problems with analog multiplexing. Signals less than 1V generally require differential low level analog multiplexing, which is expensive, with programmable gain amplifiers after the multiplexing. The use of fixed gain converters on each channel, with the signal conditioning designed for the channel differences and the use of digital multiplexing may be more efficient.

5. The location of the measurement points—Analog multiplexing is suitable for measurements at distances of a few hundred feet from the converter. Analog lines may suffer from losses, transmission line reflections, and interference. Lines may range from twisted wirepairs to multiconductor shielded cable, depending on the signal levels, distance, and noise environment.

Digital multiplexing can be used for distances of thousands of miles with the proper transmission equipment. Digital transmission systems have the noise rejection characteristics that are required for long distance transmission.

Systems with small-to-moderate numbers of channels can use medium scale integrated digital multiplexers which are available in TTL and MOS logic families. The 74151 is typical—eight of these ICs can be used to multiplex eight A/D converters with 8 bit resolution onto a common data bus. This type of digital multiplexing offers little, if any, advantage in wiring economy, but it is low cost, and the high switching speed allows sampling rates much faster than analog multiplexers. The A/D converters are required only to keep up with the channel sample rate and not with the commutating rate.

When large numbers of A/D converters are to be multiplexed, the data bus method shown in Figure 3-4f reduces system interconnections. The data are bussed onto the lines in bit-parallel or bit-serial format, depending on the converter's output.

A digital distribution system may not use a device that specifically acts as a digital multiplexer. Unlike the process of shunting analog information to many output channels or from many sources, the digital multiplexing function is often delegated to the devices being multiplexed since they share a set of common inputs via a bus.

Commands from the processor instruct the individual sources as to which must feed its data into the common bus. A variety of devices may be used to drive the bus, including open collector and tri-state TTL gates, line drivers, and opto-electronic isolators. The channel selection decoders can be built up from one of 16 decoders to the required size. This technique allows some additional reliability in that a failure of one A/D does not affect the other channels.

An important requirement is that the multiplexer operate without introducing unacceptable errors at the sample rate speed. In a digital MUX system, the speed can be determined from the device propagation delays and the time required to change the bus capacitance.

Analog multiplexers can be more difficult to characterize. Their speed is a function of not only the internal parameters but also external parameters such as channel source impedance, stray capacitance, the number of channels, and the circuit layout. The nonideal transmission and open circuit characteristics of analog multiplexers introduce static and dynamic errors into the signal path. These errors can include leakage through switches, coupling of control signals into the analog path, and interactions with sources and following amplifiers. The circuit layout can compound these effects. The user should be aware of the limiting parameters in the system in order to judge their effects on performance.

Low level multiplexers should always be differential or two wire, so the converter "sees" the difference in the errors of two identical channels. Leakage, gain, and crosstalk effects are greatly reduced, provided that matching is maintained. The magnitudes of most settling errors are decreased, although their duration remains the same.

When multiplexers must be operated in conditions with high common mode interference, two wire differential and guarded or flying capacitor multiplexers should be used. If considerable normal mode interference also exists, further steps are required. The most common techniques include filtering, digital averaging, and the use of integrating converters.

Low pass filters in the inputs of the multiplexer are one method of reducing normal mode interference. The filter characteristics may be tailored for the channel. The filters can increase the settling time, but this effect is usually small. The filter can also be placed after the multiplexer, but then each channel must charge the filter, greatly increasing the settling time. In differential systems, the filters should have balanced impedance in both inputs or be connected differentially to reduce the common mode effects.

If passive filtering of each channel is not practical, then an integrating A/D converter can be used to provide high normal mode rejection. This rejection can be obtained with a conversion time that is usually shorter than the settling time of a filter required to provide the same rejection. Rejections of normal mode interference to 70dB can be obtained with an integrating converter. Many integrating converters are also designed for floating guarded inputs.

In systems where processing time and memory are available, provided the converter can track the variations in input signal produced by interference, then software may be used to reduce the effects of interference. Multiple samples are taken for each channel and the results summed and averaged. The signal-to-noise ratio improves as the square root of the number of samples, provided that the sampling and interference frequencies are not correlated.

In unfavorable environments, the major source of noise is induced. In this case, the designer should rely on techniques like early preamplification and conversion, isolation, shielding and guarding, signal compression, and filtering along with an information rate using fast sampling or parallel paths with enough redundancy to allow the processor to retrieve data and use correlation and summation.

In favorable environments, the measurement process and the processing hardware introduce the major part of the system uncertainty. Emphasis should be placed on the measurement techniques: filtering, data acquisition resolution, and the use of digital processing for signal retrieval, including drift compensation and scale factor adjustments. When noise is likely to have large spikes as a major component, the integrating type converter can be used, but for random noise, the statistical properties of the noise allow it to be filtered by digital techniques.

SAMPLE HOLD DEVICES

The sample hold device has a signal input, an output, and a control input. It uses two operating modes: (1) sample or (2) track, in which it acquires the input signal and tracks it until commanded to hold; during this time it retains the value of the input signal. Sample holds are sometimes called *track holds* when they spend a large portion of time in the sample mode tracking the input.

Sample holds normally have a unity gain and are noninverting. The control inputs are usually TTL compatible. A logic 1 is usually the sample command and a logic 0 the hold command.

The sample and hold in its basic form consists of a switch and a capacitor (Figure 3-5). As the switch is closed, the unit is in the sampling or tracking mode and will follow the changing input signal. As the switch is opened, the unit is in

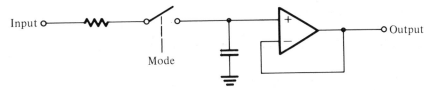

FIGURE 3-5   Basic Sample and Hold.

the hold mode where it retains a voltage on the capacitor for a period of time, depending on the capacitor size and leakage.

Most sample hold devices also use input and output buffer amplifiers along with solid-state switching circuits. The output buffer amplifier is usually a low input current FET amplifier that minimizes the leakage of the capacitor. The switch must also have a low leakage.

An ideal sample and hold, or *zero order hold* as it is sometimes called, takes a sample in zero time and then holds the value of the sample indefinitely without lost of accuracy. In practice commercial devices are usually specified in the ways that they differ from the ideal. Common parameters follow:

1. *Acquisition time* is the time from when the sample command is given to the point when the output enters and remains within a specified band about the input value. At the completion of the acquisition time the output is tracking the input.

2. *Aperture time* is the time between the hold command and the point at which the sampling switch is completely open. It is sometimes called the turn-off time. The required aperture time for a binary resolution is found by dividing the resolution by 2π times the conversion frequency.

3. *Aperture uncertainty time* is the variation in the aperture time which is the difference between the maximum and minimum aperture times.

4. *Decay rate* is the change in output voltage with time in the hold mode.

5. *Feedthrough* is the amount of input signal which appears at the output when the unit is in the hold mode. The feedthrough varies with the signal frequency and is sometimes expressed as an attenuation in dBs.

6. *Settling time* is the time measured from the command transition until the output has settled within a specified error band about the final value.

The purpose of signal sampling is the more efficient use of data processing equipment or data transmission facilities. A single data transmission link can be used to transmit many channels of information by sampling each channel periodically.

For the efficient use of processing equipment to monitor and control a process, it may only be necessary to sample the state of a process once every few minutes, perform a computation and correction, and then free the computer for the remaining time for other tasks. Continuous monitoring of a single information channel by the computer is very inefficient. In most systems it is better to use a single A/D converter which is connected to a number of information channels with sampling techniques.

In most practical devices, a sample is taken during a time period which is short compared to the holding time. During the holding time there may be some change in the output which may affect the system accuracy.

The effects on a continuous analog input signal may be determined by finding the transfer function of a sample hold. By using the impulse response of the device along with the Laplace transform, the transfer function of the ideal sample hold is found as follows:

$$G(j\omega) = \frac{1 - e^{-j\omega T}}{j\omega} = \frac{2\pi}{\omega_s} \frac{\sin \pi\omega/\omega_s}{\pi\omega/\omega_s} e^{-j\pi \left(\frac{\omega}{\omega s}\right)}$$

T is the sampling period and w_s is the sampling frequency. If the magnitude and phase of this function are plotted, then the sample hold acts like a low pass filter with a cut-off frequency fc of approximately fs/2 and a phase delay of T/2, or one-half of the sampling period.

By changing the sampling rate such that $f_s - f_c > f_c$, the result is that $f_s > 2f_c$, which is the basis of the sampling theorem.

Circuit Configurations

The type of storage element used divides sample hold devices into two classes. The more conventional type uses a capacitor for storage while the other class uses an A/D converter, a register for storage, and a D/A converter. It is more complex and costly but it has the advantage of an arbitrary hold time.

An open loop follower circuit is shown in Figure 3-5. When the switch is closed, the capacitor charges exponentially to the input voltage and the amplifier's output follows the capacitor voltage. When the switch is opened, the charge voltage level remains on the capacitor.

The capacitor's acquisition time depends on the series resistance and the current available. When the charge is completed to the desired accuracy, the switch may be opened, even though the amplifier may not have settled. This will not affect the final output value or the settling time greatly as the amplifier's input stage does not draw any appreciable current, provided that the switch is an FET and the amplifier has an FET input.

The disadvantage of this circuit is that the capacitor loads the input source which can oscillate or reduce the current required to charge the capacitor fast enough. The circuit can be modified to include a follower to isolate the source. For faster charging at close to a linear slew rate, a diode bridge can be used, as shown in Figure 3-6a. The current sources are switched on to charge the capacitor. If the bridge and current sources are balanced, current flow into the capacitor stops when the capacitor voltage is equal to the input voltage.

These circuits have the advantage of fast acquisition and settling times but they are open loop devices. When low frequency tracking accuracy is more critical than speed, a cascaded amplifier configuration will be less effective than a single amplifier configuration that provides some isolation. This can be accomplished by closing the loop around the capacitor, and using a high loop gain for tracking accuracy. Figure 3-6b shows how the input follower amplifier can be replaced with a high gain difference amplifier. When the switch is closed, the output represented by the charge voltage on the capacitor is forced to track the input, as a function of the gain and the current capability of the input amplifier. The common mode and offset errors in the output follower are compensated by the charge on the capacitor. In a modification of this circuit (Figure 3-6c), a current amplifier is used with an integrator, which allows the switch to operate at ground potential and reduces the error due to leakage. In both of these circuits, the charge is controlled by the output as well as the input.

If the basic feedback circuit is switched into hold before the output has settled at the input value, the sample may be in error. Since the loop is open during the hold, it must reestablish the input when it returns to the sample mode even if the input has not changed. This can result in a step or spike if the input amplifier has a high enough voltage gain. In the ideal sample hold, tracking is error free, acquisition occurs instantaneously with zero settling times, and hold time is infinite with zero leakage.

System Application

Sample hold devices are widely used for data acquisition. Typically the sample hold maintains the intput to an A/D converter constant during the conversion interval, while the multiplexer is seeking the next channel. When the conversion is complete, the sample hold samples a new input, and the cycle is repeated. This is known as *synchronous sampling* since the sample hold operates in a synchronous mode with the other elements of the system. If the input signals change at different rates, a programmed access scheme can be used to allow the signals with the most information to be sampled more often.

In the asynchronous mode, a number of sample holds are used which acquire and store the data at rates suited to each individual channel. These sample holds are then either sampled by analog multiplexers, or the signals are converted asynchronously and then multiplexed digitally after digital signal processing.

Sample hold circuits may be used with both A/D converters and D/A converters. With A/D converters they shorten the aperture time for the converter by sampling the input signal and then holding this value until the conversion is completed. With D/A converters they can be used to remove transients which appear at the outputs when the converters change from one analog level to another.

Fast sample holds can also be used to acquire and measure short pulses having an arbitrary occurrence and width.

In data distribution applications sample holds can also be used to smooth D/A outputs in systems that may be sensitive to spikes by sampling the outputs after the settling time is over. Sample holds may cost less than a larger number of D/A converters having comparable accuracy. Here a fast D/A converter updates a number of sample holds.

There are also a number of sample hold applications in hybrid computing and data reduction. One example is shown in Figure 3-6d. Here, a peak follower is

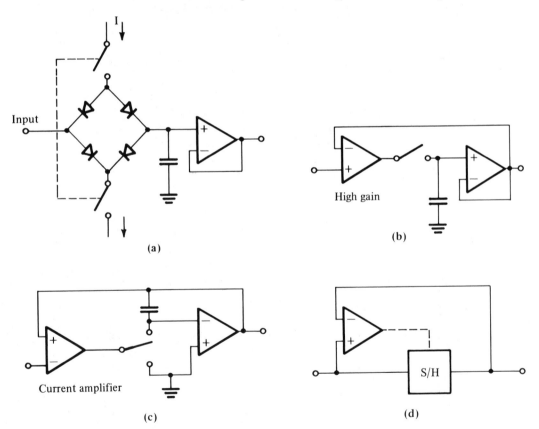

FIGURE 3-6   Sample-Hold Configurations. (a) Current Source; (b) Feedback Loop; (c) Integrator Feedback; (d) Peak Follower; (e) Counter; (f) Digital.

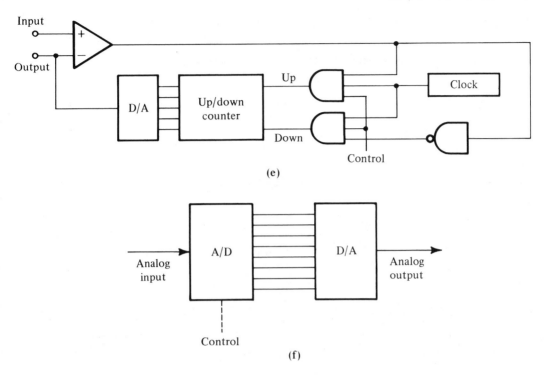

FIGURE 3-6   *continued*

made up of a sample hold and a comparator. Balancing the unknown input voltage against some internally produced reference, the comparator circuit responds the inequality between the input and the reference. The sample hold output should be biased by a few millivolts of hysteresis signal to avoid any ambiguity problems from step inputs and also to reduce false triggering from noise.

When the comparator input is greater than the sample hold output, the comparator's positive output forces the sample hold into a tracking mode. As the comparator input becomes less than the sample hold output, the comparator puts the sample hold into the hold mode until the input again becomes greater than the output. To reset the circuit, the control input is switched into sample and a low level signal is applied to the input.

Digital storage provides a means for long hold duration with no droop, no sample hold offset, no feedthrough problems, and no dielectric leakage effects. The disadvantages and cost, complexity, longer acquisition times, and the potential requirement of presampling as illustrated in the counting sample hold of Figure 3-6e.

This sample hold uses a D/A converter, an up-down counter, a comparator clock, and some logic gates. The initial acquisition time may be long, since the clock period (ts) depends on the LSB settling time of the D/A converter and the number of counts depends on the resolution. For a full scale step, the acquisition time is approximately $(2^n - 1)$ t_s. Smaller, slower changes are followed more rapidly.

The circuit can be converted into a peak follower by disabling the up count function. The range of input signal levels and polarity determines the choice of D/A converter. A BDC counter and BCD D/A converter could be used with a display for a maximum peak digital voltmeter.

Another approach is to use an A/D converter and a D/A converter (Figure 3-6f). When averaging is desired, the A/D converter must be the integrating type. The acquisition time here is approximately the sum of the A/D converter's conversion time and the D/A converter's settling time. When the D/A output of a successive approximations A/D is used, a separate D/A converter is not required and the acquisition time is then equal to the conversion rate.

DATA CONVERSION

Analog to digital (A/D) converters translate the analog data which are characteristic of most measurements in the real world to the binary words required for data processing and transmission in control systems. Digital-to-analog (D/A) converters are used to transform the data or the result of computations back to analog form for control, display, or analog processing. The analog variables are usually converted into voltages or currents. The quantities may appear as changing ac measurements, waveforms, or a combination.

Most analog variables in practice are changing dc voltages or currents. They may be scaled from a direct measurement, or subjected to demodulation or filtering techniques. The voltages and currents must always be scaled to ranges compatible with the converter input range.

Many converters use a full scale of 10 or 10.24V. For 10V, the bit values are expressed as negative powers of 2 multiplied by 10; for 10.24V, the least significant bit (LSB) is expressed in rounded numbers which are usually multiples or submultiples of 10mV.

Digital Codes

Digital numbers are represented by the presence or absence of fixed voltage levels arranged to represent the decimal fraction or weight of the bit position.

Binary is the most common code, but several other codes are used in systems, depending on the signal range and polarity, conversion technique, special requirements characteristics, and origin or destination of information.

The Gray Code is a binary code in which the bit position alone does not signify a numerical weighting; each coded segment corresponds to a unique portion of the analog range. A comparison of the Gray Code with natural binary follows:

Decimal Fraction	Gray Code	Binary Code
0	0 0 0 0	0 0 0 0
1/16	0 0 0 1	0 0 0 1
2/16	0 0 1 1	0 0 1 0
3/16	0 0 1 0	0 0 1 1
4/16	0 1 1 0	0 1 0 0

Binary coded decimal (BCD) is a code in which each decimal digit is represented by a group of 4 binary coded digits. The least significant bit of the most significant group, which is sometimes called a quad, has a weight of 0.1, the LSB of the next group has a weight of 0.01, the LSB of the next group has a weight of 0.001. Each group or quad has ten levels with weights 0 to 9. Group values in excess of 9 are not allowed.

In the Gray Code, as the number value changes, the transition from one code to the next requires only one bit change at a time.

In systems where many bits can switch at a transition, it is possible to latch in midtransition and lock in a false code. The use of the Gray Code will limit these errors to a least significant bit change.

The actual mechanization of some converters may require codes, such as natural binary or BCD, in which the bits are represented by their complements. These codes are known as Complementary Codes.

In a 4-bit complementary binary converter, 0 may be represented by 1111, half scale (MSB) by 0111, and full scale (less 1 LSB) by 0000.

In a similar manner, for each quad of a BCD converter, complementary BCD is obtained by representing all bits by their complements, thus, 0 is represented by 1111, and 9 is represented by 0110. The equivalents for 1 through 4 in complementary binary and complementary BCD (with an overrange bit) are shown:

Equivalent Binary	BCD	Natural Binary	Complementary Binary	BCD	Complementary BCD
0		0000	1111	00000	11111
1, 1/16	1/10	0001	1110	00001	11110
2, 2/16	2/10	0010	1101	00010	11101
3, 3/16	3/10	0011	1100	00011	11100
4, 4/16	4/10	0100	1011	000100	11011

If a natural binary input is applied to a D/A converter coded to respond to complementary binary, the output is in reverse order. A zero output for all 1's input results.

These Complementary Codes have involved complementing all bits and do not consider the representation of the analog polarity.

These unipolar codes represent numbers, which in turn represent the magnitudes of the analog variables, without regard to the polarity. Gray Code is an exception. Since it is not quantitatively weighted, it can represent any arbitrary range of magnitudes of either polarity.

A unipolar A/D converter responds to analog signals of one polarity and a unipolar D/A converter produces analog signals of one polarity. The signal polarity is determined from the converter reference. The conversion of bipolar analog signals into a digital code requires sign information in the form of one extra bit called the *sign* bit. This bit doubles the analog range and halves the resolution.

In some converters the sign bit is obtained by interpreting the existing middle significant bit. Then the analog range is doubled, but the resolution is twice as coarse. A 10 bit converter's resolution is 1/1024, with a +10V range. If a bipolar code of 11 bits is used, the resolution is 1/2048 and the range is ±10V.

D/A Converters

The basic D/A converter circuit uses a reference, a set of binary weighted precision resistors, and switches as shown in Figure 3-7a. In this circuit, an operational amplifier is used to hold one end of the resistors at zero. The switches are controlled by digital logic. Each switch that is closed adds a binary weight of current E_{REF}/R_j at the summing point at the amplifier's negative input. The output voltage is proportional to the total current, which in turn is a function of the value of the binary number.

In a converter which provides 12 bit D/A conversion, the range of resistance values would be 4,096:1 up to 40 meg-ohms for the least significant bit. If the resistors are manufactured from thin or thick films, or in integrated circuits, this range is not practical. If discrete resistors are used, cost, size, and temperature tracking become problems.

Resistance ladders are used to reduce the resistance range. These require a limited number of repeated values, using attenuation. One approach is to use a binary resistance quad, with four values: 2R, 4R, 8R, 16R for each group of 4 bits. An attenuation of 16:1 is used for the second quad and 256:1 for the third quad. The proper quad weights for BCD conversion may be achieved with an attenuation between quads of 10:1.

The R-2R ladder allows an even greater reduction of resistance values as illustrated in Figure 3-7b.

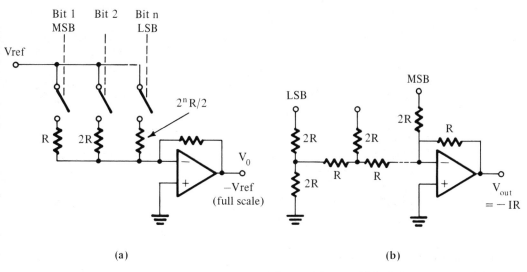

FIGURE 3-7 D/A Converter Circuits. (a) Basic Circuit; (b) R-2R Resistance Ladder.

Figure 3-7a shows a basic current weighting D/A converter. A set of binary weighted currents flows through a feedback resistor, producing an analog output voltage proportional to the sum of the currents that are allowed on by the switches.

In a converter of this type the critical parameters are the switching speed and resistor temperature matching. The range of resistance must be obtained from a consistent film, so the resistance temperature coefficients can be matched.

The switching speed of the current switch will depend upon the current available to charge the stray capacitance. The design difficulties of D/A converters increase rapidly with the resolution. Eight bit D/A converters are relatively easy to design and produce since the accuracy requirement is of the order of 0.2%. Ten bit converters are more difficult to design, since the requirement is 0.05%. As one reaches 12 bits of resolution (0.0125%), the design and tolerance control become critical.

The following general considerations should be used to define the specific requirements:

1. Resolution, or the number of bits of the incoming data word to be converted, the analog accuracy and linearity required.
2. The type of digital codes and logic levels required.
3. Output signal requirements current, voltage, full scale range. Voltage output D/A converters can be easier to use. The current output D/A converters are normally used in those applications where high speed is im-

portant rather than a voltage output. These include A/D circuits using comparators.

4. Type of reference is required, fixed, internal or external, variable, multiplying. The number of quandrants needed for multiplying.
5. Speed requirements, or the shortest time available between data changes. The allowable system waiting time for the output signal to settle after a full scale input change.
6. Switching transients or the use of filters.
7. Temperature range or ambient and internal temperature changes; the range that the converter must perform within specifications without readjustment.
8. Power supply sensitivity, or adequate stability to hold errors within specifications.

A/D Converters

There are a number of types of A/D converters. A limited number of these are available in small, modular form for incorporation into equipment or systems. The most popular types are:

1. Successive approximation.
2. Integration.
3. Voltage to frequency.
4. Counter comparator.
5. Parallel.

Each type has special characteristics which make it more useful for specific system applications.

The successive approximation converter has been in wide use for computer interfaces because of resolutions to 16 bits and high speeds in the MHz range. Conversion time is fixed and independent of the magnitude of the input voltage. Conversion is independent of the results of previous conversions, since the internal circuits are cleared at the start of a new conversion.

The successive approximation technique consists of comparing an unknown input against a precisely generated internal voltage at the output of a D/A converter (Figure 3-8a). The input of the D/A converter is the digital output of the A/D converter. The process is similar to a weighing process using a set of n binary weights.

The input should not change during conversion. If the input were to change during the conversion, the output would no longer represent the analog input. It is not unusual to employ a sample hold device ahead of the converter to retain the input value that was present before the conversion started. The status output of the converter may be used to release the sample hold from the hold mode at the end of the conversion. The sample hold may not be needed if the signal varies slowly and is noise free so that significant changes will not be expected during the conversion interval.

Accuracy, linearity, and speed are a function of the characteristics of the D/A converter, its reference, and the comparator. The settling time of the D/A converter and the response time of the comparator are typically much slower than the switching times of the digital elements. The differential nonlinearity of the A/D converter is a reflection of the differential nonlinearity of the integral D/A converter. If the D/A converter is nonmonotonic, one or more code steps can be missing from the A/D converter's output range.

The integration converter types are quite popular. These converters use an indirect conversion, first converting to a function of time and then converting the time function to a digital quantity using a counter.

The dual ramp integrator is useful for the slower responding transducers. They have been the predominant type used in digital voltmeters. Many of these DVMs use sign magnitude BCD which requires polarity sensing and switching.

The conversion circuits in this class include the single, dual, and triple ramp types and the voltage-to-frequency converters. The ramp converters use a reference voltage of opposite polarity to the signal which is integrated while a counter tracks the clock pulses until the integrator output is equal to the signal input. At this time, t, the output of the integrator is $V_R t/RC$. The number of counts is proportional to the ratio of the input to the reference. This single ramp circuit has the disadvantage that its accuracy depends on the capacitor and the clock frequency. The multiple ramp types provide cancellation compensation for these effects (Figure 3-8b).

In the voltage-to-frequency converters, a frequency is generated in proportion to the input signal and counter measures the frequency to provide a digital output proportional to the input signal.

The counter-comparator A/D converter, shown in Figure 3-8c, is analogous to the single ramp type, except it is independent of a time scale. The analog input is compared with the output of a D/A converter and the digital input of the D/A is driven by a counter. At the beginning of a conversion, the counter begins the count and continues until the D/A output exceeds the input value. Then, conversion stops and the converter is ready for the next conversion once the counter has been read and cleared.

Converters required for system applications usually receive external commands to convert or hold. For low frequency signals, the converter may be of the integrating type, which acts as a low pass filter. It is capable of averaging out the high frequency noise and nulling the frequencies harmonically related to the integrating period. The integrating period can also be set equal to the period of the line frequency, if a major portion of the interference occurs at this frequency and at its harmonics.

When the converter must respond to individual samples of input, then the maximum rate of change of the average input, the full scale voltage, and the conversion time (Tc) have the following relationship for a binary number of n bits:

$$\left.\frac{dV}{dt}\right|_{max} = \frac{2^{-n}V_{FS}}{T_c}$$

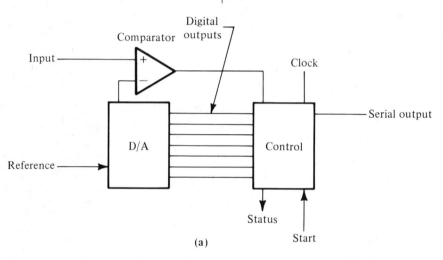

(a)

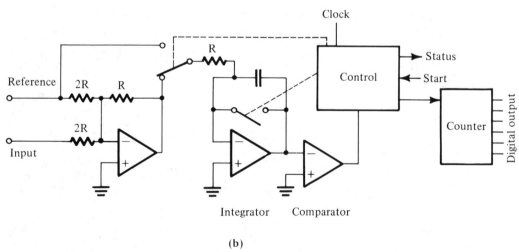

(b)

FIGURE 3-8   A/D Converter Circuits. (a) Successive-Approximation; (b) Dual-Ramp; (c) Counter-Comparator.

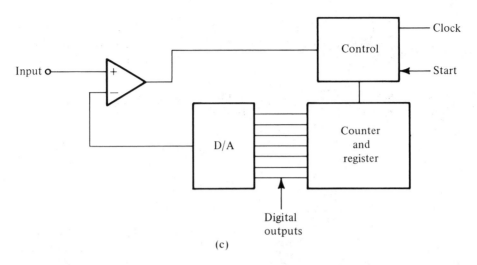

Input o—

Clock

Control

Start

D/A

Counter
and
register

Digital
outputs

(c)

FIGURE 3-8   *continued*

When individual samples are not important, but large numbers of samples will be used, as in a stationary process, the only requirement is that the signal be sampled at least twice each cycle for the highest frequency of interest.

The dual slope integrating A/D converter spends about a third of its sampling period performing an integration, and the rest of the time it is counting out the average value over the integrating period and resetting for the next sample. It will always read the average value, which results in a sample of the input waveform over the integrating window or period. Even though they are slow, the integrating A/D converters are useful for measurements of temperature and other slowly varying voltages, especially in a noise environment.

Successive approximation converters are capable of high resolutions and high speeds. If the conversion time of a typical successive approximation converter is 10us, the maximum allowable dV/dt can be 500V/s. The successive approximation converter has a weakness in that in higher rates of change, it may generate linearity errors since it cannot tolerate changes during the weighting process. Then the converted value is somewhere between the values at the beginning and the end of conversion and the time uncertainty can approach the conversion interval. When the signal is slow enough, noise with rates of change that are large can cause errors that may not be averaged. An external sample hold can be used to improve performance.

The process of selecting an A/D converter system is similar in some respects to that required for D/A converters. The following considerations include some that are analogous to D/A converters and others that are not:

1. The analog input range and resolution of the signal to be measured.
2. Requirements for linearity error, accuracy, and stability of calibration.
3. The effects on the various sources of error as the ambient temperature changes.
4. The time allowed for a complete conversion.
5. The type of reference: fixed, adjustable, or variable.
6. The system power supply.
7. The character of the input signal noise level sampled: filtered, rapid, or slowly varying. Integrating types are best for noisy input signals at slow rates, while successive approximation is best for sampled or filtered inputs at rates to 1MHz. Counter comparator types provide low cost but are slow and noise susceptible.
8. System effects—When a system is assembled in which one A/D converter is time-shared among input channels by a multiplexer and sample hold, their contribution to system errors should be considered.

SINGLE CHANNEL SYSTEMS

The simplest data acquisition circuit uses a single A/D converter, performing repetitive conversions at a free running rate. It uses an analog signal input and the outputs are a digital coded word with an over-range indication, polarity information if required, and a status output to indicate when the output is valid.

A simple example of this type of system is a digital panel meter, which uses an A/D converter and a numeric display. In this application, the purpose of digitizing is to obtain the numerical display, and the use is as a meter, rather than as a system component. Some digital meters have connections to the A/D output for digital transmission or processing.

This type of direct conversion is useful if the data are to be transmitted through a noisy environment. Consider the case of an 8 bit converter with 1/256 resolution and a 10V signal. The LSB information may be lost if the peak-to-peak noise level on the analog signal is greater than 40mV or 10V/256. Using digital transmission, the standard TTL noise immunity is 1.2V or $2.0 - 0.8V$ compared to the 40mV analog noise immunity.

Since most converters are single ended with reference to signal ground and have normalized analog input ranges of 5 or 10 volts, which are either single ended or bipolar, the signal inputs must be scaled up or down for the converter input level to allow the greatest possible use of the converter's resolution. If the signals have enough magnitude, they may have been preamplified, or they are the output from an analog power device; the scaling could be achieved with operational amplifiers using a single ended or differential configuration. When the signals are small, or have a large common mode, a differential instrumentation amplifier can be used.

In general an operational amplifier is a good choice for single ended signal inputs. Although it could be used in some cases for differential signal inputs, the input impedance and common mode rejection adjustments limit its effectiveness.

In such cases the instrumentation amplifier, or data amplifier, is the best choice. This type of amplifier uses a closed loop configuration and has the following important characteristics:

1. High impedance differential inputs.
2. Wide range of gains.
3. High common mode rejection ratio.

The actual amplifier characteristics required depend on the gain required, signal levels, and cost trade-offs. If the input signals must be galvanically isolated from the system, an optical or transformer coupled isolation amplifier can be used for isolation of the conductive signal paths. This type of isolation is required in many medical instrument applications. It is also useful in applications where large common mode spikes are encountered, applications in which the source of the measurement is at a high potential.

Sample hold devices can be used to improve the performance of single channel systems (Figure 3-9). A successive approximations type of converter can operate at greater accuracies and higher speeds by using a sample hold device at its input. Between the conversions, the sample hold acquires the input signal. Just before conversion starts it is in the hold mode and it remains here through the conversion. If the sample hold responds quickly and accurately enough, the converter will convert the changes from the preceding sample correctly, at a speed up to the conversion rate.

In practical sample holds, there is a finite acquisition time, tracking delay,

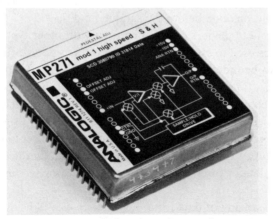

FIGURE 3-9   A sample hold module such as this can be used to improve the performance of single-channel systems. (*Courtesy Analogic Corporation.*)

and aperture time. The aperture time and tracking delay may compensate one another, or they may be unimportant if they remain consistent. Then the principal source of error is the aperture uncertainty.

The relation between the aperture uncertainty and the maximum rate of voltage change required to maintain the resolution in an n bit system is given by:

$$dV/dt \Big|_{max} = 2^{-n}V_{FS}/t_{AU}$$

for an 8-bit system let

$$f \text{ sample} = 100kHz, V_{FS} = 10V, t_{AU} = 5ns$$

then

$$dV/dt \Big|_{max} = 5mV5ns = 1.0 \text{ Vus}$$

This value is also limited by the slew rate of the sample hold.

If the successive approximations converter has a constant input applied by the sample hold, it will deliver an accurate representation of the beginning input at the end of each conversion period. Errors that are functions of time will be caused by the sample hold. This includes the acquisition errors discussed here, plus any droop during the conversion interval and linearity, offset, and transient errors. Since sample holds usually operate at unity gain, any scaling or pre-amplification should be done before the sample hold.

Sample hold devices can also be used with other types of converters. They are sometimes used to establish the timing of the signals being sampled when this is required to be independent of the conversion time. Their use is greatest when the conversion time is variable, as in the counter types of converters.

Signal conditioning in single channel systems can include a variety of techniques. The scaling of the input gains to match the input signal to the converter's full scale range is the most common example. A dc offset to bias odd ranges, such as 2.5 to 7 volts, to levels more compatible with standard converters may also be included. Isolated power converters such as shown in Figure 3-10 can also be used.

The linearizing of data from thermocouples and bridges may be performed by analog techniques, using piece-wise linear approximations with diodes or

FIGURE 3-10   An isolated power converter such as this modular unit can be used for signal conditioning. (*Courtesy Analogic Corporation.*)

smooth series approximations using IC multipliers. It can also be done digitally after the conversion, using a ROM to store the function.

Analog differentiation can be used to measure the rate at which an input varies. Integration can be used to obtain the total flow from a rate of flow. A sum and difference scheme can be used to reduce the number of data inputs for data reduction purposes.

Analog multipliers can be used to compute power by squaring the voltage or current signals. Analog dividers can be used to compute ratios, the logarithms of ratios, or square roots. These dividers can compute ratios over the wide dynamic ranges required in gas flow computations.

Comparators could be used to make decisions based on the signal levels when an input exceeds its threshold or is within a window of thresholds. Logarithmic modules are useful when range compression is required to permit the conversion of signals having greater resolutions than the converter resolution. Active filters can be used to minimize the effects of noise and undesirable high frequency components of the input signals.

In many systems all the data processing does not need to be digital. Analog circuits should be considered as candidates for performing processing or data reduction. They can be viewed as an alternate means of reducing the number of transmission channels, the software complexity, and noise.

A problem occurs with the use of analog techniques in anticipating where noise errors will appear. Digital circuits can have high noise immunity, no drift problems, high speed, and low cost; the rules for using them are simple. With the exception of pre-amplification, a great many of the functions that have been described in analog form could be performed digitally, after the conversion.

Reductions in the cost of digital circuits and the increase of chip complexity allow the development of devices that perform analog functions, but contain digital components. Examples of these include analog function generation with read only memories and the generation of arbitrary delays with timer circuits.

MULTIPLE CHANNEL SYSTEMS

In a multichannel sensor or data system, parts of the acquisition chain are shared by two or more inputs. The sharing can occur in a number of ways, depending on the requirements of the system. Some large systems can combine different types of multiplex configuration and may even use cascaded tiers of the same type.

The conventional way to digitize data from a number of sensor or data channels is to use a time-sharing process, where the input of a single A/D converter is multiplexed in sequence among the various analog sources. Data acquisition systems of this type are available as system boards, modules, or integrated circuit packages.

A typical data acquisition system board might be capable of a 110 KHz throughput rate with 12 bit resolution. In this type of system a sample hold is used with a high speed hybrid 12 bit A/D converter and a monolithic analog multiplexer. The sample hold could have a 1us acquisition time while the A/D converter does a conversion every 8us.

Many of the modular types use hybrid technology. A typical modular data acquisition system (Figure 3-11) has eight true differential input channels, 16 single ended input channels, or 16 pseudodifferential inputs for 12 bit data acquisi-

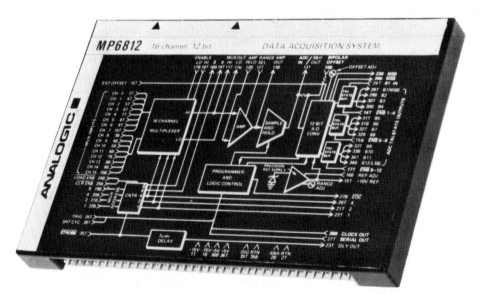

FIGURE 3-11 This data acquisition module can be used in an expanded configuration to provide 256 single ended input channels. (*Courtesy Analogic Corporation.*)

tion in a 3 x 5 inch package. Acquisition and conversion time combined are 10usec. giving a throughput rate of 100kHz. The 12 bit binary data can be expanded up to 256 single ended channels. Output coding is natural, offset binary, or two's complement.

The module shown in Figure 3-11 includes a multiplexer, programmable control logic, sample hold circuit, and a 12 bit A/D converter with three-state output buffers.

There are also single chip, 16 channel, 8 bit data acquisition systems. CMOS technology allows a 16 channel multiplexer, 8 bit successive approximation A/D converter, and the control logic to be fabricated on a single chip. This chip has low power consumption and a minimum of adjustments. There are no full scale or zero adjustments. Latched and decoded address inputs and latched TTL three-state output allow interfacing to microprocessors. The input multiplexer allows access to the 16 single ended analog input channels and provides the logic for additional channel expansion. Connection of the multiplexer output to the converter input is done externally, permitting signal conditioning such as linearization or the use of a sample and hold. The 8 bit A/D converter uses a 256R ladder network, a successive approximation register, and a chopper stabilizer comparator for successive approximation conversion. Use of the chopper stabilizer comparator makes the converter resistant to thermal effects and other long term drift problems. Speed, flexibility, and performance over a temperature range of −25 C to +85 C make low cost chips such as this the practical answer to many custom data acquisition needs.

In these conventional analog multiplexer configurations, for the most efficient use of time, the multiplexer is always seeking the next channel to be converted, while the sample hold is in the hold mode having its output converted. When the conversion is complete, a status line from the converter forces the sample hold to return to the sample mode and acquire data from the next channel. When the acquisition is complete, either immediately or on command, the sample hold switches to hold, conversion begins, and the multiplexer switches to the next channel.

In systems where the channels may be diverse rather than identical, the multiplexer could either be switching sequentially or using some random selection technique. In some systems manual operation for checkout purposes may also be desired. With the use of the random access mode, those channels with more intelligence should be accessed more frequently.

In addition to sharing the converter and the sample hold, expensive instrumentation amplifiers can also be conserved by multiplexing. With the decreasing cost of instrumentation amplifiers, plus the disadvantage of low speeds as well as the effort involved in the successful transmission and multiplexing of low level data, there is a trend to the decreasing use of this approach.

Low level multiplexing can often use programmable gain amplifiers, or automatic range switching preamps. These can allow the use of converters having medium resolutions with range switching used to obtain the additional significant bits. For example, a 12 bit converter with 32 steps of adjustable gain can provide 17 bit resolution, if the resolution is actually present in the signal and the system can operate on it without causing degradation.

More than one level or tier of multiplexers can be required when there are about 64 or more channels to be multiplexed. Here, the problems of stray capacitance and capacitance and capacitive unbalances are increased by the capacitance of the off channels on the conducting channel. For an n channel system, the capacitance is $(n - 1)C$ plus the stray capacitance. This capacitance can be reduced by using two levels of multiplexers. In a 64 channel system with 8 channels per switch, the capacitance is reduced from 63C to 14C.

In another type of system in which the number of shared elements is maximized, use a shared A/D converter, with a multiplexer at its input to switch the outputs of a number of sample holds. This configuration is useful where the sample holds must be updated rapidly or even simultaneously and then read out in some sequence. It is usually a high speed application in which items of data indicating the state of the system must do so at the same approximate time. The multiplexing can be sequential, or it might use some random addressing scheme. The sample holds should be relatively free from droop to avoid accumulating excessive errors while awaiting the readout, which will likely be longer than in the case of a parallel converter system. Increased throughput can always be obtained by adding converters if desired.

Typical applications that use this approach include wind tunnel measurements, seismographic experiments, and the testing of fire control systems. Usually, the measurement is a one-time event and the information is required at a critical time, for example, when a ground shock or air blast hits the test specimen.

Parallel conversion is becoming more practical. Since the cost of A/D converters has dropped, in some cases a multichannel conversion system can be assembled, using one converter for every analog source, at almost the cost of a conventional analog multiplexed system.

This approach has several advantages common to resolver/synchro systems above the 10 bit level. Slower converters can be used at a given throughput rate, or the converters can operate at top speed, providing a greater data flow. For a system with a constant data rate, the reduced conversion speed and the fact that each converter is receiving continuous data, rather than changing from one level to another, may allow the sample holds to be eliminated. The parallel conversion approach can also be useful in many industrial data acquisition systems, where many strain gages, thermocouples, thermistors, or other sensors are separated over a large area.

Digitizing the analog signals at the source and transmitting serial digital data can reduce noise problems greatly. Even in these systems digital transmission allows a considerable immunity to line frequency pickup and ground loop interference. The digital signals can be transformer or optically coupled for complete electrical isolation. The use of low impedance digital driver and receiving circuits can greatly reduce any vulnerability to noise.

By digitizing the sensor signals at their source, we can perform logical operations on the digitized data before it is fed into the microcomputer. Processing can then be streamlined and the test problems minimized. The microprocessor can access data from slow thermocouple sensors less frequently, while reading data from faster, more critical sources at greater speeds. Such digital subsystems can make their own decisions as to when particular data should be used. If certain signals remain constant or are within a narrow range for long periods, then changed rapidly later during the process, ignoring the data may be useful until the changes occur.

A good deal of flexibility and versatility can be gained by transferring from analog multiplexing to digital multiplexing. Decision circuits may exercise judgment on when and what data to use and this in general will improve the overall interface.

In some applications, such as when data are being transmitted from space to earth, the transmission channel is crowded, and some data compression is required to ensure that the data that does get through contains the most intelligence rather than redundant information.

SYSTEM CONFIGURATION

Computer control equipment has evolved to the configuration where a single computer performs the supervisory control and process monitoring (Figure 3-12a). The computer monitors a process by reading analog or digital data from the input/output equipment. It might change a control set point in performing a supervisory control function. The control console may consist of CRT displays, input keyboards, and printers. The peripheral equipment is connected to the I/O channel by interface circuitry. This hardware performs most of the functions required for the operations between the channel and the peripheral device, such as address detection, decoding, timing, and error detection/correction.

The single computer configuration is used in many installations and has had a major influence on the technical and human aspects of computer control, such as the development of control software and instrumentation.

Multiplex Configuration

Remote multiplexing uses a configuration similar to that shown in Figure 3-12b. The remote multiplexer units are located throughout the system and the ana-

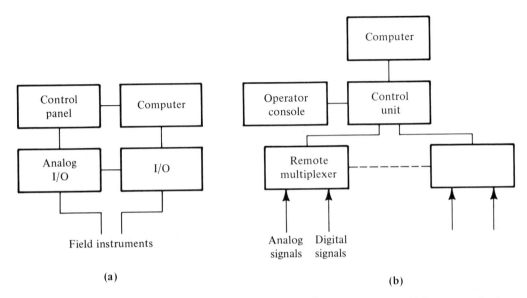

FIGURE 3-12 System Configurations. (a) Basic Computer System; (b) Remote Multiplexing.

log/digital signals are sent to the nearest remote multiplexer. A/D converters are used in the remote multiplexer to convert signals to digital words which are usually 16 bits long.

The control unit will signal the remote multiplexer to request the unit to send a particular block of words or a group of blocks. The remote multiplexer will respond by sending the data requested to the control unit. The control unit will scan the data to check that the transmission is complete. After the data are checked, they may be sent to the central computer or displayed.

The installed cost of wiring has been increasing, while at the same time installed cost of digital systems has been decreasing. The use of remote multiplexing has increased since it can reduce a large amount of the wiring necessary for an installation.

Remote multiplexing can also be used in systems without a central computer. A reduction in wiring is still achieved.

In some systems, only analog signals must be remote multiplexed. In this case a more simple multiplexer, which sends analog data, can be used.

Remote multiplexing has been mainly used for process monitoring rather than process control. However, the reliability of remote multiplexing is high enough now to be used for control signals as well.

A control system with remote multiplexing uses an expanded loop configuration similar to Figure 3-12b. This approach is sometimes called *total multiplexing*. This type of system allows both analog and digital signals to flow either to or from the control computer and the remote multiplexers.

In a flow control, flow transmitters might provide the analog signals to the remote multiplexer. The remote multiplexer then converts the analog signals—typically currents of 4–20 milliamps—to digital words from an analog-to-digital converter and transmits it back to the control computer. A control unit might convert the digital data back to a 4–20 milliamp analog signal for use in an analog instrument, such as a flow controller.

Most remote multiplexing equipment transmit data at a rate which allows the updating of each analog value once every second. In many petrochemical applications, process values which are updated each second act as continuous signals for purposes of the process control.

Although the use of electronic instrumentation predominates, many systems still use pneumatic instrumentation, especially those in hazardous locations.

Many of these pneumatic systems are computers and an interfacing method is required. Two basic techniques are used for interfacing the pneumatic instruments to the computer: One technique is to use P/I (pneumatic to current) and I/P (current to pneumatic) converters. In this system the output of the converters is continuous. In the other method a pneumatic multiplexer/converter is used in either a host or remote multiplexing configuration.

A typical pneumatic multiplexer can handle six pressure inputs per second. Each input of 3 to 15 psi is converted into an analog dc voltage. The computer controls the pneumatic multiplexer as it steps from one input to the next.

Another approach is to use a separate control device which accepts the analog signal from the pneumatic multiplexer and controls the stepping. This reduces the software required in the computer. The cost of a single pneumatic multiplexer is much less than the cost of individual P/I converters.

Computer Configuration

Early process control computers were used only for process monitoring. Then closed loop control came into use for plant optimization. The computers became larger and larger as the number of functions increased. This growth paralleled that of large mainframe computers in data processing installations.

More recently the trend is away from the use of large, single, central processing units toward the use of a number of smaller microprocessors, which can be located in some proximity and connected.

In the distributed processing approach, the processors are not usually dedicated to any single function, but are assigned tasks by one master processor which operates the network. An advantage of distributed processing is the division of labor, since the remote units offload the processor for improved performance of the total system. Time response may be increased with less overhead in the system and an improvement in the execution of functions without waiting for the availability of a central computer. Distributed processing systems have a modularity which is not present in a centralized system. Thus, as sections of a system are automated, remote units can be added. Initially, no need exists to install a system large enough for all anticipated expansions. The system must be designed to include any planned expansions.

Distributed processing can also improve system reliability and failure tolerance. The remote units can allow operation independent of the central computer for a short time, thus a short outage in the central computer or a communications link can be overcome. The central computer can also maintain control for a limited loss of remote units.

When two or more processors are interconnected, they form a multiprocessor system. Microprocessors can be configured in a typical private bus system (Figure 3-13). These can contain a variable amount of RAM and program ROM in each module, and operate in one of two ways: A system master can be used to load the programs into the module RAM, or the operating programs can be stored

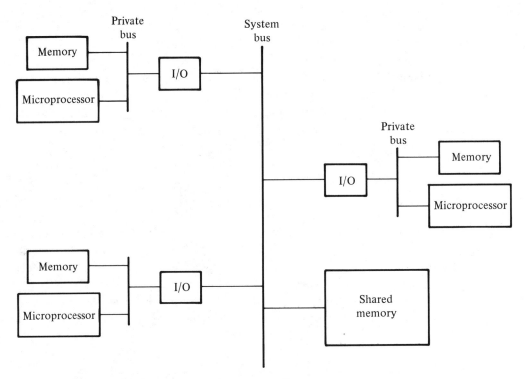

FIGURE 3-13   Private Bus Multiprocessor System.

in the module ROMs. The modules can be distributed to where the processing power is needed, such as a production line where each module is responsible for a different task.

The system bus provides the communication between modules and a system master monitors conditions and issues instruction changes.

The master may also perform diagnostics on modules and report malfunctions and their causes to the operator. If the operation requires high reliability, redundant processor modules can be added to take over if a primary module fails. The computing demands may require the use of a special unit that decides on resource allocation. This is called the system bus arbiter. It must handle multiple resource requests, independent requests and returns, and detect and prevent deadlocks.

The arbiter may be a microprocessor or specialized hardware, depending on the system tasks. A microprocessor arbiter can be tailored to the system to ensure maximum throughput and avoid deadlock. A hardware arbiter is useful mainly for granting high speed, single byte accesses to a common memory. In a large microprocessing system both hard-wired and microprocessor types may be required.

A basic allocation scheme is to match requests with available resources on a pseudorandom basis. Here any resource is assigned to the first request encountered in a linear search of the request list. Other schemes include round-robin or first-in-first-out queuing.

The system arbiter should be designed to allocate resources based on priorities assigned to the system. In operations the arbiter must grant one request while holding all others for execution. A hardware arbiter can sometimes be adapted from a priority interrupt control unit.

Duplex Operation

Any duplexed configuration allows an increase in the reliability of the system. The primary or master processor performs the control task. When a failure occurs, a backup processor takes control. The primary processor can then be taken off-line, repaired, and returned to service without interrupting control.

The duplexed processors provide twice the computing capacity, since either processor can control the process. This excess power can often be used for other optional tasks. To guarantee that the higher reliablity will be achieved, the system should be analyzed for the effects of each failure classification. The software must also be designed to achieve the higher potential reliability of the duplicated hardware.

Duplexed operation allows special characteristics to be taken advantage of in the selection of the two processors. One processor could be designed to perform fast floating point calculations, its normal function during operation of the system.

For the backup processor to keep up-to-date with the status of the program, it must have access to the current status information. If a shared file system is used, then this information is available in the shared file. In a system with several processors, requests from one processor to the other can be posted in the shared file. Then a scheduling processor periodically checks to determine if a request is waiting. The request can then be scheduled according to the operating system priority.

Disc files are serial devices which two processors cannot use at the same time. This contention problem must be solved by circuits or another microprocessor (an arbiter) which may be part of the file interface.

Another basic method of connecting microprocessors is a connection through the I/O channels. This could be either a serial or parallel data transfer. The transfer of information then uses the data rate of the slowed microprocessor in a number of operating modes. The microprocessors could cooperate in solving a problem which requires more computing speed than a single microprocessor provides. Each microprocessor could control a portion of the process. The necessary coordination can be effected through the interconnection.

Some of the sensors in the system may be connected to several process I/O subsystems and the rest to only one. The shared sensors can be those which are critical to the process. If one of the processors becomes inoperable, the other processor still has access to the data in order to maintain control.

In a duplexed configuration the designer must consider when the backup processor should be activated. One approach is to have the primary processor set a timer in the backup unit on a periodic basis. Failure to set the timer causes the backup unit to take control and disable the primary unit. The type of timer is known as a *hardware* or *watchdog* timer. The switchover from the primary to the backup computer must be designed so that it results in only a small control deviation.

Distributed Control

The increased use of computer networks in data processing, along with the increased popularity of remote multiplexing, has resulted in the wider use of distributed control in industry. Distributed control differs from distributed processing, in that it also makes use of remote multiplexing. The processors communicate with field multiplexers with each microprocessor usually dedicated to performing the same task in an on-line environment.

The two basic techniques for distributed control are the loop approach and the unit approach. The loop approach uses satellite microprocessors which perform a fixed number of functions. A single microprocessor might perform a single function for a number of different loops. If a microprocessor is dedicated to a single loop, then a failure in the microprocessor will cause the loss of only that loop.

The unit approach uses a separate control system for each unit in the system with a microprocessor assigned to each unit. For example, a microprocessor might be assigned to the control of a distillation column.

Some of the advantages of distributed control can be realized by either approach but others can only be gained by using the unit approach.

One major advantage is the reduction in wiring. With the microprocessor's I/O equipment located throughout the system, wires must be run only a short distance. This provides savings in installation and since the runs are shorter, there are fewer problems from interference on low level signals.

Also, enough CRT control consoles can be used in the system so that a sufficient number of operators can function during critical periods.

The reliability and maintainability of a microprocessor system are improved with distributed control. The cost of microprocessor hardware is low so processors in a unit configuration can be backed up with a spare (duplexed processing) so a single processor failure will not affect operations. Since microprocessors are normally on a single printed circuit board, the system is easily repaired, usually by replacing one board.

The unit approach is highly modular. Computer control can be added to the units in the system one at a time, without disrupting control of those units already on computer control. The most likely unit for computer control can be established after a system study is conducted. The unit may be a system or process bottleneck, a large energy user or one which is difficult to operate without computer control.

Networks

Computer networks are an extension of the distributed processing concept. The differences between distributed processing systems and computer networks tend to be a matter of degree.

Computer networks consist of two or more computer systems which are separate. The distance may be the number of feet within the same room or the miles between separate facilities connected by common carrier lines.

Computer networks can be organized in a number of ways. The computers are usually loosely coupled and capable of stand-alone operation.

The network configuration may be defined as master-slave, hierarchical, or peer. The configuration definition classifies how the control responsibilities are assigned among the processing units.

In the basic master-slave network, a host processor is connected by one line to a satellite processor. The communication line between the processors is referred to as a *link*. The link may be any communications channel, such as a coaxial cable or telephone line. Each device in a network is sometimes referred to as a *node*. A master-slave network with multiple slaves is also commonly referred to as a *multipoint or multidrop configuration.*

In the master-slave network, a single processor has control and it determines which slave computer shall operate on a task. Communications between the slave computers is under control of the master. After it has been assigned a task by the master, the slave operates asynchronously with respect to the master until completion of the task, or until it requires service from the master. The master-slave configuration is used mainly to provide system load sharing.

The hierarchical network uses a multiple level master-slave scheme. The various levels in the hierarchy are assigned responsibilities for certain functions. The highest level in the hierarchy makes the major decisions, while the lower levels have responsibility for the specific operations.

In contrast to the master-slave and hierarchical networks which use a top-down control philosophy, the peer network configuration uses mutually cooperating computers in which no defined master or slave relationship exists. This type of configuration requires that the operating system of each computer is aware of the status of the other computers in the network and that a scheduling program is used to provide the task distribution. When a job is passed to a computer in the network, the originating computer moves to a new task. A computer which is busy passes the task on to an available computer to execute the task.

The time response in peer networks can be difficult to predict, since one computer does not know the workload of another and no master imposes tasks. Peer networks can provide access to specialized facilities not available on the originating computer and the processors can share the computing load in a dynamic manner for a more efficient use of the total facility.

In addition to the control configuration, a number of physical connections may be used for a network. The star configuration has a host or master processor in the center of the system (Figure 3-14a) and each processor communicates with the host. Communication between these satellite processors is through the host. This is also known as a *radial* or *centralized* configuration. This configuration can be limited by the speed of the host. If more than one satellite is allowed to talk to the host at the same time, the host may be burdened with controlling the data flow between satellites.

The multidrop configuration as shown in Figure 3-14b is also known as a data bus, data highway, or multipoint configuration. The host controls the flow of data between any two nodes. Any satellite may communicate with the host or any other satellite at any time.

The loop configuration of Figure 3-14c is normally used in remote multiplexing systems. Then, if a single link breaks, the nodes can still communicate. This is also known as a ring configuration. The loop can begin and end at a loop controller, which is a microcomputer that controls the communications.

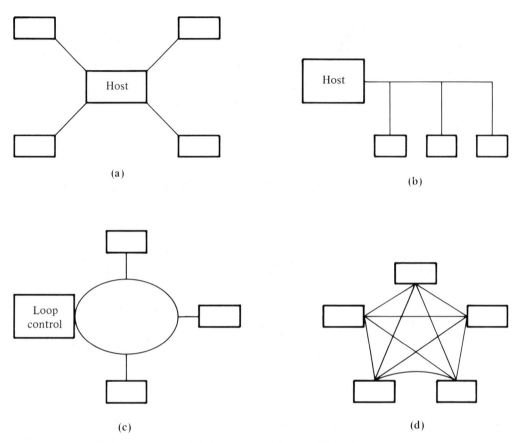

FIGURE 3-14   Network Configurations. (a) Star; (b) Multi-Drop; (c) Loop; (d) Point to Point.

Messages between computers in the loop are handled as a string of words with some bits or words containing information on the originator and addressee. When a computer recognizes a message addressed for it, it accepts the message. As a processor receives and verifies a message, the starting and ending address of that message is passed to the destination processor. This processor then encodes and retransmits the message to the next point in the network. The supervisory processor or controller maintains network information and data link assignments based on equipment conditions, message load, and the most direct route to the final destination. Loops can be difficult to control and the fact that messages must pass through the computers requires higher data rates.

In the point-to-point network, all processors have a direct access to every other processor (Figure 3-14d). For n processors, $n(n-1)/2$ interconnections are required. For a three processor system, three communication links are required.

For five processors the number of links required is 10, for a ten processor system the number of links becomes 45. This configuration allows a faster response or lower grade communications lines. The alternate paths allow messages to be forwarded even if some of the links are broken.

If microprocessors are organized into a multiprocessing system, three advantages over one larger processor exist:

1. System throughput can be increased.
2. Processor tasks can be segmented more easily.
3. The system can be expanded for more flexibility by adding modules.

Multiprocessing architectures increase computing power by allowing communications between processors. In many cases, multiprocessing produces a system solution that is less expensive and faster than a single processor. Typical applications for several interconnected processors are:

1. Remote, microprocessor controlled data terminals linked to a central or host processor.
2. Systems where a single processor offers too little computing power, but a more powerful unit offers too much.
3. Multiple, related tasks which cannot be handled efficiently by one processor.

Multiprocessing requires an evaluation of the architectures and the communication options. Of the basic methods of communications, serial links can be used when the volume of system data is small, while parallel bus techniques will handle more information. Interprocessor communication may require the passing of information between two processors located a considerable distance apart.

Communication

Any facility that allows signals to be sent from one place to another is considered to be a communications channel. This includes telephone lines, coaxial cables, and microwave links.

Communications channels for voice transmission require a bandwidth of 1000 to 3000 Hz, depending on the fidelity desired. Commercial television requires bandwidths to 6MHz. The bandwidth required for digital transmission is a function of the data rate. The digital data can be transmitted as modulated signals. The transformation of digital to modulated signals and the transformation from modulated to digital signals is performed by a modulator-demodulator, which is called a *modem*. Communication over short distances of several feet do not require modems. The digital logic signals may be sent directly over these distances.

Modem channels are divided into three types: simplex, half duplex, and full duplex. Data can be sent only in one direction on a simple channel. A half duplex channel allows data travel in both directions, but only one direction at one time. The data may travel in both directions at the same time in a full duplex channel.

A modem can use a particular type of modulation scheme as well as a particular communications protocol. The protocol defines the sequence in which the data are to be transmitted, the structure of the message, and the method of synchronization for the transmitter and receiver, as well as the error checking procedure. Some examples of protocols in use are IBM's Binary Synchronous Communications (BISYNC), DEC's Digital Communications Message Protocol (DDCMP), and IBM's Synchronous Data Line Control (SDLC). This use of communications protocols has simplified many problems of data transfer between digital devices in a network configuration.

Data communications in the network may be either asynchronous or synchronous. Asynchronous data transmission is suited for data that occur in short bursts while synchronous data transfers are more suited for longer streams. The asynchronous messages are preceded by a start signal to synchronize the transmit and receive circuits. Synchronous data will normally contain a defined bit pattern for synchronization. The pattern is defined by the protocol.

Since the data transmission can be affected by noise and signal failures, it is useful to determine when the transmitted message is received without errors. This is important in many control systems where undetected errors could cause a loss of control. A basic checking technique requires the receiving station to retransmit the message received back to the transmitting device. This requires that the message must be transmitted twice. This can be done in many applications.

Various other methods of error checking can determine if the message were received correctly. Another basic method involves adding an extra bit, called the *parity bit*, to each block or word of data. This is sometimes called *Vertical Redundancy Checking (VRC)*. This bit can be set such that the total number of 1s in the block is always odd for odd parity. If even parity is used, then the number of 1s in the block or word is always even. Parity checks do not always provide sufficient security for control applications and for that reason other methods are normally used.

Other more elaborate methods of error checking are Longitudinal Redundancy Checking (LRC) and Cyclic Redundancy Checking (CRC). A combination may be used in some systems, depending on the security requirements.

LRC adds an additional block to the message. The bits in this check block are calculated by performing an Exclusive OR operation on corresponding bits in the other blocks of the message. The receiving device compares the contents of the block with the data received.

In CRC a polynomial is calculated from the data. This polynomial is divided by a constant polynomial and the remainder or residue is used as a check block.

EXERCISES

3-1. List the application factors for selecting instrumentation amplifiers.

3-2. How is the sense terminal of an instrumentation amplifier used? Discuss how interconnections of devices can produce errors as a result of currents' impedance levels and transient overloads.

3-3. Describe some isolation techniques used in isolation amplifiers. When is guarding necessary, and when should individual channels be individually guarded? If the guard should be present up to the input of the preamplifier, how can the guard be switched?

3-4. In general, the objective of amplifier selection is to choose the least expensive device that will meet the physical, electrical, and environmental requirements for the application. In evaluating the trade-offs, discuss the three types of budgets that must be considered: cost budget, system time budget, and error budget.

3-5. Discuss how difficulties in single channel, low level circuitry are compounded by the addition of low level multiplexing.

3-6. Discuss the trade-offs in using digital multiplexing before transmission or remote A/D conversion and serial transmission.

3-7. Discuss the use of sample holds in data acquisition. What system factors affect the way they are used?

3-8. What is meant by a cyclic code? What are its advantages?

3-9. If the speed of an A/D converter is limited by the settling time of an input buffer follower, how can the throughput rate be increased?

3-10. Discuss some ways of avoiding errors when external resistors are connected to a converter. Why is it helpful to understand resistor tracking in actual applications?

3-11. Discuss under what circumstances it is more economical to choose a general purpose converter, which will meet the needs of a large number of system designs, or to go through an optimum selection process for each individual application?

3-12. Discuss the trade-offs for the different A/D converter circuits, based on the availability, cost of devices, and the cost of alternatives.

3-13. How should one treat terminals that are used to determine the signal voltage range in a converter involving analog signals to protect the low resolution levels? How should one treat cable runs that carry logic signals?

3-14. Discuss the design of a system where many digital sources must be multiplexed into a central computer or data transmission channel. Can they be tied to the computer by a common set of parallel bus lines?

3-15. What technique allows a unit to add computer control with a modest capital outlay? What must be successful in terms of the payout period before computer control can be added to other units in the system?

3-16. Discuss a system configuration where both processors are of equal importance in maintaining control of the process and both must be operating to obtain optimum performance.

3-17. Discuss the disadvantages in a configuration, such as the star network, which is typical of a small system master-slave relationship, where the master is directly connected to each of the other processors.

3-18. In a multidrop configuration all computers in the network are connected to a single communication line, which may be a multiplexed channel or a serial data highway, over which any two but only two processors may communicate at any one time. Discuss how this physical configuration can be used in a master-slave configuration.

3-19. In a system with duplexed processors what advantages are not gained automatically and must be designed into the system hardware and software?

3-20. A good example of an arbiter is found in a data communications network made of three intelligent data receiver-transmitter processors. Each processor communicates both with a high speed serial data link and with others over the system bus. What is required for this network to operate at high speeds? Should the arbiter incorporate a microprocessor or a hard-wired unit?

4

Motor Control Techniques

Although motor control is commonly considered as speed control, many other motor control functions such as switching, direction control, or programming are also motor control techniques.

Speed control is most easily accomplished on motors whose speed is a function of power input such as dc or universal brush type motors. In most types of ac motors, speed control requires varying the frequency. A variable drive can be built using a variable frequency trigger. In some controls the control range will be limited due to a decrease in torque as the frequency is reduced.

The variation of the average power input to a dc motor can be accomplished by pulse width modulation techniques. This requires a switching between the on and off states while controlling the percentage of the on time. In ac systems, this can be done using phase control to control the phase relationship between the gate trigger signal of an SCR or similar device and the supply voltage. The percentage of the on time in each cycle is controlled.

In a dc brush motor the speed of the motor can be detected by using the back EMF generated by the motor when the controlling SCR is off. In the separately excited shun field, wound and permanent magnet field motors, this EMF is directly proportional to the speed. In series motors, the field is not energized at this time and the residual magnetism must provide the back EMF used by the circuit. This residual magnetism is a function of the past history of the motor current, so the voltage the control circuit must use is not a function of speed alone.

On-Off Control

In addition to controlling or varying the speed of a motor, other control functions can be accomplished using solid-state control. One of the most basic is the use of a triac as a switch for contactor replacement. In combination with a reversing type split capacitor motor, a pair of triacs can provide a task responding, reversing motor control. The triacs can be gated by a number of methods. The use of a solid-state static switching circuit can also be used to provide motor over temperature control.

The choice of solid-state over electromagnetic relays (EMRs) should be based on life cycle expectancy, as there is a definite limitation on the life of EMRs in high cycle rate applications.

Solid-state relays (SSRs) are packaged in potted phenolic blocks with screw terminals and standard contact and coil systems. Thus, they are easily applied by a user familiar with contact relays. SSRs are available in ac and dc output versions and are not interchangeable.

Optical isolation between the input and output can withstand 1,500V ac rms and some are available to 3,750V rms. Other devices use internal reed relays for isolation. Some devices are transistor coupled with no true isolation between the input and output.

The trigger circuits can provide zero crossover switching in ac applications to reduce electrical noise generation. Output current ratings in ac or dc types range from 1A to 40A. Ac load voltage ratings range up to 480V rms and dc types up to 250V.

Some forms of SSRs allow rail and panel mounting with LED indicators such as used in programmable controllers for control outputs. Low profile printed circuit board mounting packages are also available. SSR costs may be several times that of an equivalent EMR.

A number of low profile EMRs permit printed circuit cards carrying such relays to be mounted in closed circuits. Such relays measure less than one-quarter inch high and have double pole, double throw contacts rated at 1A. Other versions pack dry reed relays into integrated circuit type dual-in-line packages and TO-5 transistor cans. Dry reed relays are pc board mounted for control signal multiplex scanners. The relays permit complete circuit isolation between field wiring and control equipment. In flying capacitor type scanners, for example, both ends of the capacitor are switched by DPDT contacts between the field sensor wires and the control input circuits.

Similar relay isolation techniques, at higher power levels, are used to isolate the field terminations at the output of some control equipment. If an ac or dc power source is connected to these terminations, the hard contact isolation will protect the control equipment; an ac triac can fail if dc is applied.

Relays are also used to insert adjustment time delays after some machine or process event. These time delay relays come in several forms: on delay, off delay, or on and off delay with some EMRs incorporating solid-state timing circuits.

Speed Control

A basic dc adjustable speed drive must have certain features for the particular application. All the basic features and functions must be identified. The common signals, mechanical interfaces, and electrical I/Os must be designated. From this information the hardware can be designed with modularity and flexibility. To meet the requirements of interfaces, all I/O and interfunction connections must be designed.

Some additional features may also be on the basic drive system:

1. Adjustable linear acceleration and deceleration controls.
2. Solid-state logic sequencing buffered from 115V operator control signals.
3. Solid-state antiplugging control.

In addition to these options, the drive can be modularized to enable it to expand with the need for new features.

The motor controller design approach should be towards the maximization of hardware accessibility and the minimization of wiring and mechanical complexity.

Subassemblies can be provided in modular form. A basic drive may consist of five subassemblies: (1) ac protection, (2) field supply, (3) ac contactor assembly, (4) power converter, and (5) regulator.

Electrical wiring and mechanical interfaces should be simplified for greater ease in maintainability.

The ac protection panel should have ac circuit breakers for all incoming line connections.

Phase Control

AC power is convenient and phase control is a convenient way of regulating this source. Thus, phase control is used to control a wide variety of motor types. Most motors were not designed for this type of operation. But they are used since they are available and inexpensive.

Pulse width modulation can be used with a variable time delay between the turn on and the turn off pulses. This allows an adjustment of the on time in each cycle. A fixed time delay between the on and off pulses requires varying the repetition rate. This produces an output with a fixed on time and a variable off time.

Many control circuits on the motor characteristics and improper motor selection can cause poor circuit operation. The control circuit is only a part of the overall system and it can be no better than the overall system design. Most motors have ratings based on operation at a single speed and they depend on this speed for the proper cooling. The use of the motor at a lower speed may cause heating problems. Odd order harmonics in a phase controlled system can produce undesirable side effects in induction motors. The speed torque characteristics of some induction motors may make them unsuitable for use with variable voltage controls. Some circuits for universal series motors depend on the residual magnetism in their magnetic structure. This is a characteristic that the manufacturer may be attempting to minimize in the motor design.

A drawback of motor speed control which is a function of the average power input is that as the input power is lessened to reduce the speed, the output torque is also reduced. This problem is overcome by using a feedback signal to advance the SCR firing angle in proportion to the load on the motor, supplying a greater amount of power as more torque is required.

The proper motor in a phase control circuit allows some versatility in application. In some control systems, a wide variety of motor sizes and speeds may no longer be required. A single motor might be used with the different requirements being compensated by means of the control system.

If the overspeed control is built into the system, there may be no need for overvoltage capability in the motor, allowing some savings in motor costs.

Contactors for motor control, usually packaged and marketed along with the motors, are actually relays with high contact ratings. They generally have life expectancies which are a function of the number of operations at rated load. These expectancies can range from 100,000 operations at full load to 10 million operations at no or low signals.

System Considerations

Many motors in industry are remotely controlled using the logical analysis of contact-closure patterns by computers and programmable controllers. These motors and their contactors are then a part of the overall dynamic control system. The simplest manually controlled machine tools use relay control schemes. The most sophisticated tools are numerically controlled by computer (CNC). Here, the trend among users of these machine tools is towards a more central data collection with the downloading of programs from control and monitoring computers. Interface problems increase as the once independent unit control systems are integrated into central supervisory control systems.

The growing use of remote multiplexing systems has increased the system sophistication in relay control schemes. In a typical application of motor operated valve is opened or closed from a computer output. Protective logic for limiting the open or closed movement of the valve takes the form of a positive torque switch on the valve. In a hard-wired system a ten conductor cable is required for the open/close/power indication. Additional two conductor cables are needed for both the valve open and overload conditions. When a solid-state multiplexing system is used to replace most of the wiring, the protective and control logic remains basically the same. This is necessary so that a failure in the remote multiplexing system does not damage the motor or valve.

The output functions can be programmed in PROMs in the form of Boolean algebraic operations as well as latched functions.

In another trend the development of high performance dc servo motors has been bypassed by a need for microcomputer systems with increased speed. These applications include printers, memories, nc machine tools, and other production equipment where computer commands must be translated into mechanical motion. A typical application is the system for a digital tape or disk drive. The design techniques may include the control of acceleration, deceleration, velocity, and mechanical drive reversal.

DC AND UNIVERSAL MOTOR CONTROLS

The shunt wound dc motor can be used with solid state speed control systems to provide a wide range of speed control. The speed of a shunt motor is almost constant with changes in torque, thus permitting speed control to be achieved by controlling the voltage applied to the armature. The use of a small compound series winding can make the speed almost independent of torque. Also, a small amount of feedback with speed information for the control that supplies the armature voltage will reduce variations of speed with torque.

Shunt Wound Motor Control

Figure 4-1 shows a simple solid-state speed control for shunt wound dc motors. This circuit uses a bridge rectifier for full wave rectification of the ac supply. The field winding is connected across the dc output of the bridge rectifier. Armature

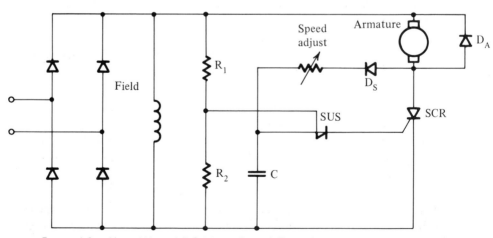

FIGURE 4-1   Shunt Wound DC Motor Speed Control.

voltage is supplied through the SCR and is controlled by turning the SCR on at points in each half cycle; the SCR turns off only at the end of each half cycle. Rectifier DA provides a circulating current path for energy stored in the inductance in the armature at the time the SCR turns off. Without DA, the current will circulate through the SCR and the bridge rectifier and prevent the SCR from turning off. At the beginning of each half cycle, the SCR is in the off state and the capacitor C starts charging by current flow through the armature, rectifier DS, and the adjust resistor. When the voltage across C reaches the breakover voltage of the SUS trigger diode, a pulse is applied to the SCR gate, turning the SCR on and applying power to the armature for the remainder of that half cycle. At the end of each cycle, C is discharged by the triggering of the SUS, register R1, and the current through R1 and R2. The time required for C to reach the breakover voltage of the SUS governs the phase angle at which the SCR is turned on and this is controlled by the magnitude of the adjust register and the voltage across the SCR.

The voltage across the SCR is the output of the bridge rectifier less the reverse EMF across the armature. The charging of C is partially dependent upon this counter EMF. If the motor runs at a slower speed, the reverse EMF is lower and the voltage applied to the charging circuit is higher. This decreases the time required to trigger the SCR, increases the power supplied to the armature, and compensates for the loading on the motor.

The energy stored in the armature inductance results in a current flow through rectifier DA for a short time at the beginning of each half cycle. During this time, the reverse EMF of the armature cannot appear, thus the voltage across the SCR is equal to the output voltage of the bridge rectifier.

The length of time required for this current to reach zero and for the reverse EMF to appear across the armature is determined by both the speed and armature current. At low speeds and high armature currents, the rectifier DA conducts for a longer time at the beginning of each half cycle. This action causes faster charging of capacitor C and provides compensation that is dependent on both armature current and motor speed.

This circuit provides a good range of speed control adjustment. The feedback signal from the speed and armature current improves the speed regulation over the inherent characteristics of the motor. The inductance of the field winding of a shunt motor is usually large, resulting in a significant length of time required for the field current to return to normal value after the motor is energized. It is desirable to prevent application of power to the armature until after the field current has reached a normal value. This soft start function can be added.

This approach is capable of speed regulation of approximately 10%. For higher performance, a tachometer feedback circuit can be substituted for the trigger circuit.

The regenerative dc servo drives using two-SCRs and one- or three-phase rectifier bridges connected across the armature terminals are limited in their response because of the maximum switching frequencies of the SCRs. Higher switching frequencies with increased bandwidth can be obtained through the use of chopper circuits using time ratio control.

To control the speed of a dc series field motor at different torque levels, it is necessary to adjust the voltage applied to the motor. For a constant voltage the motor speed is determined by the torque requirements and the speed is reached under minimum torque conditions. If a series motor is used as the drive for a vehicle, the voltage to the motor must be controlled for the different torque requirements of grade, speed, and lead. The most common method of varying the speed of the motor is inserting a resistance in series with the motor to reduce the power. This type of speed control is inefficient and wastes power, especially under high current, high torque conditions due to the I^2R losses.

Pulse Techniques

A more efficient method of controlling the voltage applied to the motor is the pulse width modulation method. A variable width pulse of voltage is applied to the motor to vary the average voltage to the motor. A diode is placed as shown in Figure 4-2 in parallel with the inductive motor path to provide a circuit for the inductive current when the switch is opened. This current path prevents any abrupt current changes and resultant high voltages across the switching device.

The shunt diode allows more average current to flow through the motor.

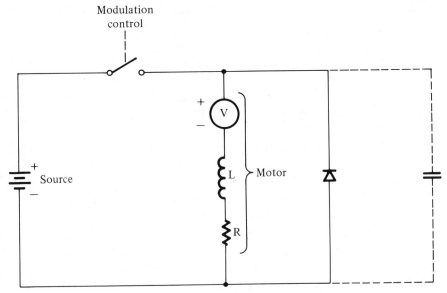

FIGURE 4-2   Basic Pulse Width Modulation Technique.

The power taken from the battery is approximately equal to the power delivered to the motor. The energy stored in the motor inductance at the battery voltage level is delivered to the motor at the same approximate current level, but at a much lower voltage level when the battery is disconnected. For smooth motor operation, the current variations through the motor should be kept to a minimum during the switching cycle. The amount of energy that can be stored in the motor inductance limits the power delivered to the motor during the off time. For low speeds the on time must be brief with a rapid switching rate.

Pulse frequency modulation can offer not only the same advantages as pulse width modulation, such as ease of computer programming and the use in open loop speed controls which are sticky around null, but it will also save control energy. Thus, when the energy source is limited, PFM should be a potential candidate.

Series motors operating as tractive motors require high current levels depending upon the application and condition of operation. For fork lifts, carts, and other small electric vehicle applications, a pulse width modulated motor control may be required to handle current levels of 400–500 amps unless current limiting circuitry is used. Even with current limiting circuitry, the torque demands may require switching currents from 250–300 amperes.

The maximum current level that any transistor can handle in any particular circuit is dependent upon the switching load line and the safe area of operation. For the pulse width modulator the possibility of two different load lines depends upon whether a capacitor is used to tailor the load line. One can either parallel the motor with the high capacitance necessary or add additional transistors in parallel to meet the load line safe area requirements. With each transistor operating at a lower current level and higher gain, the total drive current requirements for the transistor switch are generally reduced. Because of the variation of V_{BE} (sat) for each transistor, collector current levels of each transistor in the parallel combination can be different unless the drive requirement is equalized. The usual method is to insert a resistance in series with each transistor emitter.

The power loss in the drive circuit can be high. A 200 ampere motor requirement requires a drive current of at least 10 amperes, assuming a forced gain of 20.

If a saturated Darlington circuit is used, the drive current contributes to the load current. The .5 volt drop across the saturation resistor in series with the collector can dissipate 95 watts at the maximum current level of the motor. The power loss in this resistance will be in proportion to the peak motor current demand, while the peak power dissipation in the emitter follower circuit will be at a constant and will not depend on motor demand. Because the drive transistor in the Darlington circuit will begin to turn on before the final transistor, unless the input to the driver is restricted, this transistor may use a load line that is beyond its safe area. The driver transistor should be selected with a safe area that will not be reached by the available input to the driver.

Since most high current transistors have slow switching characteristics, they tend to match in load line excursions closer than the faster switching transistors. The slow switching transistor power dissipation is greater and if switching rate is rapid, reverse bias may be necessary to reduce switching power dissipation by decreasing the switching off time. This reverse bias can be produced by a separate transistor converter circuit or by charging a capacitor during the on drive and discharging it in the reverse direction to drive off the output transistors.

One of the circuits for producing a variable pulse width at a constant switching rate is shown in Figure 4-3. The unijunction transistor is charged through the potentiometer and the side of the flip-flop which is in the off condition. When the capacitor CA is charged to a level where the emitter voltage reaches the peak point voltage, the emitter to B1 junction of the unijunction will

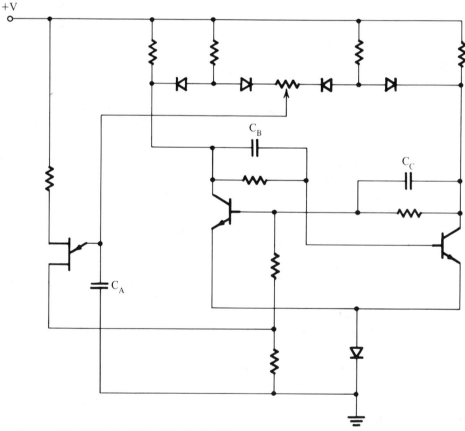

FIGURE 4-3   Variable Pulse Width Oscillator.

turn on and supply a trigger to the flip-flop. The off side of the flip-flop will now turn on and the on side will turn off. The capacitor now charges from the opposite side of the potentiometer. This charging of the unijunction capacitor from alternate sides of the potentiometer gives two different time constants which determine the on and off times. Diodes are used to disconnect the charging potentiometer from the divider current that could flow through the conducting transistor and to eliminate the effect of transistor bias current levels on the charging current through the potentiometer.

Because of the high motor current levels that can be encountered in the driver circuit under stall or starting conditions, a current limiting circuit becomes desirable. The motor current should be limited to a value that will only meet the maximum torque requirements to reduce the motor temperature and the total number of power transistors. The current limiter senses the total current through the transistors and, when a set current level is reached, the drive circuitry is turned off. The current can be sensed by the drop across a resistor common to all transistors or sense the voltage drop across an individual transistor emitter resistor in a matched parallel combination. The limiting should be as fast as possible and repeatable at the set limit under the conditions of operation. The circuit in Figure 4-4 can be used for this application. The current level at which the transistor will switch on is determined by a calibration to the characteristics of the tunnel diode. If current limiting action is sustained for any length of time, the power transistor

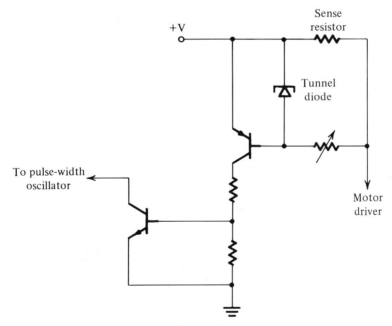

FIGURE 4-4   Motor Driver with Current Limiting.

may attain a high temperature unless reverse bias is used to decrease switching power.

Permanent magnet motors behave in a similar manner to shunt wound motors, since they also have field strengths independent of armature current. Thus, the circuit in Figure 4-1 can also be used with a P–M motor. Some types of P–M motors, such as the so-called *printed circuit motors,* do not lend themselves readily to phase control. The effect of the low inductance, low resistance, and low operating voltage of these motors makes the peak-to-average current ratio excessive using phase control, causing high RMS currents in the line and possible demagnetization effects in the motor field magnets. This type of motor is more suitably controlled by means of a chopper type of circuit.

Figure 4-5 shows a chopper type motor drive. SCR_1 is the load SCR and must be capable of carrying the maximum load current. The function of SCR_2 is to commutate SCR_2. For a speed control circuit to perform a high current function, SCR_1 may have to be capable of carrying 200 amperes. This might be accomplished by parallel operation with the necessary precautions for current sharing and triggering. The value of the commutating capacitor C_C can be approximated from the following equation:

$$C_C = 1.4 \frac{t_o\, I_A}{V_C}$$

Where t_o is turn-off time of the device, I_A is the anode current before commutation, and V_C is the voltage across C before the start of commutation.

If SCR_1 is a parallel combination of devices, rather than a single device,

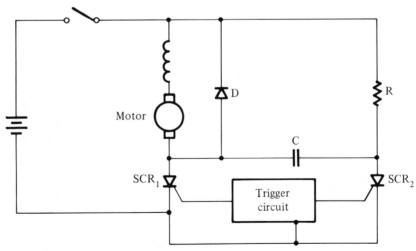

FIGURE 4-5   Chopper Motor Driver with Resistive Charging.

there will be a decrease in the required value of C_S. For a current of 200 amperes and a supply voltage of 36 volts, the commutating capacitor is 90 microfarads. The minimum value of R depends upon the minimum time required to charge C with a repetition rate of 30 cycles per second and a minimum on duty cycle of 10%. The minimum charge time is 3.33 milliseconds. Setting this time equal to three time constants, we obtain:

$$3 \text{ R}(90 \times 10^{-6}) = 3.33 \times 10^{-3}$$
$$\text{or}$$
$$\text{R} = 12 \text{ ohms}$$

The dissipation in R based on a worst case 90% duty cycle is approximately 108 watts.

Another approach that tends to eliminate that dissipation as well as an additional reduction of C is illustrated in Figure 4-6. In this circuit SCR_3 is gated on at the same time as SCR_1. This allows the resonant charging of C by L to twice the supply voltage and permits the reduction of C by a factor of two. In high current applications the saving in the cost of the commutation capacitor may offset the cost of the additional SCR. The value of L determines the charging rate and charging for C. The maximum value of L should be based on a resonant frequency whose half cycle time is equal to 3.33 milliseconds of 150 cycles. With a commutating capacitor of 45 microfarads, the inductance is approximately 25 millihenries. Since the commutating capacitor is charged to twice the supply voltage, the SCR must have a reverse voltage capability of this same amount. A smaller value of inductance can be used to decrease the charging time. However,

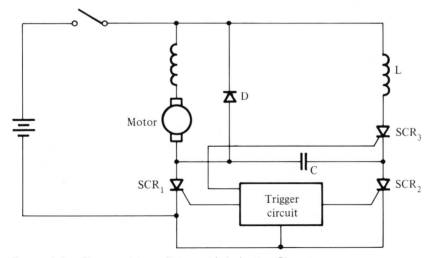

FIGURE 4-6   Chopper Motor Driver with Inductive Charging.

as the charging time decreases, the peak charging current increases, and to charge to twice the supply voltage, the Q must be high.

Comparing SCR motor control with transistor control, the transistor version can have a lower voltage drop. At 200 amperes, the drop across a transistor control may be about 1.3 volts and that across an SCR version would be about 1.5 volts. Losses in an SCR control are approximately 33% resistive while those in the transistor control are 75% resistive.

The SCR control could be more resistive to damage by fault conditions than a transistor control. A short across the motor can cause the current to rise too rapidly for current limiting to save the transistors. The same overload could be withstood by the SCR control, until the circuit breaker tripped.

Nonregulating Control

In some cases speed regulation is not required. If the load characteristics are relatively fixed, or the motor drive is part of a larger overall control system, a nonregulating control circuit may be desired. These nonregulating circuits can provide a considerable cost savings. Figure 4-7 shows one of the simplest half wave circuits. It uses one SCR with a minimum of other components. A divider network supplies a phase shift signal to a neon bulb which triggers the SCR. By varying the setting of the potentiometer, the gate signal of the SCR is phase shifted with respect to the supply voltage which turns the SCR on in the positive ac half cycles. The capacitor fires the neon bulb on both positive and negative half

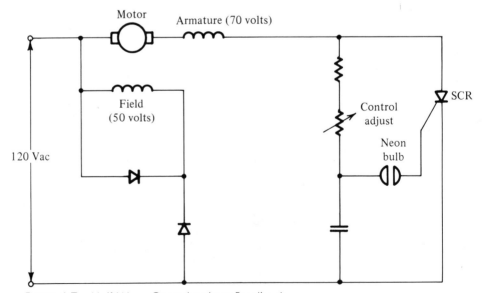

FIGURE 4-7　Half Wave Control without Feedback.

cycles. The negative half cycles are disregarded by the SCR. Replacing the neon bulb with another trigger device, such as a Diac or a Silicon Unilateral Switch, the performance and reliability of the circuits are improved because these trigger devices have a longer life and a more stable firing point than neon bulbs. Because of their lower trigger voltage, these trigger devices give a wider control range.

Reversing Control

Sometimes the direction or time of operation of a motor must be controlled as well as the speed. An example is an automatic mixing machine. An SCR control could periodically reverse the motor, which drives the agitator to give the required agitator cycle without the use of a mechanical transmission. Another example would be an automatic tapping machine. Best results are obtained when the tap is advanced several turns and then partially withdrawn so that it can clean the threads. Speed adjustment is required for different taps and material. This can be combined in an SCR control. An example may be found in the automotive field. For example, automobile windshield wipers should be operated at reduced speed when the precipitation is light. However, we do not want the wiper blade to move slowly across the windshield and stall part way across. We would prefer to have the blade move across the windshield at full wiping speed, but less frequently. The circuit of Figure 4-8 will do this. The unijunction trigger can give a broad range of repetition rates. When the SCR is triggered on, the motor begins to operate and closes S_2. S_2 shorts the SCR, turning it off, and carries the motor current. S_2 remains closed for a complete wiping cycle and then it opens. Since the SCR has

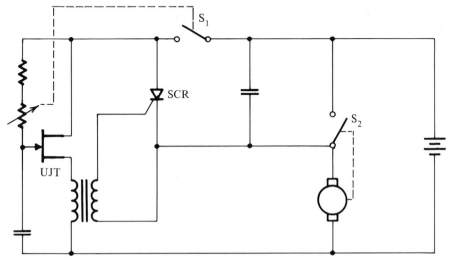

FIGURE 4-8   Windshield Wiper Type of Control.

been turned off, the wiper remains in the rest position until another trigger pulse initiates another wiping cycle. The capacitor protects the SCR and the switch contacts from the voltage transients due to the motor current.

It may be desirable to use a separate negative terminal so that the SCR can be connected to the negative side of the motor, permitting the elimination of the transformer by connecting the UJT base to the gate of the SCR. The switch would also have to be on the negative side of the motor.

SCRs can be used for supplying both armature power and field excitation to dc machines. A full wave reversing control or servo as shown in Figure 4-9 can be designed with four SCRs which are controlled by two UJTs. A transistor clamp synchronizes the firing of one UJT to the anode voltages across one pair of SCRs. A potentiometer is used to regulate the polarity and the magnitude of output voltage across the load. At the center position, neither UJT fires and no output voltage

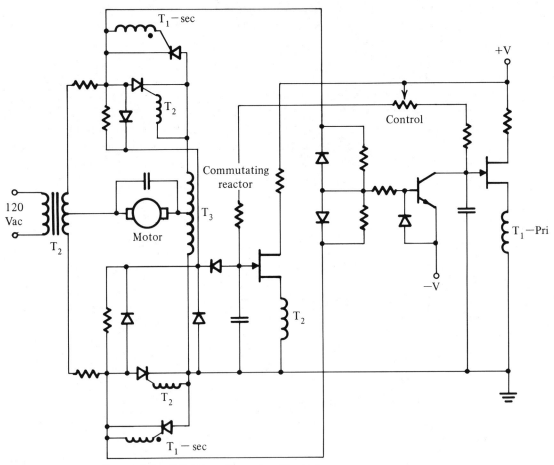

FIGURE 4-9   Full-Wave Reversing Control.

appears across the load. At the extreme left-hand position, full output voltage appears across the load. As the resistor arm is moved to the right of center, a similar action occurs except the polarity across the load is reversed.

For a dc motor, plugging action occurs if the resistor arm is moved abruptly to reverse. Resistors are used in series with each end of the transformer to limit fault currents if a voltage transient should fire an SCR pair simultaneously. A commutating reactor and capacitor limit the dv/dt which one pair of SCRs can impress upon the other pair.

Half Wave Drive

A phase sensitive servo drive supplying reversible half wave power to the armature of a small permanent magnet or shunt wound motor is shown in Figure 4-10. The circuit uses two half wave circuits back-to-back fired by a unijunction transistor. It fires on either the positive or negative half cycle, depending on the imbalance of the control bridge. The controlling element can be a photoresistor, thermistor, or the output from a D/A converter.

The universal series motor is used in a wide variety of consumer and light industrial applications such as blenders, hand tools, and mixers. The control circuits to be described provide, in effect, a variable tap on the motor.

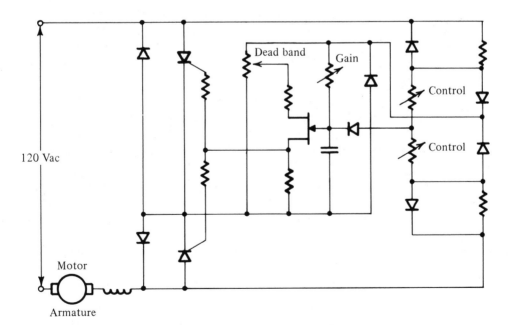

FIGURE 4-10   Half-Wave Phase-Sensitive Reversing Control.

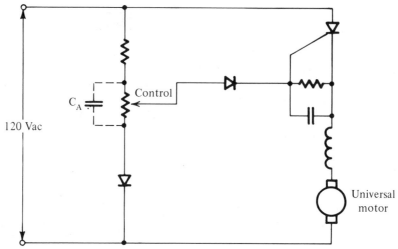

FIGURE 4-11   Half-Wave Universal Series Motor Control.

The half wave circuit of Figure 4-11 supplies half wave dc to the motor. To have full speed operation, the motor must be designed for a normal voltage of about 80 volts for operation on 120 volt ac lines. Brush life of a motor driven by a half wave supply tends to be shorter than for the same motor on full wave ac. The half wave circuit shown employs residual back EMF feedback to provide increased motor power as the motor speed is reduced by loading. This back EMF voltage is dependent on the residual magnetism of the motor, which is a function of the magnetic structure and the characteristics of the iron. The motor used must have sufficient residual magnetism.

The circuit operates by comparing the residual back EMF of the motor with a circuit generated reference voltage. In the divider network with the potentiometer, current flows only during the positive half cycle due to the diode. The voltage at V_A then is a half sine wave. If the residual back EMF is greater than this value, then the motor is going faster than the selected speed and the SCR will not be triggered to supply power to the motor during this half cycle. As the motor slows and its back EMF drops, current will flow through the gate of the SCR_1, thus triggering it. The speed at which this occurs may be varied. The smallest impulse of power that can be applied to the motor is one-quarter of a cycle, since the latest point in the cycle that the SCR can trigger is at the peak of the ac line voltage.

If the motor is loaded so that its speed and back EMF continue to drop, the SCR will trigger earlier, supplying more power to the motor. If the motor is lightly loaded and running at a low speed, one-quarter cycle may change the speed by a considerable amount. When this happens, a number of cycles may be required to return to the speed at which the SCR will again trigger. This causes a hunting effect which may be accompanied by mechanical noise. To overcome this problem, the smallest increment of power must be reduced from a full quarter cycle to that required to compensate for the motor energy lost per cycle. To accomplish this, the capacitor C_A is added to the circuit. The capacitor voltage becomes sinusoidal in shape during the positive half cycle. This voltage is phase shifted by an amount determined by the circuit time constant and an exponential decay during the negative half cycle. Two main effects occur: The first is that the latest possible triggering point is delayed, thus reducing the smallest increment of power. The second is that the amount of change required to go from minimum power to full power is reduced, providing a more effective control. Increasing C_A even more, the triggering point comes still later and the required voltage $\triangle V$ becomes still smaller. Care must be taken not to go too far, because increasing C_A decreases $\triangle V$ and increases the loop gain of the system, which can lead to instability and hunting.

The impedance level of the divider network should be low enough to supply the current required to trigger the SCR without extra loading. The current available for triggering from this network approaches a sine wave.

In some cases, low speed operation without a tight specification on gate current to fire requires such a low impedance network that the high power ratings of the resistors and the large capacitor size become expensive. In these cases, a low voltage trigger device such as an SUS can act as a gate amplifier. Use of an SUS allows a much higher impedance divider network with smaller size and lower cost components. Here the reference voltage must exceed the back EMF by the breakover voltage of the SUS, which is about 8–10 volts. When the SUS triggers, it discharges a capacitor into the gate supplying a pulse of current to trigger the SCR. This also eliminates the need to select SCRs for the gate trigger current.

Full Wave Drive

The circuit in Figure 4-12 also derives its feedback from the lead current and thus does not require separate connections to the field and armature windings. The diode bridge applies a full wave rectifier voltage to the SCR. When the SCR conducts, the bridge appears as a low resistance in series with the motor. The line voltage, less the drop across two of the rectifiers, the SCR, and its resistor is applied to the motor. By delaying the firing of the SCR until a later portion of the cycle, the voltage applied to the motor is reduced and its speed is reduced proportionally.

The capacitor phase delay of the SCR firing is obtained by the charging of the speed control from the voltage established by the zener diode. When the charge reaches the firing voltage of the unijunction transistor, it triggers the SCR.

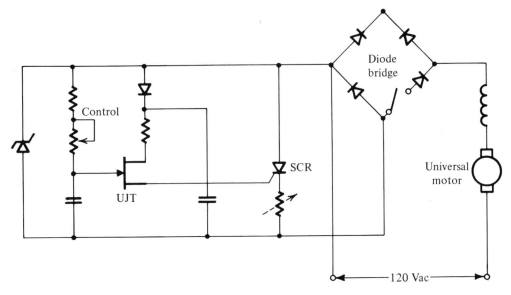

FIGURE 4-12   Full Wave Universal Series Motor Control.

During discharge the emitter current of the unijunction transistor drops below the holding current level and the unijunction stops conducting. While the SCR is conducting, the voltage drop on the SCR leg falls below the breakdown voltage of the zener diode. Since the SCR leg voltage is a function of motor current, its capacitor is charged during the conduction period proportional to the motor current. The amount of charging required to fire the unijunction transistor is decreased by an amount proportional to the motor current. Thus, the firing angle at which the unijunction transistor will fire has been advanced in proportion to the motor current. As the motor is loaded and draws more current, the firing angle of the unijunction transistor is advanced even more, increasing the voltage applied to the motor and its available torque. Since the firing voltage of the unijunction transistor depends upon the Base 2 to Base 1 voltage, it is necessary to support the Base 2 voltage using the capacitor and diode, during the conduction portion to prevent the feedback voltage from firing the unijunction.

The SCR current limiting resistor depends on the motor characteristics. It can be an adjustable wire wound rheostat which can be calibrated in terms of motor current. If the value is too high, feedback can be excessive and surging or loss of control may result. If the value is too low, a loss of torque can occur.

Figure 4-13 shows a circuit of a full wave series motor speed control with feedback which requires that separate connections be available for the armature and field.

In this circuit, firing is accomplished when the voltage on the wiper arm of

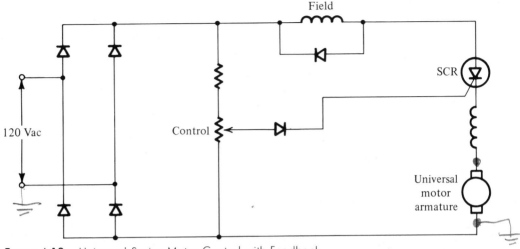

FIGURE 4-13   Universal Series Motor Control with Feedback.

the potentiometer rises to a high enough value to forward bias the series diode, allowing the gate current to flow. Since the counter EMF tends to reverse bias this diode, the firing point depends upon the counter EMF. The counter EMF, in turn, is a function of speed. As the motor is loaded, reducing the speed and the counter EMF, the series diode becomes forward biased earlier in the cycle, thus triggering the SCR earlier in the cycle and supplying the motor with greater power to offset the loading effect.

The full wave bridge supplies power to the series networks of the field, SCR, and armature. This circuit works on the same principle as that in Figure 4-11, using the counter EMF of the armature as a feedback signal. As the motor starts running, the SCR fires when the reference voltage exceeds the forward drop of the series diode and the gate-to cathode drop of the SCR. The motor then builds up speed and as the back EMF increases, the speed of the motor adjusts to the setting of the potentiometer in the same manner as the circuit of Figure 4-11.

One of the drawbacks of this circuit is that at low speed settings, because of the decreased back EMF, the anode-to-cathode voltage of the SCR may not be negative for a sufficient time for the SCR to turn off. When this happens, the motor receives full power for the succeeding half cycle and the motor starts hunting. This circuit is also limited by the fact that the SCR cannot be fired consistently later than 90°. A capacitor on the control arm is not a cure because there will be no phase shift on the reference due to full wave rectifier charging.

In general, for most universal motors, the feedback voltage is low; the residual field is low. Usually it is in the 1–5 volt range. For maximum feedback effectiveness, the change in counter EMF between zero and full load on the motor should approach the peak-to-peak voltage of the ramp generated by the trigger circuit. Because of the voltages involved, low cost breakover type devices are ruled out and the trigger circuit is limited to the types of circuit shown in Figure 4-13.

When the field can be kept at a high value during the off time of the SCR, as would be the case with a permanent magnet motor, this same technique could be applied with a breakover type of trigger circuit.

Since the universal series motor is generally designed to run on the 60 Hz ac line, the simplest approach to a nonregulating control is the full wave phase control circuit of Figure 4-14. The operation of this circuit is a function of the control resistor since the motor feedback does not come into play. The capacitor and resistor across the TRIAC act as a dv/dt suppressor.

INDUCTION MOTOR CONTROL

A variety of induction motor types exist, and within these are a wide range of possible characteristics. Some of these characteristics make a given motor type unsuitable for phase control. The major difficulty is that induction motors tend to be more frequency sensitive than voltage sensitive and phase control is a variable voltage, constant frequency technique. If the motor is not designed for phase control, the motor should be as voltage sensitive as possible. Variable voltage drive of induction motors is a compromise, usually dictated by economics but it can be satisfactory in the proper application. A variable frequency drive is superior but more expensive than phase control. Variable frequency inverters that can be used for drives include the class A, B, D, and E types as well as the cycloconverters.

Certain types of single phase induction motors—the split phase and capacitor start motors—require a switched start winding. Since there is a torque discontinuity when the start switch cuts in or drops out, controlling the speed of the motor at these points is not possible. Thus, if the higher starting torque of a switched start winding is required, the motor must be designed so the switching

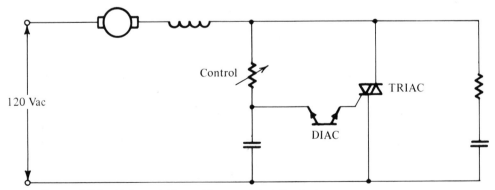

FIGURE 4-14   Universal Motor Nonregulated Control.

point is below the range of speed over which the phase control is desired. Another consideration is the power factor of the motor. A highly inductive motor can require a complex control to avoid the problems associated with the phase control of inductive loads. This is because the synchronization must come from the supply and the trigger must be continuous for a good part of the conduction period.

Unlike the brush type motors, induction motors produce no convenient electrical indication of their mechanical speed. For applications like fixed fan loads, direct voltage adjustment with no feedback is satisfactory. An example is the circuit of Figure 4-14 with a permanent split capacitor motor or with a shaded pole motor. The proper motor load combination can be shown by speed torque curves. In the case of a low rotor resistance, varying the voltage of the motor will produce very little speed variation, while a higher rotor resistance motor is satisfactory.

Speed Control

The problem of speed regulation of induction motors can sometimes be overcome by referring to the complete system. Consider the control of the speed of a blower in a hot air heating system in response to the temperature of the air. The prime interest is the temperature of the air, not the speed of the motor. This system can use the circuit illustrated in Figure 4-15. In this circuit the thermistor acts in response to air temperature to control the power supplied to the motor. The phase control network serves to set a minimum blower speed which provides the air circulation.

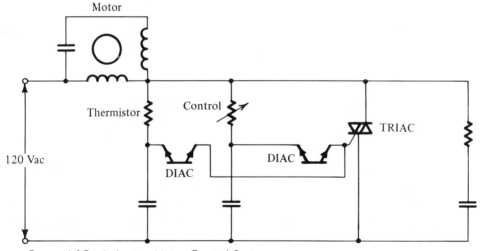

FIGURE 4-15   Induction Motor Control System.

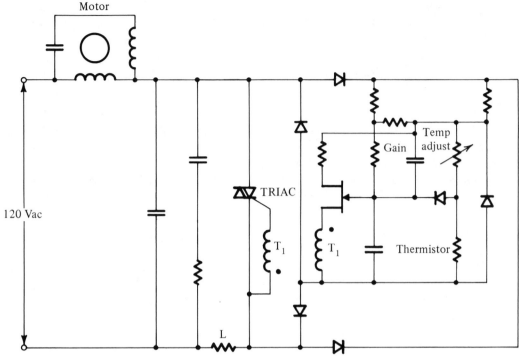

FIGURE 4-16   High Gain Induction Motor Control System.

Figure 4-16 gives an example of a more complex control system which is capable of a higher control gain. This can be used to control a blower motor in response to cooling coil temperature to prevent air conditioner freeze ups.

This circuit uses a ramp-and-pedestal system for the control of fan or blower motors of the shaded pole or permanent split capacitor type in response to the temperature of a thermistor. The circuit includes both RF noise suppression and dv/dt suppression.

To regulate the speed of an ac induction motor by means of phase control, it is necessary to provide speed information to the circuit with a small tachometer generator. Such a generator can be inexpensive since high precision is not required.

The speed torque characteristics of the motor should be voltage sensitive. A motor with low rotor resistance can be difficult to control in a stable manner since the open loop system characteristics are highly nonlinear through the controlled speed range. The drop-out point of the start switch, if used, should be below the lowest desired controlled speed.

Figure 4-17a shows a general block diagram of an induction motor phase control system. A ramp and pedestal control circuit, as shown in Figure 4-17c, can

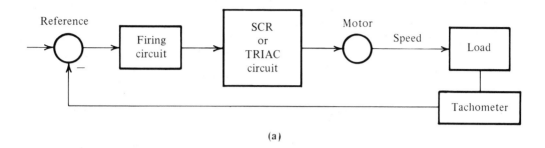

(a)

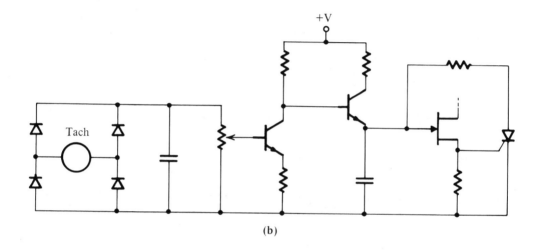

(b)

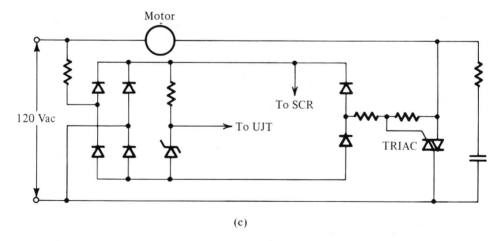

(c)

FIGURE 4-17   (a) Induction Motor Phase Control System; (b) Tachometer Circuit; (c) Phase Control Circuit.

be combined with the tachometer input connection shown in Figure 4-17b. The connection is shown for an ac tachometer. The filter time constant should be chosen to give adequate filtering at the lowest desired speed and tachometer frequency, consistent with the system stability requirements. The system is also useful for multiphase controls as well as dc motor drives.

In some cases, a capacitor start or a split-phase motor must operate where there is a high frequency of starts, or where arcing of a mechanical start switch is undesirable, for example, where explosive fumes could be present. In these cases, the mechanical start switch can be replaced by a triac. The gating and dropout information may be presented to the triac in several ways. The simplest form of connection is to use a conventional current or voltage sensitive starting relay at the pilot contacts for a triac static switch.

In another method, shown in Figure 4-18, the triac is gated on by the motor current transformer. As the motor speeds up, the current drops and can no longer fire the triac. A variation is to replace the current transformer with a small pickup coil, mounted near the end windings of the motor. If a tachometer generator is already sensing the speed of the motor, this same signal can be used to control a firing circuit for the triac. With this arrangement the dropout speed can be set to be outside the desired speed control range.

A pair of triacs can be used to provide a reversing control for a split capacitor motor as shown in Figure 4-19.

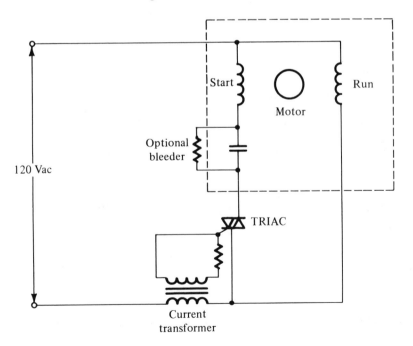

FIGURE 4-18 Motor Starter.

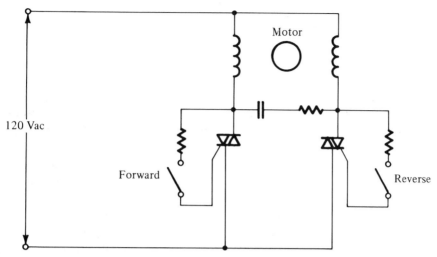

FIGURE 4-19   Reversing Control.

Variable Frequency Drive

Variable frequency ac drives (VFD) are used in an increasing number of applications where an ac motor offers advantages over the dc motor. VFDs can substitute for traditional methods of varying the speed of an ac motor. The variable frequency drive offers excellent reliability and allows the use of standard NEMA-B ac induction motors. The low cost and reliability of the ac motor make it well suited for industrial applications. Now, it can also be adapted to variable speed application.

Operation of the VFD is typically from a 3 phase, 60Hz, 480V or 240V ac supply. The VFD frequency and voltage output is controlled or modulated to achieve the desired variable speed ac motor operation. The variable frequency voltage output can be achieved in three ways: pulse width modulation, voltage source, or current source.

In the case of a current source VFD, the control system consists of a converter, series inductor, and inverter. The converter section is a conventional phase controlled, full wave dc drive. The converter feeding the series inductor functions as a controlled, adjustable current source. The inductor feeds current to the motor by operating the inverter thyristors as switches and steering the current into the motor phases sequentially in steps. Changing the thyristor switching rate varies the frequency of the motor current. The waveform is suitable for ac motors and results in only about a 5% additional current than when operated directly off the ac line. The motor speed varies directly with frequency. The motor voltage is also varied with frequency to achieve a high and consistent flux in the motor. Varying the frequency and voltage in a constant ratio allows the motor to be operated with relatively constant torque characteristics over the speed range. The wide speed range with constant torque capability is about the same as would be found with a dc motor drive.

In applications in which alternatives of ac or dc motors are available, a comparison of initial capital cost will usually favor the dc drive system. Although the ac motor is less expensive than the dc motor, the higher cost of the VFD makes this combination more expensive. However, initial VFD controller cost is only one factor in choosing such systems. Depending on the system, the following features could influence what system to use:

1. Applications where multiple motors are run from one control, hence the lower cost of the motors can offset the higher single drive price.
2. AC motors offer higher speeds for a given horsepower than dc. This allows faster processing without the need for a gear increaser.
3. Lower rotor inertia with no commutation problems during starting with ac motors allows fast acceleration and deceleration.
4. No brushes or commutator allows less maintenance for ac motors.
5. The lower weight and smaller frames of ac motors per horsepower simplifies the motor mounting and allows more flexibility in motor placement.
6. Process changes from constant speed to variable speed may favor a VFD if the existing ac motor can be used.
7. AC explosion proof motors are often the only types available.
8. Higher reliability requirements can be met with an ac motor in those applications which allow use of a constant speed bypass if a control system fails.
9. The ac motor allows higher reliability, lower replacement cost, and quicker shipment time.

Rising energy costs can also be a factor in the increased use of VFD. Traditional methods of varying the speed of an ac motor tend to be inefficient compared to VFD. The conventional drives dissipate the slip energy in heat. Since VFD controls both the frequency and voltage, the motor dissipates little additional heat than when operated on the line. Losses are primarily from the voltage drop across the thyristor and the inductor losses. Power cost savings with the VFD may more than offset the additional VFD cost. A power cost comparison can illustrate the power cost saving resulting from the higher VFD efficiency. A constant torque load where horsepower decreases directly with speed can result in substantial power savings.

Savings can also be achieved by modifications of the current source circuit. Replacing the converter with a rectifier/chopper can yield a 0.95 power factor over the complete speed range; the use of solid-state power factor correction can eliminate the need for power correction capacitors. Many electric utilities assess a penalty based on power factor. One method is to increase billed kW demand by the deviation of actal power factor. A constant power factor can thus yield a considerable savings.

STEPPING MOTORS

A stepping motor takes a single step due to the excitation of its stator winding and the rotor translates to a new position. The motion of the stepping motor rotor in a single step mode is similar to that of a torsional pendulum. The excitation through the stator flux and the rate of increase of flux determine the maximum kinetic energy input to the rotor. The load friction is in the form of damping and the system inertia consists of both the rotor and load inertias. The effects of the system dynamics on position error can be large.

As the acceleration increases, the available torque becomes less, because during acceleration, the rotor-stator lag angle is greater than the angle at steady state conditions. This additional lag causes a loss of torque.

The acceleration performance of a stepping motor is a function of driver, load friction, inertia, ramp time, starting frequency, and final frequency. A digital positioning system can generate motion pulses having a linear velocity ramp such that the last pulse of motion and zero velocity occur simultaneously. Arriving at the final position at too high a speed, or undershooting and using too much time by creeping to this position can occur with exponential and linear voltage systems.

The motor is a transfer device between the electrical information presented by the driver and the mechanical motion delivered to the load. The functions of the stepping motor system are dependent both upon the load and the driver.

Many systems are difficult to characterize since the load cannot always be analyzed readily. The practical approach is for the designer to understand the general parameters and problems involved, make a reasonable choice of driver and motor, and then test the combination. This technique has been more useful in the past than attempts to thoroughly quantify and analyze the complex system dynamics.

Permanent magnet (PM) stepping motors contain a stator with a number of wound poles. Each stator pole may have several teeth for flux distribution. The rotor is cylindrical and also toothed. PM motors use permanent magnets and most have the permanent magnet in the rotor assembly. The magnet is usually axially charged except for the smaller motors which may be radially charged.

The PM motor operates due to the interaction between the rotor magnet flux and the magnetomotive force generated by an applied current in the stator windings. If the pattern of winding energization is fixed, then a series of stable equilibrium points are generated around the motor. The rotor will move to the nearest of these and remain there.

When the windings are excited in sequence, the rotor will follow the changing equilibrium points and rotate in response to the changing pattern. The use of a permanent magnet causes a small detent torque in the motor even when the windings are not excited. This torque is normally a few percent of the maximum torque.

The variable reluctance (VR) motor has a stator with a number of wound poles. The rotor is cylindrical with teeth that have a relationship to the stator poles. This relationship is a function of the step angle required. When a current flows through the windings, a torque is developed which tends to turn the rotor to a stable position of minimum magnetic reluctance. This is not an absolute position since there are many stable points in the motor. If a different set of windings is energized, this minimum reluctance point switches to another set of poles and rotor teeth, causing the rotor to move to a new position.

When the proper energizing sequence is used, the stable positions rotate about the stator poles producing a rotational speed. If the energizing is fixed, the rotor position is also fixed, thus the shaft position can be stopped by changing the energizing pattern. In a VR motor, the rotor teeth have only a small residual magnetism compared to the permanent magnet motor, thus there is no torque when the stator is not energized. The VR motor behaves in a similar manner to an ac electromagnet, in which a magnetic attraction occurs regardless of the direction of the magnetic flux.

Linear devices also exist in the form of linear PM or VR stepping motors. Their operation is the same except that the magnetic paths are along a linear axis.

Other stepping motor designs use internal torque amplification along with the stepping motor. One technique is to use internal gearing to reduce the step angle with no power gain. Another approach is to couple a low torque stepping motor with a linear power amplifier. In the torque amplification technique a harmonic drive with a large gear reduction can be used.

The power amplification technique uses a smaller stepping motor with a hydraulic amplifier. The stepping motor drives a control device in the hydraulic motor. When the cost of the system and the hydraulic supply can be justified, the power gain is considerable. Hybrid motors of this type can provide several horsepower.

Compared to VR motors the PM types are more efficient with better damping ratios and they are available with higher power outputs. The VR motors are simpler in construction, with a low rotor inertia, and when they are lightly loaded, they can provide high speeds.

System Considerations

The general nature of the power output relationships takes the following form: at very high speeds, zero torque and zero power are delivered; at zero speed, the torque is high but power delivered is zero. One should normally operate at the peak of the power output curve to obtain the maximum performance from a given motor-driver combination.

The operating point selection and motor driver choice must also consider the thermal conditions in the motor. Since the motor thermal time constant may be 20 minutes or more, it is possible to allow the high motor dissipation for short periods of time, depending on the duty cycle. The major criterion is the final temperature rise which is a function of the ambient temperature, the motor heat sink, and the operating times. The main limitation on the power output of most stepping motors is the temperature rise. More torque can be obtained at higher than rated current provided that the average motor temperature is held within its rating.

The designer can determine if a motor will overheat by measuring the winding resistance before and after it has operated at the duty cycle for 3 to 4 hours. The long period is required since the thermal time constant of most motors can be more than 20 minutes. If the winding resistance is measured when the motor is at room temperature and again at the end of a period of operation at the worst duty cycle, the temperature rise of the winding is approximately 2.5 times the percentage resistance increase from the room temperature resistance. This temperature rise should be limited to the value recommended for the motor based on the maximum ambient temperature expected. Heat radiators or fan cooling are required in some applications. But in general, most control systems have only limited periods where high speeds are required and heating is not a severe problem.

Among the driver-motor combinations, resistance limited drivers frequently offer good performance when used with a suitable motor. However, consideration must be given to the power dissipated in the limiting resistance and the effects of the resulting temperature rise on the remaining system components. In many systems, the limiting resistor dissipation can be a significant percentage of total system power. Thus, a more efficient driving system can be considered, even though the system drive requirements can be met adequately by the resistance limited drive.

Another trade-off arises when the stepping motor system is used as a high speed positioning device and angular accuracy is critical. Here, the start-stop speed and damping characteristics are important. Motor drives that provide high start-stop rates tend to be poor in settling characteristics. Reverse pulse damping or a current drive can provide a solution for the damping problem. Mechanical dampers can be used, but their life span may not be satisfactory in some applications.

Microprocessors are used in many stepping motor controls. The microprocessor may be used with some hard-wired logic to advance the motor by a step when an output appears from a state generator. The logic sequence can also be stored in memory with 1 bit sent out at a time. A step counter may be used to ensure that the motor goes through the correct number of steps. Two modes of operation can be used: constant speed operation or automatic acceleration and deceleration. The constant speed mode moves the motor through the steps at a constant rate. In the acceleration-deceleration mode, progressively decreasing or increasing time delays are used between the steps to increase or decrease the stepping rate. The basic control algorithm appears in Figure 4-20.

STEPPING MOTOR DRIVES

In a stepping motor system the overall performance is heavily dependent upon the drive system which provides the power delivered to the load.

The stepping motor drive system accepts the drive input signal and converts it to the proper format for driving the motor windings. A power amplifier is used to drive the current through the windings and a power return system is used to remove the current from the windings at the completion of a step. Since the motor rotates in response to the changing patterns between the rotor and stator magnetic fields, a state generator must be used to create the proper sequence and pattern of states. The state generator is usually controlled by a serial pulse train.

Drive Techniques

Two major types of sequences are used to cause the motor to step. In the wave drive scheme, only one set of stator poles are energized at a time (Figure 4-21a). The two phase drive energizes both sets of poles simultaneously as shown in Figure 4-21b. Either one of these causes an N step motor to step by increments of N, but there is an N/2 displacement of stator and rotor between the two sequences.

The wave drive currents are shown in Figure 4-21c. The A1 current energizes the phase A poles to create a North Pole at the stator pole teeth. The B_1 current generates a North Pole at the phase B pole teeth. The A_2 current generates a South Pole at the pole A teeth while B_2 creates a South Pole on the B phase poles. As a result of this sequence, the rotor advances to align the rotor and stator teeth. The logic used to generate these states can be placed in memory and used to control a microprocessor.

The wave drive is a common technique, but it is not preferred for all two

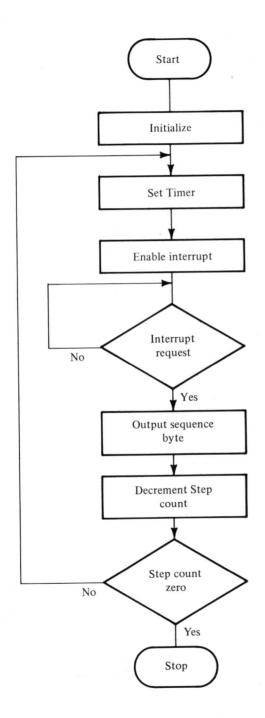

FIGURE 4-20   Microprocessor Stepping Motor Control Flow.

phase stepping motors since the ampere-turns on the stator poles are 40% higher if both of the phase A and B poles are driven using the four windings A_1, A_2, B_1, and B_2 two at a time. A net gain in torque per watt is approximately 20% with the two phase sequence of currents shown in Figure 4-21d. The current in A_1 creates a North Pole in the A stator poles and the current in A_2 generates a South Pole. The B_1 currents generate a North Pole in the B stator poles and B_2 creates South Poles.

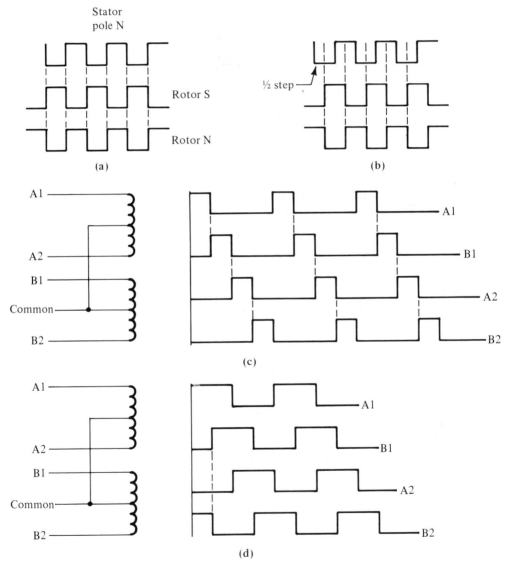

FIGURE 4-21   (a) Wave Drive; (b) Two Phase Drive; (c) Wave Drive Currents; (d) Two Phase Drive Currents; (e) Half Step Drive Currents.

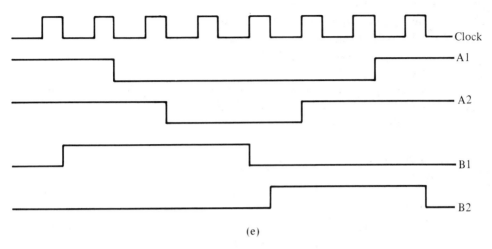

(e)

FIGURE 4-21 *continued*

The four combinations of current in the two windings will produce four motor steps and the pattern repeats every four steps. The logic required is slightly simpler than the wave drive. For this reason and for the increased performance, the two phase drive is more commonly used over the wave drive.

When four windings of a stepping motor are energized with the wave drive using the tooth alignment shown in Figure 4-21a, the successive steps have a spacing of 1.8°. The two phase drive with the tooth alignment of Figure 4-21b also has 1.8° steps. When the half step drive is used, it alternates between the wave type and two phase drive as shown in Figure 4-21e and the step is one-half the normal step, or .9°.

Of these three drive sequences, the two phase drive is most widely used since it is more efficient. The half step drive is useful in some applications for reducing resonance problems. A difficulty with the half step drive is that the two types of steps may have somewhat different characteristics due to the different magnetic alignment.

It can be used with either bipolar or unipolar drivers. With unipolar drivers, A_1, A_2, B_1, and B_2 represent the individual winding drivers. With bipolar drivers, a positive voltage at A_1 drives a positive current into the A-phase winding pair and a positive voltage at A_2 drives a negative current into the winding pair. This convention also holds with B_1 and B_2 with respect to the B phase winding pair.

In using a half step drive system with a constant current per winding, in certain cases the two windings are energized. The motor power is at a maximum and the available torque is at a maximum. When only one winding is energized, the motor power is reduced by half. This results in a strong step and a weak step.

In terms of positional accuracy, if the load is less than about 40% of the maximum torque, the stiffness of either step is about the same around the origin. Above this torque the weak step has a difference in stiffness that appears as a loss of stepping torque at low frequencies, but gives about equal performance at high frequencies. The difference between the strong step and weak step can be reduced by changing the current in the windings to drive the motor at rated power on every step. The weak step will not increase enough to equal the strong step, but the difference will be reduced from 30% to about 15%.

Power Amplification

Stepping motor drive requirements can vary from a few volts at about 50 ma to 100 volts or more at 20 amperes. The requirements on a driver depend not only upon the stepping motor involved, but also on the speed and torque requirements of the system. The standard motor windings represent a reasonable match to available semiconductors.

The current rate of rise in a winding is proportional to V/L, so that higher voltages can give better high speed performance up to a point. It is usual for the source voltage for pulse initiation and termination to be 10, 20, or 30 times the steady state motor voltage.

The stepping motor winding can be considered as an inductance in series with a resistance. A parallel resistance appears across the winding representing the load power, but it generally has a negligible effect on the driver design. The time constant of the winding, L/R, is typically 10 milliseconds.

If a voltage source is impressed across this winding, approximately 95% of full current is reached in three time constants, or 30 milliseconds. The drive system will apply current to each winding at one-quarter of the input pulse rate, so each winding receives 95% of full current at an input rate of $4/(30 \times 10^{-3}) = 133$ steps/second. In many applications this limitation on motor speed may not be acceptable. A typical 200 step/revolution motor can deliver 15,000 steps/second slew speed, when properly driven.

Waveform Generation

Thus, the objective of the drive circuitry for high speed operation is to provide a high voltage to move the current in and out of the winding at the pulse transition times and a low voltage to sustain the correct current during the steady state portion of the current pulse. The required waveform can be generated in four ways:

1. A series resistance is added to the windings.
2. A chopped waveform is used in which a high voltage is applied to the windings at the beginning of the step to allow the current to build up and then it is chopped at its proper value.

3. A bilevel drive uses a high voltage during pulse initiation and termination and a low voltage during the constant part of the pulse.
4. A programmed voltage source with phase controlled SCRs or constant current transformers is used to deliver the voltage to maintain a constant regardless of the pulse rates.

Unipolar resistance current limiting is probably the most widely used method, in which resistors are added in series RS with the motor windings and raise the supply voltage to increase the motor current under steady state conditions. A typical circuit is shown in Figure 4-22. The winding time constant, L/R_M, is reduced to $L/R_M + R_S$ on both pulse initiation and termination. The actual current rate of rise also depends upon the back EMF generated and thus the rotor dynamic response.

Unipolar resistance limited drives are simple and reliable and they can be adequate where speed and torque are low. However, following are several disadvantages:

1. The system efficiency can be low. A 10 watt motor may use a driver of 200 watts at 20 times overdrive.
2. The windings in the motor are not all used, since the current duty cycle is only 50%.
3. The voltage is decaying exponential and the current is a rising exponential. The current rate of rise is less than in systems where the full supply voltage is available until the desired current is reached.

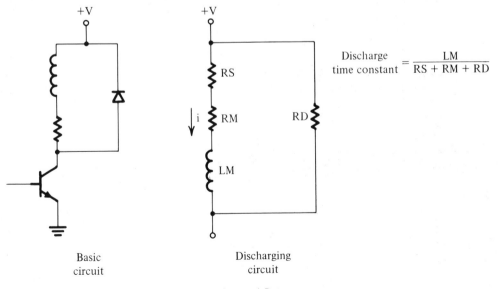

$$\text{Discharge time constant} = \frac{LM}{RS + RM + RD}$$

Basic circuit

Discharging circuit

FIGURE 4-22   Unipolar Resistance Limited Drive.

A large improvement in performance occurs between wave drive and two phase drive when two of four windings are used instead of one of four. If all four windings are energized all the time, a further increase in performance can be achieved. The full 40% increase in torque is not reached because of the nonlinearity in the B-H curve of the iron. A 25% to 30% improvement in torque can be reached. If the four windings are connected as shown in Figure 4-23, currents in A_1 and A_2 have the same sign as the A phase stator poles. When the windings are driven with the bidirectional currents, alternate magnetic poles are created in the stator similar to a unipolar drive, except the two phase windings aid each other. The logic required is the same unipolar two phase logic.

An improved motor efficiency may be realized as increased torque and speed.

One method of obtaining the required drive wave is the split supply system shown in Figure 4-24. Equal positive and negative supplies are used. The overall complexity of the system is approximately the same as the unipolar driver. The power switches are in series with the supply voltage so a delay must be incorporated to prevent one switch from being turned on while the other is already conducting. The power resistors may also be split, as shown in Figure 4-23, so that overlaps in switch construction will not be harmful.

Another technique for bipolar drive uses full bridge operation, as shown in Figure 4-25. This circuit requires only a single supply voltage obtained at the expense of doubling the required number of power switches to eight per motor. The switch timing concerns apply as in the split supply case and the use of a delay network can be avoided by using more power resistors. Due to the operation of the bridge, the total peak-to-peak voltage on the motor windings is twice the supply voltage. Thus, the full bridge can drive a winding at twice the voltage of the split-supply bipolar or the unipolar circuits.

A level drive system can obtain a high current rise time. The circuit uses two supplies: a high voltage supply for pulse initiation and termination and a low voltage unit to supply the sustaining current during the pulse duration. One circuit

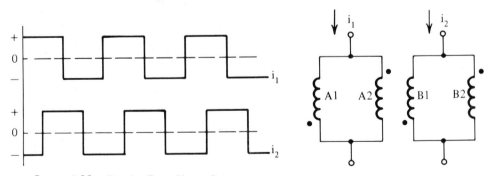

FIGURE 4-23   Bipolar Two-Phase Drive.

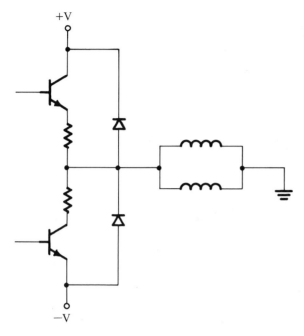

FIGURE 4-24   Split-Resistor Resistance Limited Bipolar Driver.

technique is shown in Figure 4-26. The high voltage supply is turned on until the motor current reaches the operating level. Then the high voltage switch is turned off. At the end of the pulse, the lower switch is turned off and the current in the winding is returned to the high voltage supply. This results in rapid pulse initiation and termination. The current rise time is higher than in the resistance limited

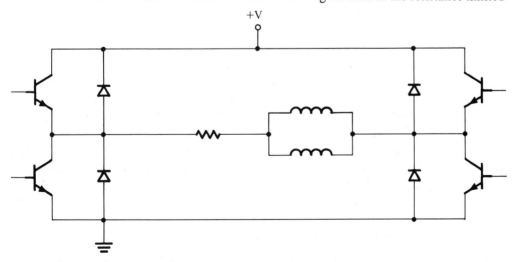

FIGURE 4-25   Bridge Type Resistance Limited Bipolar Driver.

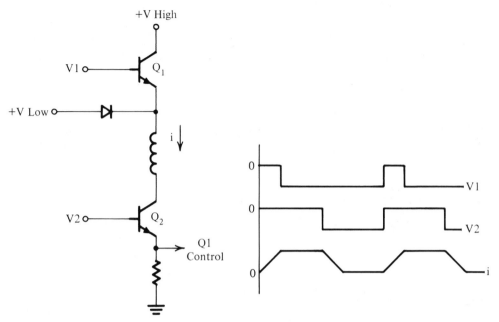

FIGURE 4-26   Bilevel Drive.

current case where the voltage is exponential. With either drive system, unipolar or bipolar, higher driver efficiency is obtained if the voltage source can be programmed to deliver the required current at all stepping rates. Programming the voltage efficiently can be done in the initial conversion from the power line, using methods such as ferro-resonant transformers, saturable reactors, and SCR phase control. These methods have response times of 50–200 milliseconds.

A block diagram of an SCR constant current system is shown in Figure 4-27. This mode of operation allows good resonance control. Only enough voltage is applied to force the rated current through the stepping motor winding so the problem of rotor overshoot and resonant modes is reduced. Another result of this mode is that high overdrive does not exist at low speeds. The average current to the drive system is regulated. As the frequency is increased, the average voltage is increased, forcing the set level of current through the motor winding until the voltage limit of the supply is reached. At high speeds, a large fraction of the motor current is returned to the supply by the recirculating diodes, but this is ignored by the system which continues to adjust the voltage on the basis of the average drain from the supply. The result is that the high speed torque is enhanced, but the RMS current through the windings, the RMS voltage, and the motor dissipation are increased during the high speed operation. This type of operation produces the greatest power output from a given stepper motor, but thermal factors, which

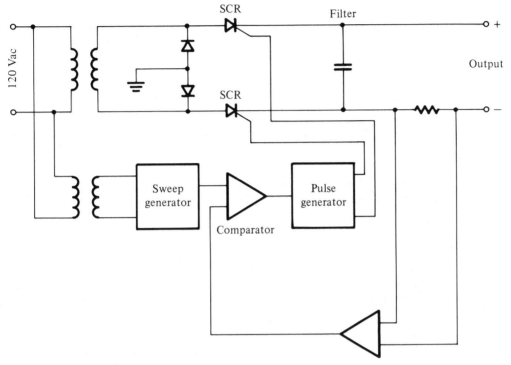

FIGURE 4-27 SCR Constant Current Control System.

include duty cycles, heat sinking, and ambient temperatures, must be considered in the design.

The long response times tend to limit the start-stop performance. When the system is at rest, a low percentage of the supply voltage is applied to the driver. In a system that will deliver 100 volts at high speeds, the voltage may be 3 volts when the motor is stopped. Since the voltage must rise only at a relatively slow rate, the start-stop response is limited also. This limitation may be allowable in a ramped system where the motor is accelerated to a high speed after starting from a low speed. Here the time to accelerate to high speed after starting is affected little by the driving system, since the response of motor and load limits the acceleration.

Care must be used in system design to prevent excessive current flow in the switching system. This can occur if the system is suddenly switched to a low speed from a period of high speed operation. Here, the system voltage is high to allow the high speed, thus the filter capacitor is fully charged. As the motor is suddenly switched to low speed operation, the capacitor charge is pumped through the switches, causing possible excessive current flow or switch dissipation. In a normal ramped operation, the motor speed is decelerated gradually and this problem does not occur.

Another system that offers certain efficiency advantages in stepping motor driving is the chopper system with current limiting using voltage modulation. The full high voltage supply at 10 to 20 times the motor voltage is applied to the motor winding until the correct current level is reached (Figure 4-28). The voltage is then switched off and the current circulates in the motor winding. When the current decay reaches a predetermined lower level, the voltage is applied to drive the current back to the correct upper level. This cycle is continued through the driving pulse time. At the end of the driving pulse, the winding current is recirculated to the high voltage supply. Thus, a high voltage is used to initially charge the motor winding, then a low average voltage is achieved by time modulation to sustain the current and a high voltage return to discharge the motor winding inductance. The effective motor voltage is increased at high frequencies, since the voltage is switched on to the motor for a high percentage of the time. The current sensing system as used in the current fed system does not sense the current circulating in the motor winding.

The bipolar chopper is a constant current driver that uses a switching regulator to control the motor current. The winding is connected by a 4-transistor bridge and the winding current is sensed in the common side of the bridge. The bridge is turned on until the current reaches the rated value and then is turned off.

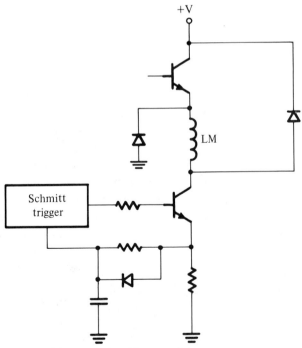

FIGURE 4-28   Unipolar Chopper Drive.

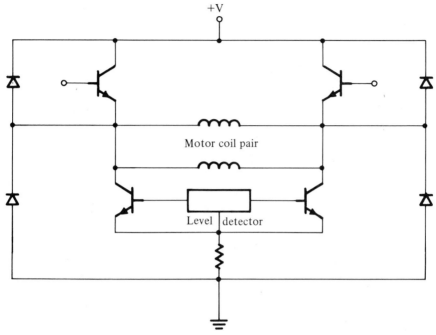

+V

Motor coil pair

Level | detector

FIGURE 4-29   Reversing Bipolar Chopper Drive.

During the off interval, the winding current is recirculated to the power supply through the diodes. The basic driver is shown in Figure 4-29.

Chopper operation is well suited for rapid start-stop applications. The high voltage is available to accelerate the rotor rapidly with a maximum of low frequency torque.

In ramped applications, resonance must be considered and the resonant regions avoided. It may be possible to accelerate through resonance regions, but it may be difficult to achieve stable steady state operation in such regions.

A bilevel drive can give approximately the same results. A current fed driver is smoother in operation, but has poor start-stop characteristics. All three types of drives give similar results at high stepping rates of 1.5KHz. They present identical waveforms to the motor in this region, where they become voltage sources switched by a transistor bridge.

Heating Effects

High efficiency drives result in an increase in RMS motor current and voltage. A large fraction of the power increase is delivered as output power to the motor, but there is a considerable increase in motor heating losses at high speeds under high overdrive conditions. An increase in motor dissipation also occurs in less efficient drives such as resistance limited systems. This dissipation is the result of the overdrive required for high speeds.

If the current or voltage ratings are used at low frequencies and low ratios of power supply voltage to motor voltage, the motor temperature will be safe. However, if large overdrive ratios are used, the input power to the motor can rise above the rated power. In the unipolar resistance current limiting case, the power input to the motor is equal to or less than the dc power input for overdrive ratios up to about 8. For ratios over 8, the motor power increases with frequency. For a ratio of 16, the power at 1500 steps per second (sps) is 1.8 times the low frequency value. Thus, overheating could be a problem if the motor runs continuously at this frequency. This is because, although the current is decreasing with frequency and the I^2R heat is dropping, the stator and rotor switching losses are increasing with frequency. These represent the power dissipated in the motor. Also the voltage across the winding increases with frequency, increasing the core losses. The switching losses increase with increasing overdrive since the voltage across the motor windings increases at high speeds.

The bipolar resistance current limiting method of driving motors gives higher performance by driving all the windings with bidirectional currents, or two phase bipolar driving can be used. With a bipolar R/L driver, the motor power increases rapidly with frequency. The maximum occurs when the I^2R losses are still large and the core losses are increasing. Eventually, the I^2R losses decrease faster than the switching losses. At high frequencies, the losses are almost all due to core loss. For frequencies above 10K sps, the motor power continues to rise. Bipolar switchng doubles the effect of the power supply voltage since the peak-to-peak swing across a motor winding is twice the supply voltage. Also, the rated motor voltage is less when two windings are connected in parallel because the resistance is halved, but the current in the pair only increases by $\sqrt{2}$.

Similar increases in motor power occur with a constant current drive. The constant current source adjusts the supply voltage upward as the motor current tends to increase with speed. The motor dissipation can rise to more than 400% of the rated motor power at constant high speeds. The increase in stepping motor performance is achieved at the cost of heat in the motor.

The motor ratings must always be considered with the type of driver used. When high performance drivers are used, the duty cycle of the application must be considered.

The simplest and least expensive drive circuit is the resistance limited unpolar circuit. However, it does not deliver the maximum output available from a given stepping motor and is inefficient from the standpoint of power requirements and heating.

The bipolar resistance limited system requires more complexity but it considerably improves the results over the unipolar circuit. The improvement can be 20–40% in torque, 30–50% in speed, or 50% in lower power. The high efficiency systems such as bilevel, current fed, or chopper also allow dramatic improvements in motor performance and efficiency, but at the expense of complexity and cost.

STEPPING MOTOR APPLICATION

Some stepping motor applications are straightforward. If the load is light and well controlled, the speed requirements low, and the environment constant and moderate, the main concerns tend to be price and a package that will fit in the space allocated. However, if a larger load must be driven as fast as possible in an application where time is valuable, the motor and driver choice can be critical. Usually the designer must make a choice based on a comparative study. The load requirements should be studied so the load parameters are as clear as possible. The designer should also consider modifications to the load in order to optimize the system.

Friction and inertia, combined with the required load speed and acceleration, determine the basic requirements for the stepping motor system. Friction determines the power output and inertia required. The required speed sets the amount of kinetic energy that must be put into the system on starting and removed on stopping.

Most stepping motor systems are more sensitive to inertial loads than to friction loads. If fast response and quick settling times are desired, the best load configuration will probably be the one that minimizes the inertia at the motor.

The stepping motor used with an appropriate driver can offer an infinite speed control range with the lower limit being as low as desired. The output speed is not affected by the load up to the torque limit of the motor and the speed is easily programmed or controlled from an external source, such as a microprocessor control. The variable speed capability can be used to run two motors at identical speeds in remote locations or the two motors can run at selected ratios of each other, independent of load. The block diagram of such a variable ratio system is shown in Figure 4-30.

Shaft position encoders can be used with stepping motors for improving the acceleration and confirming the motor position. Acceleration improvement may be achieved if the shaft position is determined accurately as a function of time. It is then possible to control the drive pulse timing to obtain maximum acceleration. The encoder must allow the determination of direction such as the class of encoders with quadrature tracks. An overshoot of shaft position could otherwise be interpreted incorrectly by the decoding circuitry. An encoder can provide an acceleration improvement that is difficult to obtain in any other way. Step confir-

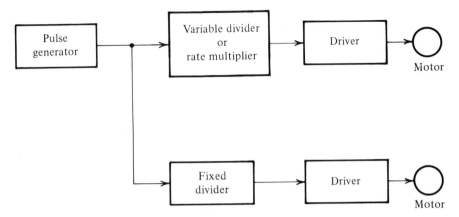

FIGURE 4-30   Variable Ratio Drive.

mation is achieved using an encoder with two quadrature tracks which act to de-code direction information. The encoder output can then be used to ensure that all will command steps are stepped through by the motor. This will guard against transient load conditions or other momentary malfunctions in the system. A step can be repeated if a malfunction occurs.

A common type of operation for stepping motors is the line operated synchronous mode. The driving waveform is made up of two sine waves which are 90° out of phase.

The synchronous drive is similar to a bipolar drive at a fixed frequency. The motor will step when either one of the inputs reaches either a positive or a negative maximum. The stepping rate is four times the line frequency. At a line frequency of 60Hz, the rate will be 240 steps/sec. Thus a 100 step/rev motor will operate at 240/100 rev/sec, or 144 RPM. The 90 waveform can be generated with an RC network. When three phase power is available, a Scott T transformer can be used, especially if several motors or high power motors are to be driven.

The components in the RC phase shifting network may have some effects on motor performance and the designer may have to optimize some parameters by adjusting some components from nominal values. In some cases, audible noise can be reduced, at the expense of some torque, by adjusting the component values. The starting characteristics under inertial loads will also be affected by changes in component values.

A synchronous control system that uses a stepping motor appears in Figure 4-31. The synchronous mode requires a minimum of extra parts to provide a reversible drive. This mode can provide a low synchronous speed when operated directly from the line frequency: 144 RPM for a 100 step/rev or 72 RPM for a 200

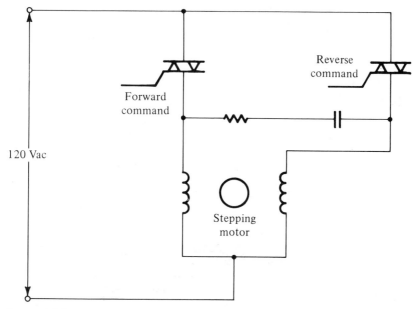

FIGURE 4-31   Synchronous-Mode Control.

step/rev motor at 60Hz. The low speed and a relatively high torque are provided without a gear box. The motor will be well damped with positive starts and stops and it may be stalled indefinitely without damage.

As with digitally operated stepping motors, a synchronous motor system can have differing starting and running characteristics as a function of the inertial load. Design curves should be used to determine if a proposed control system is within the capabilities of the selected motor.

In some applications, open loop stepping motor systems can be used in place of a closed loop control system. In other systems stepping motors are used as the mover in the system as shown in Figure 4-32. Here, the stepping motor is used in a digital stepping mode. This type of system is limited to the resolution represented by a single step.

The system operates in a bang-bang mode where an error signal produces a quantized response and the dead band required for stability is a function of the distance between the steps. The resolution is a function of the step size as seen by the load through any gearing. If a holding torque is applied to the load, then almost full torque can be generated by the motor with a shaft position error of less than the dead-band, and there will be no system error signal.

The speed of response is independent of the load characteristics. The load is

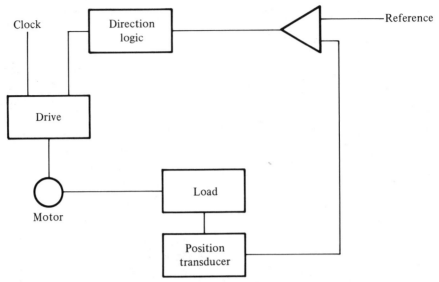

FIGURE 4-32   Stepping Motor in a Closed Loop System.

driven to the null point at the same speed regardless of the magnitude of the error. The overshoot problems are greatly reduced, since the system controls the overshoot. The stepping motor does not have the brush contact problems of dc motors and will also operate over a wider frequency range.

EXERCISES

4-1. Discuss the advantages and disadvantages of the system designer being the original specifier of both motors and motor drives.

4-2. What are some characteristics of motor drives and drive systems that should be included in specifications?

4-3. In the manufacturing industries, specialized machines and robots are a major application area for both motors and motor drives. Discuss the characteristics of a standard drive for some typical microprocessor system applications in this area. What protective features might be required?

4-4. Why must the operating ac or dc input voltage ratings be chosen with EMRs? How does the hard contact relay differ from SSRs in voltage and current ratings and contact isolation?

4-5. Discuss some methods of controlling motor speed with SCRs. Which techniques are also applicable to motor position controls? Give some examples of SCR systems which can be used to control other parameters, such as heat and light.

4-6. Consider a motor control system that uses tachometer feedback phase control through a D/A converter from a microprocessor. What are some potential advantages and disadvantages of this system?

4-7. The speed torque curves of stepping motors with half step logic using a bipolar drive show that the reduction in resonances is worth the increased logic states required. Discuss the operation of a system which operates on the strong step with two pulses per step using a microprocessor for the state generator.

4-8. For applications with ramp times of less than 0.5 seconds, a half step motor drive may deliver more torque than full step drive. With longer ramp times, full step drive delivers more torque because of its higher magnetic efficiency. Describe how a microprocessor system could be used to provide dual ramp control. What additional components might be required?

4-9. Describe some techniques of increasing stepping motor response through modification of the control pulse?

4-10. What is the most practical way to apply stepping motor systems with regard to changing load conditions? Many loads can change during the system life. How can one handle transient load conditions that occur infrequently?

4-11. What are the disadvantages of unipolar resistance limited drive?

4-12. Describe how a circuit with chopper switching frequencies of up to 1kHz can operate at a drive motor bandwidth of 25–35Hz. Compare this to the upper bandwidth for conventional SCR motor drives.

4-13. Discuss how the dynamic response of a motor control system can be measured by applying a signal of increasing frequency and monitoring the controlled quantity for a phase lag. How would the phase lag be produced?

4-14. Describe some applications where an open loop stepping motor system would be chosen over a closed loop encoder feedback system. What system characteristics would require the use of position feedback?

5

Control Systems

CONTROL SYSTEM ANALYSIS

Modern technology provides a variety of components for processing digital signals. Analyzing the performance of a control system with digital components can be similar in many respects to analyzing analog control loops. All regulating systems are closed loop systems. Closed loop systems are common in our everyday lives (Figure 5-1a,b). For example, in the automobile, the driver controls the degree of pressure on the steering wheel by considering such factors as road surface and curvature, speed of the automobile, and objects on the road. The driver's eyes monitor these factors for the proper control to allow safe travel. This is also a closed loop operation.

If the driver's eyes were to close, we would have an open loop system. An open loop system is one which does not allow for unpredictable conditions.

CLOSED LOOP SYSTEMS

Three terms have been used to describe closed loop systems: regulator, servomechanism, and servo. A regulator is a closed loop system that holds a steady level or quantity such as voltage, current, or temperature, and often uses no moving parts. A servomechanism is a closed loop system that moves or changes the position of the controlled object in accordance with a command signal. It may include some moving parts such as motors or solenoids. The term *servo* has been adopted to mean either a regulator or a servomechanism and will be used as such in this section. A servo has an advantage over an open loop system. It can accept a remote, low power signal and precisely and quickly control a large amount of power. Because of the self-checking action, it decreases the error which can result

FIGURE 5-1a   The General Motors Computer Command Control emission system uses this Electronic Control Module (ECM) which houses a microcomputer. Functioning as the on-board computer it receives data from engine-mounted sensors at a rate of up to 160 times per second. (*Courtesy General Motors.*)

from external or internal disturbances. Its action can be quite smooth and it allows unattended control. To obtain these advantages, the servo design must prevent such problems as instability and slow response. In general, these same principles apply to hydraulic or mechanical systems.

As shown in the block diagram of Figure 5-2, a closed loop system may contain the following features:

1. The applied control or reference is a device giving an input signal, R, which sets the level or position the servo is to maintain.
2. The controlled quantity is the level position, temperature pressure, or other variable being controlled by the system.
3. The feedback element is generally a voltage or current sample used to obtain an accurate indication of the controlled quantity and to process

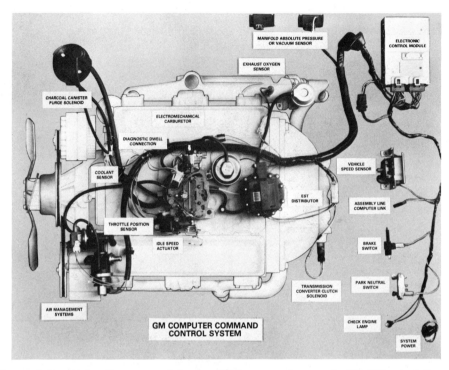

FIGURE 5-1b   Using the sensor data, the ECM may perform up to 350,000 calculations per second. It commands the fuel mixture solenoid in the electromechanical carburetor to make compensating rich or lean adjustments in order to send the optimum air/fuel mixture to the engine's cylinders. The ECM functions as part of a closed loop system to provide electronic spark timing and idle speed commands, as well as emission and fuel economy control. (*Courtesy General Motors.*)

that quantity in a suitable form for comparison with the reference signal.

4. The error detector can be any device which is capable of comparing the reference and feedback signals, which may differ in magnitude and phase, and to provide a resulting error signal.

5. The power control is the device or devices which are used for receiving the error signal and amplifying it. It may be one or more active elements. This element includes the output device which produces or changes the controlled quantity.

6. External power is the power from a source which is controlled proportionally to the error signal.

7. The disturbance block represents the external disturbances acting upon the loop. They might be the load or torque requirements on the motor or external disturbances.

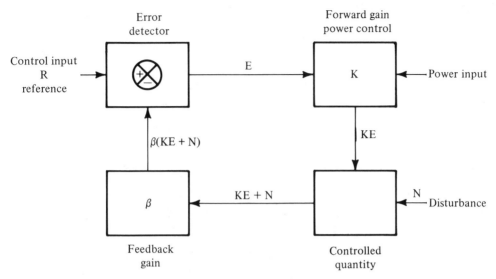

FIGURE 5-2   Basic Closed Loop System.

The closed loop operation of Figure 5-2 is given as follows: The error signal E passes through the power control system resulting in KE. KE+N is modified by the feedback element and then sent to the error detector for comparison with the reference.

Many systems have minor loops of feedback around an active element in addition to the major loop for the control of the intended variable. We will discuss the major loop even though the theory applies to the minor loops as well.

Feedback

A system can be regenerative, such as a positive feedback system used for oscillators, or it can be degenerative, such as a negative feedback system normally used for most control applications. The negative feedback systems as usually employed in control loops have self-correcting features. Let reference R be constant, which results in a constant controlled quantity, if all other factors remain unchanged. Let N cause an increase in KE+N. This increases $\beta(KE+N)$, decreases E and KE, and returns KE+N close to its original level. Thus, a closed loop system with negative feedback suppresses both internal and external disturbances and tends to maintain the controlled quantity at a level requested by the reference. It can be shown that the degree of maintaining the controlled quantity becomes a function of β, K, and the time lags in the system.

The output signal of the power control system is KE while its input is E. We define the forward gain as +K. Similarly the output of the feedback elements is

$\beta(KE \pm N)$ while its input is $KE \pm N$. Therefore the feedback gain is $+\beta$. Since we are applying the convention of $+\beta$ for negative feedback, an increase in β results in a decrease in error signal E. The product βK is known as the loop gain. The loop gain determines the overall gain of the system, or $(KE + N)/R$. Since

$$E = D - \beta(KE + N)$$
$$R = E + \beta(KE + N)$$

solving for the ratio of controlled quantity to the reference signal

$$\frac{KE - N}{R} = \text{overall system gain} = \frac{KE + N}{E + \beta(KE + N)}$$

And letting N approach zero, we get

$$\text{Overall system gain} = \frac{K}{1 + \beta K}$$

If $\beta K \gg 1$, the overall system gain $= \dfrac{1}{\beta}$. Thus, in negative feedback systems where $\beta K \gg 1$, the overall system gain is determined by the feedback and not the forward gain. The operational amplifier uses this principle. Its system gain is dictated by the feedback elements. In most servos, a useful property is $\dfrac{KE + N}{N}$, which is the system's ability to compensate for disturbance with a constant or zero R. Letting R approach zero in the loop gain equations and solving for the ratio of controlled quantity to disturbance, the effect of disturbances $= \dfrac{1}{1 + \beta K}$.

For example, if $\beta K = 20$, the effect of a disturbance is $\dfrac{1}{21}$ of what it would have been without the feedback loop. This is sometimes referred to as the *stiffness* of a servo.

In the ideal theoretical closed loop system, the feedback signal is produced instantaneously with a change in the reference signal. In a practical servo the energy storing elements in the system, such as inductance, capacitance, and inertial elements, delay the control and feedback signal. This can be considered as a phase lag.

The block diagram of Figure 5-2 shows that the function of the error detector is to subtract the feedback signal from the reference signal to give a resulting error:

$$E = R - \beta(KE + N)$$

When the reference signal has a high rate of change, there may have to be a sufficient phase lag between R and $\beta(KE+N)$ so that an addition will occur in the error detector. Thus, if the loop gain is large, a controlled quantity will be produced by $\beta(KE+N)$ without a reference signal. This is the positive feedback condition which is the basis of instability and self-oscillation. This is a useful property in oscillators. Fortunately, the loop gain decreases with increasing phase lag due to the shunting effects of capacitors and the frequency response of the active elements. But most servos, if not properly designed, have a tendency to be unstable when excited at some higher frequency for which $\beta(KE+N)$ has a sufficient phase lag and the loop gain is large enough to overcome the phase lag system losses.

Thus, any closed loop system is primarily concerned with obtaining the maximum sensitivity (gain) while maintaining the loop stability. The higher the gain of the system, the faster the transient response and the closer the control can be.

Frequency Response Characteristics

The analysis of closed loop control systems requires knowledge of the frequency response (gain and phase) of each component of the system. If this information is not known, frequency response tests may be conducted, usually at the component level. A useful test is to subject the system or component to an input with one of the following characteristics: step, ramp, or parabola. The response of the system to this input is recorded and compared with the response of systems with known characteristics to these displacement, velocity, and acceleration inputs. Computer programs are available for this purpose. When the characteristics are determined, then the maximum gain for the loop stability may be calculated from the gain phase or root locus relationship as discussed later.

The dynamic response of a first order type of control element to a step input change is represented by:

$$T \frac{d\theta}{dt} + \theta = \theta_f$$

where

θ = value controlled by the element instrument:

θ_f = final steady value

t = time

T = a time constant, RC seconds or minutes in an equivalent resistance circuit.

For given initial conditions this linear first order differential equation has the following solution:

$$\frac{\theta}{\theta_f} = 1 - e^{-t/T}$$

This represents a single exponential response. As the time constant, which is the measuring lag, becomes larger, the response, while maintaining the same shape, becomes proportionately slower. The time constant T is the time required for 63.2% of the complete change.

Process Effects

In automatic control systems two basic effects are exhibited by every process: (1) load changes caused by conditions in the process and (2) process lag, which is the delay in time required by the process and process variable to reach a new value after a load change occurs. Process lag can be caused by capacitance, resistance, and dead time. One or more of these characteristics can be involved.

The load for a process can be considered as the total amount of corrective control element required at any one time to maintain the established process level. Since a load change can call for corrective action either above or below the selected set point, the control element must be capable of furnishing the entire span. If a factory room is to be maintained at a constant temperature, the load can request heating or cooling. Thus, both a heating and an air conditioning unit are required as corrective control elements and each unit must be capable of maintaining the temperature of the room. Load changes can occur gradually over a period of time or they may change quite rapidly.

In addition to process lags, dead time must also be considered in the process reaction. This is especially true in those continuous processes where a fluid flow is transferred over a distance, since it takes a finite time for the fluid entering the process to flow through the process steps, and it takes the process time to adjust to the energy change. During this dead time the control element cannot take corrective action.

A first order system can be described by the time constant, just as first order elements were. A basic equation is:

$$\frac{\text{Output}}{\text{Input}} = \frac{1}{Ts + 1}$$

This equation describes a single capacitance and a single resistance type of first order system. This form of expression is often called a *transfer function*. It is also the Laplace transform of the differential equation describing the system. Two systems that have the same transfer function have the same response with respect to time. Thus, any single capacitance, single resistance system has this transfer function.

To describe a process with a mathematical expression, output must be related to input and how one varies with respect to the other with time must be shown. This gives a differential equation differentiated with respect to time. The equation

$$A - B = C\frac{dH}{dt}$$

might say that, if the outlet volumetric rate B is subtracted from the inlet rate A, the difference equals the area of tank bottom C in head dH during a change in time dt. Stated in differential terms

$$(\text{Rate A} - \text{rate B}) \times \text{time dt} = \text{volume change dV}$$

we can use a hydraulic analog for flow through a resistance. Let B, the outlet rate, be equivalent to the electrical current produced by a potential across a resistance. Then we rewrite the equation as:

$$A = B(CsR + 1)$$

where

$$H = BR$$
$$\frac{d}{dt} = s$$

Placing this in the output/input form:

$$\frac{\text{Output}}{\text{Input}} = \frac{B}{A} = \frac{B}{B(CsR + 1)} = \frac{1}{CsR + 1}$$

$T = RC$, so this equation is the same as for the basic first order system.

The time constant of the single time constant process is determined by finding the time required to reach 63.2% of the final change in the output following a step change. The step change in input to a single time constant process might be a sudden increase in flow rate, a sudden decrease in flow rate, or a step change in the input voltage which produces a change in the output response. The equation of the response curve is

$$1 - e^{-t/T}$$

This equation can be used in plotting the percent incomplete of $e^{-t/T}$ on logarithmic paper.

Many systems have additional lags because of a mass that must be accelerated, or more than one element of fluid capacity is involved. These are second order type systems.

Most mechanical elements have some mass which must move during operation.

These second order systems possess a dynamic response to a step change which can be described by a second order differential equation:

$$a_2 \frac{d^2\theta}{dt^2} + a_1 \frac{d\theta}{dt} + a_0\theta = a_0\theta_F$$

where

θ = value controlled by the system

θ_F = final steady value

t = time

a_2, a_1, a_0 = constants

The constants in this equation must be derived by analysis.

Gain and Phase Plots

The response and stability of either an open loop or closed loop system can be analyzed by plotting the gain and phase change versus the frequency. In Figure 5-3,

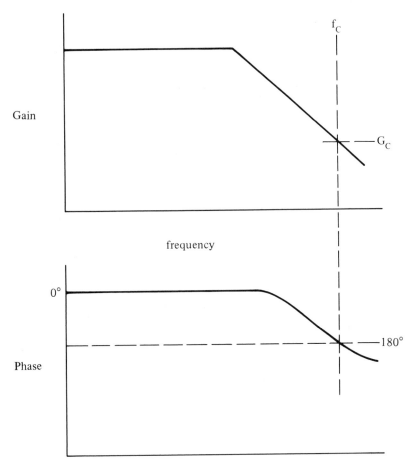

FIGURE 5-3   Gain Phase Plot.

the gain and phase of open or closed loop systems are shown as a function of frequency. The frequency at which the phase shift is $-180°$ is called the critical frequency (f_c). The gain at this frequency (G_c) must be less than one for a stable system. The effect of any energy storing element in a closed loop may be shown as a decreasing slope of 6 db/octave. This slope will eventually result in a $90°$ phase lag between R and B(KE+N).

The Bode plot is one in which the slope is calculated for each main energy storing element in the system. These points are then placed on a base line, usually the steady state loop gain of the system, with a -6 db octave line drawn from each point. The resulting points are then vectorially added. A smooth curve may then be drawn to pass approximately 3 db inside the corners or breakers. This curve represents the system response. Figure 5-4 shows a Bode plot of a system with

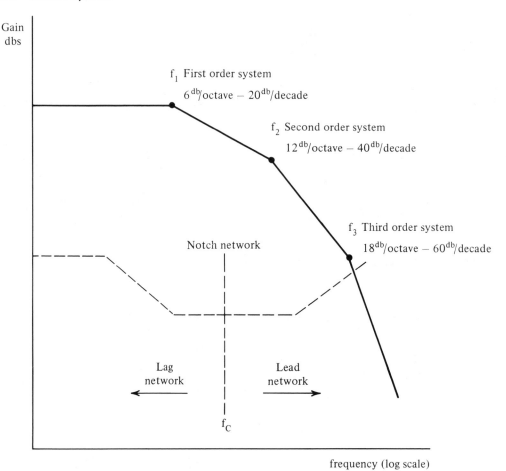

Gain
dbs

f_1 First order system

$6^{db}/octave - 20^{db}/decade$

f_2 Second order system

$12^{db}/octave - 40^{db}/decade$

f_3 Third order system

$18^{db}/octave - 60^{db}/decade$

Notch network

Lag
network

Lead
network

f_C

frequency (log scale)

FIGURE 5-4   Bode Frequency Plot.

three energy storing elements. A 6 db/octave slope (first order system) will ultimately result in a 90° phase lag between R and B(KE+N) and a 12 db/octave line
(second order system) will ultimately result in a 180° phase lag. The system can
oscillate where the smooth curve has 12 db/octave slope or greater and a gain
greater than zero db (point of unity gain). The system shown may oscillate since it
has sufficient gain concurrent with a 12 db/octave slope. A complete analysis
would require a plot of the phase lag.

Compensation

It is necessary to reduce the loop gain to prevent the system from becoming unstable or self-oscillating. We have previously established that a high loop gain may
be desirable. Thus, it is often necessary to install corrective networks in the system

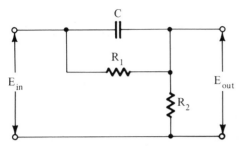

FIGURE 5-5   Lead Network.

to allow a high gain, stable system. These networks are usually placed in the low power circuitry where they can be small but still have an overall effect on the system. Since instability is caused by phase lag along with the efficient loop gain, we can counteract the lag using a lead network. A typical lead network is shown in Figure 5-5.

At low input frequencies, the output is dictated by the ratio of R_1 and R_2. As the frequency is increased, the capacitor tends to reduce the series impedance and increase the output voltage. If this network is placed so that it cancels enough of the system lag, a stable high gain system can result. This is seen on the frequency plot by summing a 6 db/octave increasing slope with a 12 db/octave decreasing slope, which results in a 6 db/octave decreasing slope with a phase lag of 90°.

Another technique is to place a phase lag corrective network in the system. A typical network is shown in Figure 5-6. At low frequencies the output nearly equals the input. As the frequency is increased, the capacitor reduces the shunting impedance until e_{out} is determined by R_1 and R_2. The lag network results in a lower gain at higher frequencies. But the total phase lag created must not result in a 12 db/octave slope prior to crossing the zero db line.

Combining the lead and lag networks results in a lead lag or notch network which uses capacitors in both legs: the R1—C of the lead network and the R2—C of the lag network. As in a lag network, at low frequencies e_{out} nearly equals e_{in}. When the frequency is increased, e_{out} decreases due to the action of C_1 and C_2. The components must be chosen so that a reduction in e_{out} does occur at some fre-

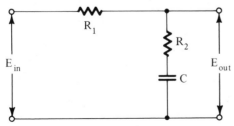

FIGURE 5-6   Lag Network.

quency. As the frequency is further increased, e_{out} then increases due to the shunting effect of C_1 on R_1. Networks of this type are used where additional sloping is required over that which can be obtained by a lead or lag network. The results of such a network are also shown in Figure 5-4. For simplicity, the final smooth curve is not shown.

Stabilization

Simple systems in which the controlled quantity is a motor speed are usually not difficult to stabilize. But systems in which the controlled quantity is a position produced by a nonstepping motor are more difficult to stabilize. This is due to the 90° lag while the motor is establishing its position. The system starts with a unit slope and may require corrective networks to avoid a 12 db/octave unit slope. A more difficult case involves those systems which use one positioning motor to position another. These have an immediate 12 db/octave slope during positioning and always need corrective networks. These problems are compounded since one usually attempts to obtain maximum response with minimum position overshoot.

A frequency exists above which the accumulative phase lags may cause instability. In some systems a lower frequency exists below which KE+N is not regulated. These upper and lower limits specify the frequency range of R for which KE+N will be controlled.

Also, due to dead zones and saturation effects from small and large amplitudes of R, respectively, the response of KE+N is not proportional for all values of R. The frequency and amplitude thus dictate a range of R for which KE+N changes continually and proportionally. This range is called its *linearity*. A system which operates within its linearity is considered to be linear. The range of amplitude and frequency values within the domain are functions of each other. A linear system does not have an infinite linearity range and a linear range can often be found for a nonlinear system, however small. The frequency approaches are designed to analyze linear systems within their linear range. Other techniques must be used for systems which operate outside their linear range. These include the phase plane method which is a graphic technique for solving the nonlinear differential equations and the describing function technique which allows sinusoidal inputs.

The Root Locus Technique

The root locus method uses a plot of the roots of the characteristic equation or transfer function of the system in the complex frequency or s-plane. The equation is the Laplace transform as a function of s, the complex frequency, with real and imaginary parts of the transfer function. The location of the roots is plotted for increasing values of the factored gain of the open loop equation. The value of the gain when the roots enter the righthand side of the plane (to the right of the half plane line) is the maximum gain for a stable system. This corresponds to the point where the loop gain increases above unity at the 180° phase change point in the gain phase plot.

The location of the roots for a fixed gain can also be used with the input response method to show that any increasing response (gain) with respect to time represents an unstable condition when the roots are located in the righthand plane. Figure 5-7 shows the responses to a step input and the root locus plots for each response. A conditionally stable state is not acceptable in most systems because some margin of safety away from the half plane line is required. Some gain margin and phase margin can always be maintained by adjustment of the system parameters or adding corrective networks.

Mode Control

Automatic control indicates that the control occurs as a corrective measure in response to a signal and that it is accomplished without human intervention to effect the control action. A basic characteristic of the system is the manner in which the

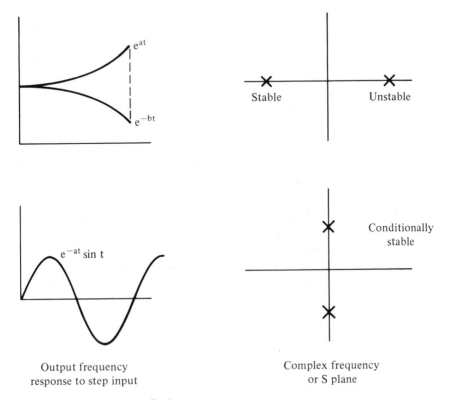

Output frequency response to step input

Complex frequency or S plane

FIGURE 5-7   Root-Locus Technique.

controller acts to restore the controlled variable to the desired value. This is the mode of control. Some common modes of control consist of proportional, proportional plus reset, and rate. A control system may use one or more of these modes.

Another test method involves adjusting the gain while the system is in operation. This method can be applied only when the system allows close observation and shutdowns can be accomplished quickly and without detrimental effects. Tuning the controller involves adjusting the gain for optimum performance and must often be done dynamically in the real application.

The three mode controller permits higher overall gain since it limits the gain to prevent instability at frequencies near the characteristic or critical frequency and yet permits higher gain at frequencies not near the critical frequency. The response characteristics for gain vs frequency of a three mode controller are similar to the notch network shown in Figure 5-4. The controller may have high gain at all frequencies other than near the critical frequency. The reset mode allows high gain for low frequencies; the rate or derivative mode provides rapid response from high gain to a disturbance of high frequency.

Tuning the controller involves setting reset and rate gains so that the critical frequency is in the center of the proportional or flat part of the frequency band.

The nature of proportional control is to realize the magnitude of the error between the actual and desired levels of KE+N and adjust the power to the load accordingly. Thus, KE+N tends to approach its final level at a lower rate than it had when the error signal was large. This produces a smaller overshoot and quicker recovery.

Some steady state error must exist to supply enough power to the load to maintain KE+N at the desired level. The magnitude of this error is directly related to the gain of the system. If only a small error is necessary, the system can have high gain. As the gain is increased, a point is reached when the proportionality feature is eliminated and the system reverts to on-off control. This produces larger overshoots but has the advantages of being able to hold KE+N with less error when subjected to internal and external disturbances.

The difference in error signal between voltage application and removal is known as *hysteresis*, which reduces the regulation accuracy of an on-off control.

Other considerations are overshoot and speed of response. If large overshoots cannot be tolerated, then a proportional control should be used. As a result, the speed at which KE+N will respond decreases. Many systems cannot use a pure on-off control. This is especially true of systems where a transport lag is present. Transport lag differs from overshoot in that KE+N is not realized until sometime after the initiation of an error signal. This results in a time delay as opposed to a phase delay associated with an overshoot. If an on-off control were used in a system with a transport lag, KE+N can vary between its maximum and minimum limits continuously due to the time lapse. A proportional control would allow the desired intermediate level to be maintained. Generalization is difficult when recommending proportional or on-off control. Each application should be analyzed considering such factors as discussed previously.

DIGITAL CONTROL SYSTEMS

Digital control systems are part of a general class of systems known as sampled data systems, for which techniques such as the Z transform have been developed. The Z transform is analogous to the s or Laplace transform of analog systems. A sampled data system is one whose output occurs periodically at some sampling rate or frequency. Some digital components have a steady state output for a steady state input and do not have to be characterized by Z transforms, e.g., position encoders or indicators at rest and hard-wired logic gates.

An analog signal may be converted into a sampled data signal by sampling. Sampled digital systems require one or more clock periods before the processing is completed and the output corresponds to the input. During this period, system outputs are usually provided to indicate that data updating is taking place. Figure 5-8 shows a signal before and after sampling. The Laplace transform of the analog signal is given by:

$$Lf(t) = \int_0^\infty f(t)^{-st} \, dt$$

This transform is in a form that can be manipulated and analyzed, using tables of Laplace transforms. The Laplace transform of the sampled function is given by:

$$L \text{ sampled } f(T) = \sum_0^\infty f(nT) \, e^{-nsT}$$

If we let $Z = sT$ in this expression, we have the Z transform:

$$L \text{ sampled } f(T) = \sum_0^\infty f(nT) \, Z^{-n}$$

This is the Laplace transform of the sampled data function with $Z = e^{sT}$, where T is the sampled period. If the sampling frequency is less than twice the highest frequency of the signal, as shown in Figure 5-7, some frequency components will overlap and information will be lost. The sampling frequency should be about ten

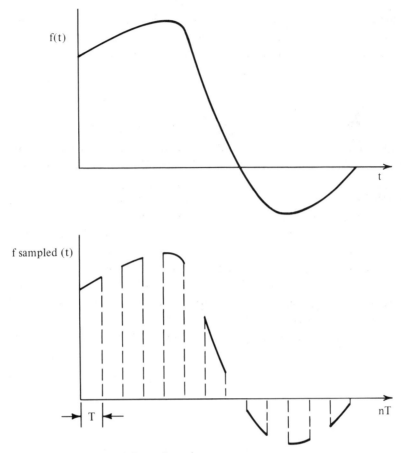

FIGURE 5-8   Sampled-Data Signal.

times the critical frequency; if it is much higher, the system will not provide good low frequency and steady state response.

A digital system can be tested for frequency response or input response in much the same manner as an analog system, depending on the equipment available and the type of input control desired. The test configuration can take a variety of forms. A sinusoidal generator might be used with an analog-to-digital converter if the system control input is digital. If the system input is parallel digital, then the step input test could be simulated by switching the most-significant bit line to a 1 and a ramp input could be obtained by mechanically driving a position encoder or indicator. If a minicomputer or microprocessor system is available, then the desired input function can be simulated by the computer and used as the input to the system under test.

The basic analog control loop consists of a summing network, controller or control system, actuator or controlled quantity, and some form of feedback sensor (Figure 5-2). A digital control loop has a similar form, except that summing is now done with digital components, either hard-wired or programmed logic. Devices for converting between analog and digital formats are required if digital components are not used throughout the system.

Digital control allows a large variety of equipment to be used in the input command mode or for monitoring and storing performance data of the system. A variety of equipment is available for use in digital systems, as shown in Table 5-1.

TABLE 5-1   DIGITAL COMPONENTS

Function	Comments
Input Command	
Switches	Programmed or nonprogrammed
Keyboard	
with CRT	
with Readout Display	LED, incandescent, liquid crystal
Magnetic Tape/Disk	Erasable, nonvolatile
Punched Card	Easily changed, nonvolatile
Paper Tape	Permanent, nonvolatile
Character Readers	Optical or magnetic
Summing and Processing	
Hardwired Logic	Volatile
RAMs, PROMs	Volatile, erasable
Microprocessors	Volatile, erasable
Minicomputers	Volatile, erasable
ROMs	Nonvolatile
Sensing	
Rotary Position	Frequency dependent
Linear Position	Nonvolatile
Pressure	Nonvolatile
Temperature	Volatile
Actuators	
Solenoids	Frequency dependent
Stepping Motors	Frequency dependent
Digital Valves	Parallel format
Data Retrieval	
CRTs	Monitoring, volatile
Readout Displays	Monitoring, volatile
Magnetic Tape/Disk	Storage, nonvolatile
Punched Cards	Nonvolatile, permanent storage
Paper Tape	Nonvolatile, permanent storage
RAMs	Temporary storage
Printers/Plotters	Nonvolatile, permanent storage

A programmed switch can be a multipole switch with poles connected either as 1s or 0s in a parallel format, while a nonprogrammed switch could be an array of single pole switches which can be configured at will.

Erasability implies that the data can be erased at will and new data put in its place; *volatility* means that a power interruption can cause the data to be lost. *Permanence* implies that the data are neither erasable or volatile. Punched cards (although each is permanent) do allow a program to be easily changed by replacement of a few cards. Read only memories (ROM) and some types of programmable read only memories (PROM) are hard-wire coded and, therefore, are nonvolatile and not erasable or changed after the initial coding.

Many systems use analog as well as digital components because a wider variety of analog components are available and presently in use. The judicious selection of components can limit the use of analog-to-digital or digital-to-analog conversion, for these components add nothing to the performance of the system and can degrade it. The proper system mechanization and modernizing techniques can retain expensive long-life components (such as large control valves) while improving the data input and output capabilities of the control system.

CONTROL SYSTEMS APPLICATIONS

Some typical applications being served by control systems are:

Parameter Being Controlled	Application
Speed	Pumps, fans, blowers, web drives, traction drives, conveyor belts, extrusion machinery, rolling mills, printing, silage unloading, drills, blenders, saws, knives, and mixers
Position	Welder controls, antenna drives, gun controls
Temperature	Ovens, heating, chemical processing, extrusion head controls, diffusion ovens, dryers, dryness controls, water heaters
Light	Illumination, street light controls, photo-projectors
Voltage and Current	Battery chargers, dc to ac inverters, dc converters, ac to dc power supplies, ac to ac converters

Many other applications are presently being satisfied by other systems than these. This list is presented to give some indication of the variety of tasks which can be performed by a servo system. In the future, the applications will increase.

We will consider some examples of control systems utilizing SCRs and triacs for controlling power to a load. The approach will be to briefly explain the basic control loop of these systems and to point out their features.

Temperature Regulation

Figure 5-9 shows a temperature regulator which uses a ramp and pedestal phase control. A triac supplies phase controlled ac power to the heater load. This increases KE+N, which is the temperature at a rate slower than the switching frequency of the triac. Thus KE+N has a long averaging time constant. The feedback element, a thermistor, senses the temperature; this is compared by the unijunction error detector with the reference signal to control the triac which regulates KE+N.

Phase control methods control the power to the load by controlling the time within the half cycle at which the thyristor will turn on. Since the thyristor is a latching device, it stays on for the remainder of that half cycle. A large error turns the thyristor on early in the half cycle and a small error results in a later turn on or possibly none at all. The result of the magnitude of the error is the point within the half cycle at which the pulse turns the thyristor on—not the presence of the pulse. Figure 5-9 shows how this is accomplished with the ramp and pedestal. As the magnitude of the ramp and pedestal rises sufficiently to trigger the unijunction, a pulse is delivered which turns the thyristor on.

The control must be stable when subjected to variations in factors like voltage and ambient temperature. The error remains constant for constant R and β(KE+N) throughout these variations. This circuit uses common voltages and the stability of the unijunction to maintain the accuracy of the relationship.

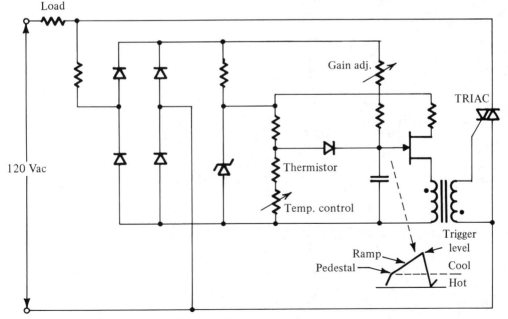

FIGURE 5-9   Ramp-and-Pedestal Temperature Regulator.

The overall accuracy of a temperature regulator is a function of the rate of change and magnitude of the disturbances as well as the capacity of the heater when compared to the volume of the controlled chamber. A control of this type is capable of controlling temperature to within ±2% of the thermistor resistance over a ±10% variation in line voltage.

Figure 5-10 shows a temperature regulator which uses half wave, zero voltage switching, on-off type control to control a hot wire relay. The relay activates a heater or other load operating directly across the power line. KE+N is the temperature and the thermistor is the feedback element. Two SCRs form the error detector. The output SCR operates on the positive half cycle only. The decision to allow this SCR to turn on is made during the previous negative half cycle, which is possible because the leading voltage is supplied by the capacitor. If the second SCR is off when the line starts to go positive, the output SCR will turn on. If the second SCR is on, the output SCR cannot turn on for the remainder of the cycle. This is an example of a control SCR being used to provide zero voltage switching of a power controlling SCR. The decision to allow the output SCR to turn on is made by the feedback element, and reference and error detector.

The SCR pair forms a major part of the error detector. They subtract $\beta(KE+N)$, as derived by the thermistor from the reference level. This is accomplished by comparing the magnitudes of the voltages at each SCR gate. As the two

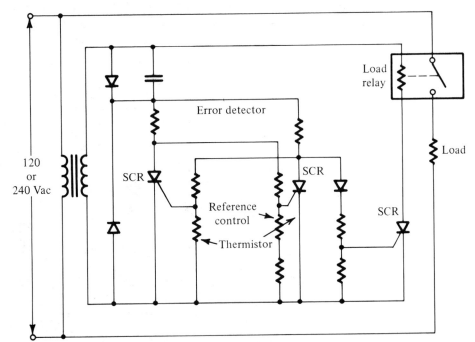

FIGURE 5-10   Relay-Controlled Temperature Controller.

SCRs are cross-coupled, only one can be on during each half cycle. The first one to turn on stays on for the rest of the negative cycle and entire positive half cycle.

The stability for this type of control for voltage and temperature changes is achieved by the circuit symmetry. A control of this type can be capable of regulation within an accuracy of $\pm 0.2\%$ of thermistor resistance and over a $\pm 10\%$ variation in line voltage. The use of a relay with SCRs allows the control of a large amount of power with relatively small SCRs.

POWER DRIVE CONTROL

Increasing interest has been placed on the control of analog power drives through positioning and tracking commands from digital computers. The more powerful computational aspects of small digital computers have been accompanied by the problems in providing satisfactory dynamic control for inputs which are in the form of discrete signals. Recent trends are towards the development of distributed systems and the use of microcomputers as control elements. This has resulted in more freedom with a growing number of control system designs.

Modern digital control systems have evolved from programs stored in a time-shared general purpose medium sized computer, followed by the use of minicomputers; then as special purpose hard-wired SSI and MSI logic and currently compact designs incorporate microcomputers.

A certain amount of commonality can be found in the hardware configurations and computer programs used in the earlier designs. The basic dynamic characteristics of the power drives were usually modeled after a Type 1 servo: a double integration with velocity feedback and velocity and acceleration limiting. Electrohydraulic power drives generally fall into this class as well as most electrical power drives.

The design approach usually was to make the model of the controlled system appear to have the same form from the control computer's point of view. This was done mainly through the selection of velocity loop gains. The resulting control was applicable to a broad class of power drives. Early control hardware used a variety of MSI and LSI elements. More recent designs are implemented using microprocessors.

These microprocessor system designs use primarily four types of functional sections or modules: interface module, analog output module, status/test module, and processor module. The interface module allows many different types of systems to be controlled. These systems may have differing characteristics such as acceleration limits; unique coefficients for difference equations and the solution of various polynomials may be required. The values of these coefficients are stored in ROMs. The signal interfacing requirements of a particular system and computer are also handled by the interface module.

The analog output module contains the buffer amplifiers and D/A converters for conversion of the digital commands from the controller into analog signals required by the system. Each output of the D/A converters is available for connection to a separate summing amplifier so that it can be summed with the output of the feedback device. Some D/A converters may be used for other functions such as monitoring position and velocity errors.

The status test module is used to generate input commands for test purposes. A number of standard commands, such as steps, ramps, and sinusoids may be generated from this module. The coefficients for these functions are stored in ROMs.

The processor module provides the arithmetic capability, temporary storage for calculations, and the mode select and timing and control functions. All calculations are usually performed in this module. The use of microprocessors with their significant space savings has resulted in the reallocation of some functions which were previously performed elsewhere.

In many distributed systems, a major goal is to avoid commands from the master computer at an excessively high rate. The high update rates can saturate the bus bandwidth, generally limiting the sample rate to about 20 to 30 samples per second. At a low rate such as this, to maintain smooth system operation may be difficult. A conventional analog system drive at 20 samples per second through a digital-to-synchro converter can result in rough, uneven operation of power drives. This can lead to premature system component deterioration resulting in poor response. The usual design approach to digital controls of this type is to use an interface that involves some form of data extrapolation for the digital-to-synchro (or resolver) conversion. The objective is to modify these discrete low sample rate commands so that they can be sent to the analog loop in the form of a quasicontinuous signal. If the control loop is closed digitally and the digital error signal is used, a sample rate of 20 samples per second can result in acceptable performance with position feedback data from a shaft encoder or synchro/resolver.

Stepping and servo motors are used in applications which require precise motion to reach a digitally defined position. One approach involves an acceleration to the operating speed and then a deceleration to the final programmed position.

Some systems use digital interpolation techniques to generate command pulses for a linear velocity ramp to minimize for travel time to the desired position. With this approach, the last motion pulse and zero velocity can occur simultaneously. The commands are supplied to force a lower final speed by creeping to the commanded position, but this can consume excessive time.

Another approach to stepping motor control is to input a linear voltage ramp to the oscillator producing the command pulses. The positioning time is reduced compared to the exponential method, but some creeping is required. This type of system uses these components:

1. A constant frequency pulse generator provides an output frequency proportional to the desired velocity. Each pulse represents one increment of motion.

2. An acceleration parameter data store is used to contain a digital number which defines the desired acceleration. The store technique could be ROM, thumb-wheel switches, or hard-wired registers. Storage is also required for the digital number which represents the distance of motion required.

3. An acceleration pulse generator is used to generate pulses which represent the commanded position increments. The frequency of the pulses increases linearly for an acceleration signal and decreases linearly for a deceleration signal.

4. Finally, a microprocessor system monitors the generated command pulses, determines the acceleration and deceleration phases, and decides when the programmed distance is reached.

Usually the commands are supplied to a stepping or servo motor in a closed-loop configuration to reduce the lag between the command and actual positions for faster positioning.

Another method to control position is to use an exponential velocity change for the acceleration and deceleration. A closed loop system with velocity and position feedback can be used to produce the exponential output velocity change from a velocity step input command. In a stepping motor system this can be achieved with a voltage controlled oscillator which is coupled to a pulse generator to produce either an exponential rising or falling voltage for the acceleration or deceleration. In a servo motor system velocity feedback with an exponential characteristic produces a lag between the command and actual positions. This lag provides a deceleration to avoid overshooting the commanded position.

In an open loop system aging and temperature effects in the analog circuits can cause variations in the speed and the exponential time constant. If the deceleration is started at a fixed distance from the final position, the part being positioned may stop short of the destination or arrive at too high a speed and overshoot the position.

Some power drives are characterized by torque or acceleration limiting and other nonlinearities. A dual mode of operation can then be used. A coarse control mode is used for large values of servo error, such as might be experienced during slewing commands. A linear algorithm is used for relatively small errors.

A desirable feature of many control applications is to achieve optimal time response. The goal is to obtain the fastest output response possible without exceeding physical constraints placed on the system. This is the same philosophy used in relay or "bang-bang" control. In a relay servo, maximum torque is applied to drive the error toward zero. Then, to prevent overshoot, full torque of the opposite polarity is applied to decelerate the output to rest with zero error.

Relay control is not satisfactory for many applications. However, good time response can be accomplished for slewing commands with the use of a switching curve to decelerate the power drive at about 75% of the maximum available acceleration.

The implementation of the switching function can be accomplished by a table lookup.

Operation in the large signal control mode can be studied with the aid of a phase plane plot. Consider an initial error resulting from a positive step function input. The initial velocity error is at a maximum positive value. If the gain factors are chosen to be sufficiently large, the output acceleration will reach a limiting value, causing an acceleration along the phase plane trajectory which is parabolic. The sharpness of the trajectory and the degree to which the trajectory follows the switching curve is determined by the magnitude of the gain.

The system can be made to follow the switching function of any similar single valued function of error as long as the acceleration limits are not exceeded. No abrupt change in gain should be made at the time of crossover from the large signal mode to the linear control mode. The choice of linear control depends chiefly on the system specifications, including the bandwidth requirements.

An improvement in system accuracy without affecting stability can be accomplished by using additive (feedforward) compensation. It can achieve low steady state error-to-ramp input commands and it generally reduces steady state errors for more complex inputs.

The time required for a linear servo to move from one point to a final position at another point is referred to as the *settling time.* It is a basic response parameter of a position servo. The settling time of a position servo may be reduced by nonlinear feedforward techniques without increasing the bandwidth or noise gain. Reducing the settling time is important since fast response and accurate position servos are widely used. Production rates can often vary directly with a servo's settling time. Thus, to increase production it is important to reduce the settling time to a minimum.

In a linear system the settling time is inversely related to the bandwidth, which usually must be held to some specified maximum frequency because of structural resonances or sampling rate limitations. Structural resonances can affect the stability if their natural frequency resides close to or within the servo's bandwidth. Also, noise from transducers, EMI-induction, converters, or sampling may cause amplifier saturation, limit cycles, or phase errors. These effects can degrade the system accuracy.

Nonlinear Feedforward Control

Nonlinear feedforward offers a way to reduce settling time at a modest cost when compared to other techniques, such as bigger power amplifiers.

A nonlinear feedforward transfer function can generally be placed in parallel with a servo's lead lag compensation (Figure 5-2). Tachometer feedback systems can also be used by feeding the position error around the tachometer signal summing junction. The reference position can be determined by a microcomputer and then entered as an offset into the error channel. The improvement in settling time is accomplished by feeding the error signal around the compensation networks and through a nonlinear gain block.

Figure 5-11 shows a circuit for generating the nonlinearity. Schottky diodes should be used for clipping because of their turn on characteristics. They also allow K2 to be greater than 1. The resistor between the K1 and K2 op amps should have a low value for complete shut off with zero gain around 0. The diode's smoother transition as they start to conduct limits noise.

Zero steady state error for ramp inputs can be achieved for a Type 1 servo by inserting a signal proportional to the rate of change of the sampled input function into the control loop. In this additive compensator section of the controller, the rate of change of the input command is approximated by the first backward difference from the present and past values.

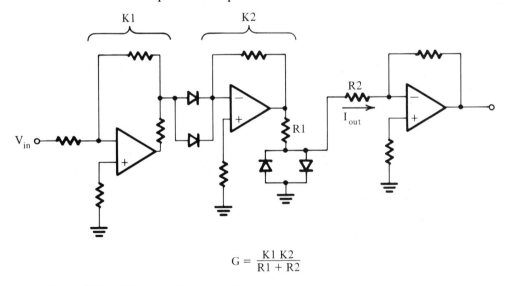

$$G = \frac{K1\ K2}{R1 + R2}$$

Figure 5-11   Nonlinear Feedback Network.

In some applications, a more reasonable estimate of input velocity is available as the byproduct of calculations performed in the master computer. Manual computer is an auxiliary means of computer in an open loop using a joystick. It consists of inserting a variable dc voltage into the summing amplifier input and is useful as a troubleshooting aid.

The digital tachometer generally provides a slight improvement in response to inputs such as sinusoids. It can be implemented as a first backward difference of the output. The complete digital system with dual control modes, feedforward compensation and velocity, and position and tachometer feedback is shown in Figure 5-12.

In many control designs the source of system control energy is assumed to be unlimited. Thus, none or very little consideration is given to control systems designs that minimize the control power. But there are an increasing number of applications where minimum energy use could make a difference.

Pulse Frequency Modulation Control

By using pulse frequency modulation (PFM) to control such a system, there can be a 40% reduction of control energy expenditure compared to a continuous unmodulated system.

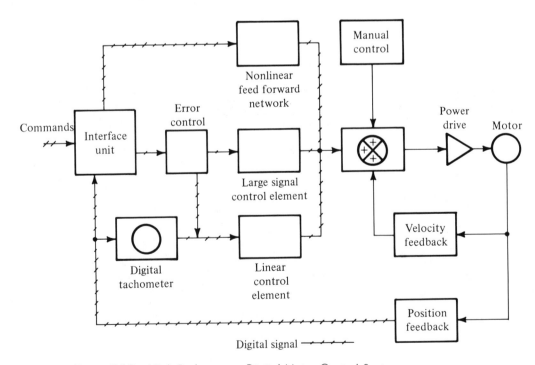

FIGURE 5-12   High Performance Digital Motor Control System.

PFM, which is a close relative of pulse width modulation (PWM), offers many of the advantages of PWM, but without the risk of oscillation at the sideband frequencies. A pulse width modulator produces sidebands which can trigger system oscillations at sampling frequencies which are multiples of the system frequency or close to it. Thus, when the sample time approaches the system natural period, the tendency is towards oscillation at the system frequency. An explanation of how these limit cycles appear is based on the system damping characteristics. A lightly damped system has a resonant frequency where it magnifies a sinusoidal input. This frequency is near the system's natural frequency. If any of the sidebands should correspond to any multiple of the natural frequency, the system tends to reinforce the oscillation with feedback around the loop rather than attenuating it.

Pulse frequency modulation converts a continuous analog signal into a pulse train. The pulses have the same width and their amplitudes are constant. The polarity is the same as the analog signal. Positive pulses are generated from a positive signal while negative pulses are generated from a negative signal. These pulses are not evenly spaced in time. Their spacing is a function of the magnitude of the continuous signal such that the repetition signal is faster for a large signal and slower for a small signal. Since it preserves the polarity and magnitude of a continuous signal, pulse frequency modulation can be used for the control in a feedback system whose plant excitation is the difference or error between the system's reference input and the controlled output.

PFM falls into the category of sampled data controls which have been discussed. The most basic of these is pure sampling of a continuous signal to yield a train of weighted and equal width pulses having the polarity of the continuous signal with equal spacing. Each pulse amplitude corresponds proportionally to the magnitude of the continuous signal. The pulse width and the pulse-to-pulse spacing are constant.

When the pulse width is very short compared to the pulse spacing, PFM approximates an impulse train or a train of delta functions. Because of the properties of an impulse function, this is a linear process. Linear processes can be studied with tools such as the Laplace and the z transforms and Bode and root locus plots. This cannot be done with pulse frequency modulation because it is a nonlinear process.

The most common type of pulse modulation is the sample and hold. It uses a sampler to provide the pulse amplitude modulation and a hold for reconstruction of the pulses. This has always been popular in digital systems because of the ease with which it can be programmed and the simplicity of its analog components.

The next most common technique is pulse width modulation, which is used in open loop speed controls and closed loop controls which would otherwise tend to be sticky around null. Modifications of these specific purposes, such as adaptive sampling, smooth the effects of disturbances by varying the sample time. The

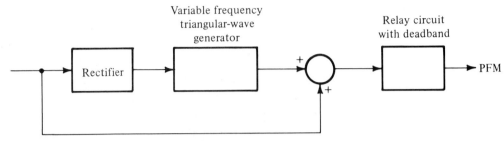

FIGURE 5-13   PFM Generation.

sampling frequency is typically five times the highest frequency of interest with all higher frequencies in the signal filtered before sampling. The sampling rate can be made low to save computer processing time in handling a large number of signals. But if a large load disturbance is encountered, the sampling rate may be too slow unless it can be increased through the adaptive process.

A PFM circuit converts a continuous signal into a train of constant amplitude equal width pulses with a repetition rate proportional to the magnitude of the continuous signal. One circuit that can be used to obtain PFM pulses is shown in Figure 5-13. The continuous signal is first rectified and then used to control a variable frequency triangular wave generator. The output of the generator is then added to the original signal and passed through a deadband circuit or clipper to shape the pulses. The deadband must be the same as the triangular wave amplitude. The method is similar to some methods of obtaining PFM. PFM can result in 40% less energy in some applications since the time response of a typical system has very little or no overshoot.

EXERCISES

5-1. Discuss the problems of a control for precision distance in computer disc memories, automated machine tools, and wafer probe testing.

5-2. Why does on-off control give improved regulation over a given proportional control in all applications?

5-3. Describe how the feedback elements' reference signal and error detector work together in a control system.

5-4. A control system has a loss through the feedback elements of 4/1 or a gain of 1/4; the closed loop gain is 20. What gain is measured by disconnecting the output of the feedback element?

5-5. Discuss the stability of a computer system for the following three cases: the condition in which the roots of the auxiliary equation are conjugate complex with negative real parts; the roots are negative, real, and equal; and the condition in which the roots are negative, real, and unequal.

5-6. Describe how control system analysis techniques depend on the equipment or tools available. What techniques may require simple equipment and paper calculations and what others require computer evaluations?

5-7. Discuss how time constants can be used in creating an analog of a system and in selecting the type of controller and optimum control settings necessary for the successful operation of a process.

5-8. How can controlled quantities with long averaging time constants help in reducing instability resulting from the energy storing elements in the system? Does the possibility of instability still exist?

5-9. Give some examples of sampled data digital formats and discuss their basic characteristics.

5-10. Discuss the problem of how to produce the best response from a servo with limited bandwidth, without sacrificing accuracy or increasing system noise.

5-11. Consider a nonlinear feedforward circuit as essentially a voltage to current conversion circuit connected in parallel with a lead network. Draw a possible schematic for this circuit.

5-12. For a power cost comparison, consider a VFD versus a hydraulic coupling. Use 0.40 kW input/hp for VFD and 0.51 kW input/hp for the hydraulic coupling and a power cost of $0.04 per kWhr. Find the annual power cost savings for a 100 hp application.

6

Temperature Sensors

TRANSDUCER SELECTION

In the following chapters of the book we will discuss typical transducers or sensors which might be used in a microprocessor system. A transducer is any device which can be used for determining the value, quantity, or condition of some physical variable or phenomenon which must be monitored.

The transducer usually measures the magnitude of some particular phenomenon for control processing purposes. The measurement consists of an information transfer with an accompanying energy transfer. Since energy cannot be withdrawn from a system without changing it in some way, the energy transfer should be kept small so the measurement does not affect the quantity being measured. Transducers use a number of techniques to produce the information using this energy transfer.

Piezoelectric energy can be converted into an electrostatic charge or voltage when certain crystals are mechanically stressed. The stress can be from compression, tension, or bending forces, which can be exerted upon the crystal by a sensing element or by a mechanical member linked to the sensing.

Energy can also be converted into a change in resistance of a semiconductive material due to the amount of illumination on the semiconductor surface. In some light sensitive transducers the change in illumination is controlled by a shutter or mask between a light source and the photoresistive material. The shutter may be mechanically linked to a physical sensing element such as a pressure diaphram or a seismic mass.

Energy may also be converted into a change in voltage due to a junction between dissimilar materials being illuminated. This principle is used in some light meters.

Transducers with a mechanical means to change the capacitive coupling between elements use energy which is converted into a change of capacitance. The capacitor consists of two conductors or plates separated by a dielectric. The change of capacitance occurs if a displacement of the sensing element causes one conductive surface to move relative to the other conductive surface. Some transducers use the moving plate as the sensing element, in others, both plates are stationary and the capacitive change occurs due to a change in the dielectric.

Energy can also be converted into a voltage induced in a conductor due to the change in magnetic flux. This type of transducer does not require any excitation and is self-generating. The change in magnetic flux can be accomplished by the movement of a magnetic material and the conductor. Energy can be converted into a change of the self-inductance of a coil due to displacement of the coil's core. The core can be linked to a mechanical sensing element.

To select a transducer, consider the following:

1. Purpose of the measurement.
2. Treatment of overloads before and during the time the measurement is taken.
3. Accuracy of the measurement.
4. Lower and upper limits of the frequency response needed.
5. Maximum error that can be tolerated during static conditions and during and after exposure to transient environmental conditions.
6. Limitations on excitation and output.
7. Power requirements.
8. Transducer's effect on the measurand.
9. Cycling line or operating line.
10. Failure modes of the transducer and hazards due to a failure in a components to other portions of the system.
11. Human engineering requirements.
12. Environmental conditions.
13. Data transmission technique to be used.
14. Processing system to be used.
15. Type of data display to be used.
16. Accuracy and frequency response capabilities of the transmission, processing, and display systems.
17. Signal conditioning required.
18. Transducer excitation voltage which is available.
19. Current drawn from the excitation supply.
20. Load on the transmission circuit due to the transducer.
21. Filtering required in the transmission or data processing systems.
22. Requirements for the detection and compensation for errors.

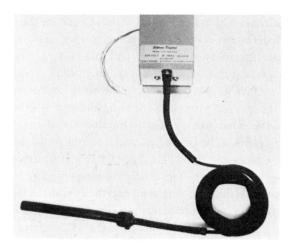

FIGURE 6-1   This temperature transducer converts the thermal expansion of a liquid into a binary output using a mechanical linkage. (*Courtesy Siltran Digital.*)

Transducer operation can be specified as continuous or as cycles where the total number and the on-time duration is stated.

The mechanical characteristics of the transducer may include specifying certain case materials and the exact nature of the sealing required for the environmental effects of corrosive or conducting fluids.

The most common transducers are used to detect either temperature or pressure changes. Some temperature transducers may convert the thermal expansion of a liquid or solid into electrical signals through a mechanical linkage (Figure 6-1). Others may use resistors, diodes, thermistors, or thermocouples to sense temperature variations by monitoring the changes in parameters such as conductivity or voltage. This second class of transducers is more relevant for microprocessor systems.

For particular temperature measurement application, one type of sensor may meet the performance and economic requirements better than all others.

In the design and development of control systems, an increased accuracy of measurements can result in more successful and economical system operation. When computer control or remote readouts are used (Figure 6-2), the long lines required can often lead to difficulty using the low output of some temperature sensors without amplification. This is where resistance thermometers appear to be best used, since they have a relatively high output and good linearity.

Resistance Temperature Detectors

Resistance temperature detectors (RTDs) use the temperature dependence of the resistance of a material to electric currents. The resistance of metals increases with

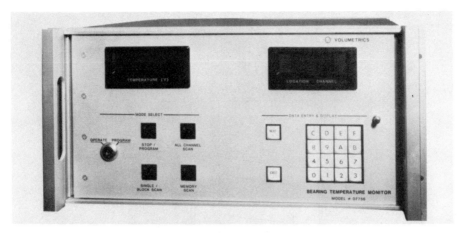

FIGURE 6-2 This 250 channel temperature monitor can be used for thermocouples or resistance temperature detectors. (*Courtesy Volumetrics.*)

increasing temperature, while most semiconductor materials decrease in resistance.

One part of the total resistivity is due to the impurities in the metal, known as the *residual resistivity*. It is lowest for pure metals. The residual resistivity may change if the detector is used at too high a temperature, or if the wire is contaminated by the environment or materials in contact with the wire. This change is relatively independent of temperature. It can be recognized if the resistance difference $(R_T - R_O)$ between two temperatures remains constant while the ratio (R_T / R_O) decreases. Changes of this type are irreversible. Another part of the total of resistivity is due to deformation and it will depend on the physical state of the metal. In a well annealed metal with low resistivity, as the crystal structure is stressed, the resistivity increases. The resistance can be increased by mechanical shock or vibration, thermal shock, or nuclear radiation at low temperatures. A difference in thermal expansion coefficients between the wire and its supporting structure (Figure 6-3) can cause stress on the wire. The RTD will act as a strain gage as well as a temperature sensor. This effect is reversible and the original resistance may be restored by re-annealing the element. Another effect sometimes encountered in high temperature applications is change in wire dimensions due to evaporation. In this case, R_T/R_O is constant but $(R_T - R_O)$ increases.

These effects must be considered in both the design and use of an RTD. The elements must be designed such that the wire remains annealed and strain free for the operating temperature range. By minimizing strain, the R vs T characteristic is similar to that of the wire alone and repeatability can be better than 0.1°C for industrial units and better than 0.001°C for platinum wire temperature standards. Strain free designs can be produced by matching expansion characteristics or by allowing the wire to expand or contract freely as temperature changes. Materials

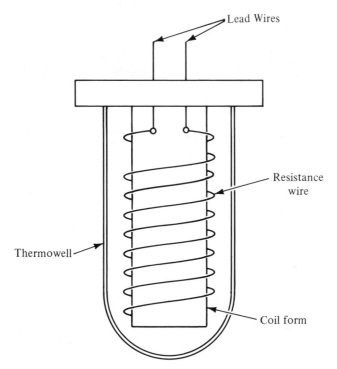

FIGURE 6-3   Typical RTD Construction.

are used which are compatible with both the environment and the sensing element. This includes thermal shocks and mechanical vibration. In some applications, the thermometer will tend to age in use until the stresses are equalized.

Copper is an inexpensive material for temperature sensing and is the most linear of metals over a wide temperature range. It has a low resistance to oxidation at moderate temperatures but poorer stability and reproducibility than platinum. The low resistivity of copper is also a disadvantage. Nickel has been widely used as an RTD element over the temperature range from −100°C to +300°C, because of its low cost and high temperature coefficient. Above 300°C, the R/T relation for nickel changes greatly. Nickel is susceptible to contamination by certain materials, notably sulfur and phosphorous, and its R/T relation is not as well known nor as reproducible as that of platinum. The resistance temperature characteristic of tungsten is also not as well known as that of platinum. Fully annealing pure tungsten has been impractical due to the brittleness caused by recrystallization; thus, tungsten sensors tend to be less stable than well made platinum sensors. Tungsten has good resistance to high nuclear radiation levels along with platinum. The mechanical strength of tungsten allows fine wires to be made and the resulting sensors can have higher resistance values.

Platinum resistance thermometers can be used with high accuracy from the triple point of hydrogen (13.81 K) to the freezing point of antimony (630.74°C). The resistance/temperature relationship of platinum is well known, reproducible, and linear over this temperature range.

Platinum is chemically inert, not easily contaminated since it does not readily oxidize, and can be used up to 1500°C. Platinum RTDs are generally more expensive than other resistance temperature sensors, although some industrial types are competitive.

Thin film platinum temperature sensors use a deposit of a thin film of platinum on an insulating substrate. The deposition technique can produce high resistance in a small sensing package with a fast response time. The fabrication techniques are similar to the ones used to manufacture ICs. Platinum wire RTD elements can be delicate and thermal cycling may cause aging. The thin film deposition technique with laser trimming can result in a small, more rugged sensor at a lower cost. However, difference in applying these high purity films to the substrate can cause the resistance/temperature characteristic to behave in a different manner from pure, annealed, strain free platinum. The temperature coefficient at room temperature is usually from 30–80% of pure platinum. The residual resistance tends to be much higher.

The variation of the resistance temperature characteristic between specimens is another potential problem. These thin film detectors can yield an accuracy and repeatability of less than 0.01°C. The small sensing tip produces a more rapid response time than thermocouples. A glass encapsulated platinum layer is normally deposited on a ceramic substrate which typically measures 10 x 3 x 1 mm. The rapid response times result from a large surface area to volume ratio and the use of the thin ceramic substrate which improves the thermal conductivity.

THERMISTORS

Thermistors are mixtures of semiconductor material with a resistance that varies rapidly with temperature. These materials are usually sintered mixtures of sulfides, selenides, and oxides of metals such as nickel, manganese, cobalt, copper, and iron. These are formed into small glass enclosed beads, disks, or rods as shown in Figure 6-4.

A high resistivity and high negative temperature coefficient is characteristic of these sensors. The resistance vs temperature characteristic is nonlinear. Maintaining narrow resistance tolerances is difficult in manufacture.

The devices are relatively inexpensive and are available in small sizes with high resistance values. With burn-in calibration, the stability is similar to the best wirewound elements.

Because of the nonlinear relationship, numerous calibration points are necessary and can be a major part of the cost for a thermistor application. To produce

Rod

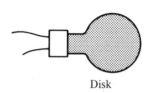

Disk

FIGURE 6-4   Thermistor Construction.

thermistors with good interchangeability is difficult, but selected thermistors or matched pairs in series can be obtained which are reasonably interchangeable. A single thermistor is usually unsuited for wide temperature spans because of its resistance characteristics. Several thermistors are used to cover a wide temperature span. Composite thermistor resistor assemblies can be used to provide linear response curves. External linearization is eliminated. Composite thermistor sensors are available in standard ranges: −5° to 45°C, −30° to 50°C, and 0° to 100°C. The networks generally display greater sensitivity than thermocouples. Components are also available as composite networks of thermistors and resistors packaged as a single sensor. These chip thermistor networks can be bonded directly to hybrid substrates, for the temperature sensing suitable for compensation of critical systems with LSI packages.

Another approach is the use of resistor pastes for direct screening onto alumina ceramics. Pastes with both positive and negative temperature coefficients are fired to yield temperature sensitive resistor elements with relatively linear responses. High and low resistivity formulas are available. Resistivity ranges from 100 ohms/square to 1.1 megohms/square.

OTHER SEMICONDUCTOR SENSORS

Germanium crystals have been used primarily for temperature measurements in the 1 to 35° K range. A single crystal of germanium with controlled impurities is used. Repeatability can be 0.001°C near the helium boiling point and 0.05°C near the hydrogen boiling point. The resistance vs temperature characteristic shows a nonlinearity similar to thermistors. Calibration interchangeability has been difficult to obtain in production.

Carbon resistors have been used as temperature sensing elements in the region below 60°K, where their behavior is similar to silicon semiconductors. A high resistivity and negative temperature coefficient of resistance are characteristics of this sensor. Disadvantages encountered are a sensitivity to pressure variations when used in pressurized applications, long term resistance drifts, short term drift after cycling between low and room temperatures, and a long response time.

Small crystals of silicon with controlled doping have the proper characteristics as resistive temperature sensors. Their temperature coefficient of resistance is positive above −50°C and their resistance vs temperature characteristics are close to linear, particularly over the most usable range, −50°C to +250°C. Below −50°C the temperature coefficient of resistivity becomes negative and the slope of their resistance vs temperature characteristic increases greatly. One type, constructed of heavily doped p-type silicon, offers a positive temperature coefficient of 0.7%/°C. Unlike common thermistors, linearity is within ±0.5° without a linearization network. Over the range of zero to 100°, linearity can be within ±0.025°. Resistance values begin at 10 ohms and progress through 40 values up to 10k ohms.

Since these resistance sensors do not produce voltages, they must be energized from external power supplies. This is complicated by the fact that the resistance cannot assume an arbitrary zero value at any temperature, but changes value from R_{T1} to R_{T2}. To obtain a zero based output signal, the sensor is used in conjunction with a resistance bridge network.

The Wheatstone bridge (Figure 6-5a) can be used as the starting point in studying resistance bridges, but it can only be used when the sensor's resistance value is high enough to mask the small additional resistance due to the leadwires joining the bridge with the element. This is generally true with thermistors.

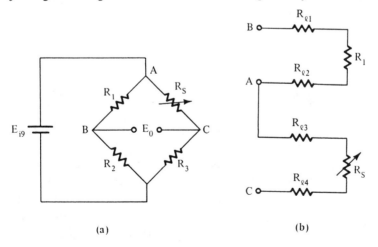

(a)   (b)

FIGURE 6-5   Resistance Sensor Bridge Techniques. (a) Wheatstone Bridge; (b) Compensating Loop.

Cable runs from a few feet to several miles may be required to connect the sensor to the bridge. If we mount the bridge with the sensor, the bridge must endure a more rugged environment and we may not gain any electrical advantage. Since we must use temperature sensitive wire to connect a temperature sensitive resistor, we cannot ignore one and measure only the other. Several methods can be used for bridge excitation.

Either ac or dc bridge excitation may be used depending on how the output signal is to be used. AC excitation has been used in some systems because of the type of recording and control instrumentation used. Its main advantage is the elimination of thermal emfs in the sensor leads. Reactive effects such as phase shifts from cable capacitance and noise interference due to cable coupling can be potential problem areas. Thermal emfs are usually insignificant.

DC excitation has been popular since unwanted noise may be filtered out. The accuracy and stability of the power supply can directly affect the accuracy and stability of the bridge output signal. Power supplies with stabilities to 0.01% may be required in high performance circuits, although 0.1% is usually adequate.

A compensation loop for reducing the effects of lead wires is shown in Figure 6-5b. Another circuit uses the three lead technique. Here, one lead is connected in series with the power supply where any resistance changes have negligible effects on the voltage. The two remaining leads are placed in series with an opposite leg of the bridge where their effects tend to cancel.

Transistors are also used as temperature sensors. Most operate off the change in emitter-to-base voltage with temperature. An accurately characterized silicon transistor may produce a linear change in voltage over the $-40\,°C$ to $150\,°C$ range. The correlation between the temperature and the transistor base-to-emitter voltage is approximately 2 mV/°C. Most sensors have a 400 mV output change. Accuracy is $\pm 2\%$ or 5 C. The thermal time constant for some sensors is 3 seconds in flowing liquid and 8 seconds in moving air. A stable constant current source is required.

IC temperature transducers are also available that contain the complete circuit required along with the sensor itself. Many of these use laser trimming to produce a precalibrated temperature transducer for applications below 150°C.

These operate as a current source and an output of one microampere per degree Kelvin is typical. Trimming of internal calibration film resistors may be performed during manufacture to produce $\pm 0.5°C$ accuracy and linearity within $\pm 0.3°C$ over the full range. Output impedance may be greater than 10 Mr which provides good rejection of hum, ripple, and noise. The high output impedance is useful in remote sensing systems with lines that may be hundreds of feet in length. These sensors can also be powered directly from 5-volt logic, to allow simple multiplexing schemes in most systems.

Also some ICs output temperature changes as a frequency variation. A voltage-to-frequency converter allows them to operate with a digital pulse output. They can be trimmed for Celsius, Kelvin, or Fahrenheit measurements using external components. The pulse train output is proportional to the substrate temperature.

THERMOCOUPLES

A basic thermocouple circuit (Figure 6-6) consists of a pair of wires of different metals joined or welded together at the sensing junction, and terminated at their other end by a reference junction which is maintained at a known temperature called the *reference temperature.* The load resistance of signal conditioning or readout equipment completes the circuit. When a temperature difference exists between the sensing and the reference junctions, a voltage is produced which causes a current to flow through the circuit. This thermoelectric effect is caused by the contact potentials at the junctions. As it was discovered by Seebeck, it is known as the *Seebeck effect.* The wires between the sensing junction and the reference junction must be made of the same material. The connecting leads from the reference junction to the load resistance are usually copper wires. They must be copper whenever the associated wiring in signal conditioning or readout circuits is made of copper so as to avoid junction potentials due to different metals.

Related to the Seebeck effect are the Peltier and Thomson effects. When current flows across a junction of two dissimilar conductors, heat unrelated to I^2R heating is absorbed or liberated as a function of the direction and magnitude of current. This is the Peltier effect. When a current flows through a wire along which a temperature gradient exists, heat unrelated to I^2R heating is absorbed or liberated from the wire. This is the Thomson effect.

The magnitude of the thermoelectric potential produced depends on the wire materials selected and on the temperature difference between the two junctions. A digital voltmeter designed for thermocouple measurements is shown in Figure 6-7.

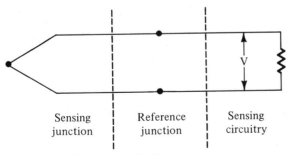

FIGURE 6-6 Thermocouple Circuit.

FIGURE 6-7   This thermocouple meter can also be used as a calibration source.
(*Courtesy Analogic.*)

The most common thermocouple materials are chromel-alumel, iron-constantan, copper-constantan, and chromel-constantan. Tables showing the thermal emf vs temperature, based on most of the standard materials, have been developed by the National Bureau of Standards, the Instrument Society of America, and many thermocouple manufacturers. Most are based on a reference temperature of 0°C.

The basic laboratory method uses an ice bath for the reference temperature, but it has limitations whenever measurements must be frequent or continuous. Some other techniques which can replace the laboratory ice bath are automatic icepoint reference; automatic references at temperatures other than 32°F, usually 150°F; and electrical compensators which use an ambient temperature couple in a bridge circuit with ambient sensitive elements.

The electrical compensator as shown in Figure 6-8 is an economical method for single channel systems with typical accuracies of ±¼°F. The oven type of reference is useful as a reference for a number of thermocouple channels. The oven type references, as well as the icepoint references, have an economic advantage wherever a number of reference couples are required.

If the leads from the thermocouples are brought to the switch box, the contact resistances of the switches may cause problems. Any nonconducting film on the contacts can cause errors since only a few millivolts are available to penetrate the film.

Some precautions when installing thermocouples include:

1. Do not attempt to locate the thermocouple near or in a direct flame path.
2. Always locate the thermocouple in an average temperature zone.
3. Preferably locate the thermocouple where the hot junction can be seen from an inspection port in high temperature zones.

4. The thermocouple should be entirely immersed in the medium whose temperature is to be measured.
5. All connections must be clean and tight to avoid high resistance contacts.
6. Avoid cable runs parallel to or closer than 1 foot to ac supply lines since induced currents will cause errors.

The use of a pocket or well will impede the transfer of heat to the sensing element and reduce the speed of response. Agitation of the fluid around the sensing element will reduce this effect and improve the speed of response.

Another problem occurs when there are two bodies and one is hotter. Then a net transfer of energy will occur from the hotter body to the colder one. When a number of bodies are all at the same temperature and enclosed in a space insulated to heat, then each body is considered to radiate energy into the surrounding medium and continuously absorb energy at the same time. Since there must be equilibrium, the processes balance one another, and the temperature of each body remains constant. This is known as the *Prevost theory.* In the case of a chamber where the walls are at a lower temperature than the internal hot gases, a thermocouple placed in the hot gas stream will be exposed to the heat of the hot gases; it is at a higher temperature than the walls so it radiates more heat than it receives and may measure a lower value than the true temperature of the gases. This phenomenon is important when locating temperature sensing devices in similar situations.

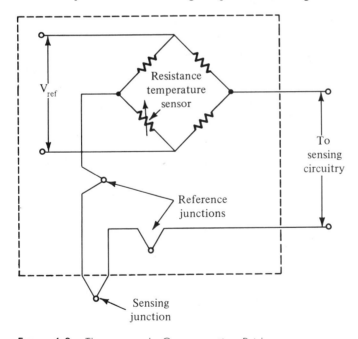

FIGURE 6-8   Thermocouple Compensation Bridge.

A thermal lag error will exist if the standard and the unknown respond to a changing temperature at different rates or if the bath is not stable with time. A self-heating error will exist if excessive power is dissipated in either the standard or the unknown. This can cause an error by heating the resistive element. Stem conduction errors may be due to heat transfer along the stem of the standard or the unknown. The thermocouple should be immersed for 10 inches in a liquid. For devices with a short stem, the stem conduction error can be determined by measuring the temperature at a number of depths.

There may also be some uncertainty of temperature in an unmonitored bath. For an ice bath made with tap water, the temperature can deviate from 0°C by as much as 0.5°C in some regions. The use of an ice bath after too much ice has melted will also cause temperature errors.

Automatic ice point devices which maintain the thermocouple at the ice-point, as opposed to the introduction of an equivalent voltage, are the most accurate references. They are available in both single and multiple channel units, and have typical accuracies of ±0.1°F. Long term stability is excellent since they operate with the reference thermocouples located within an hermetically sealed cell in which the ice water equilibrium is maintained automatically.

The thermopile is a combination of several thermocouples of the same materials connected in series. The output of a thermopile is equal to the total output of the number of thermocouples in the assembly. All reference junctions must be at the same temperature as shown in Figure 6-9.

Modern thermocouples can react more rapidly, due to their small thermal mass. A typical response is in milliseconds. Wires of 8 ten-thousandths of an inch in diameter are used to form the junction. These are inserted into a quartz insulator which in turn is assembled into a 0.008 inch O.D. metal sheath or probe that also anchors the insulated leads. The small diameter junction of less than 0.002

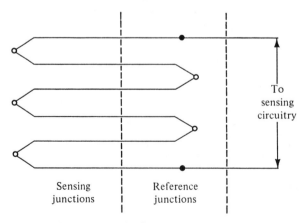

FIGURE 6-9   Thermopile Circuit.

inch is responsible for the faster response. Errors due to radiation and conduction are also reduced. But traditional drawbacks remain. Voltage output shift is small and nonlinear. Absolute measurements require a known reference. Expensive cabling and connectors are required to avoid unwanted junctions.

The major competition for resistance thermometers has been thermocouples. In the final analysis, selection boils down to an accuracy and cost comparison for the total system including sensors, signal conditioning, and readout circuits. The advantages of resistance thermometers over thermocouples include the following:

1. Higher output voltage, typically a factor of 500 or more.
2. Recording, controlling, or signal conditioning equipment can be simpler, more accurate, and less expensive because of the higher output signal.
3. The output voltage per degree for resistance sensors can be chosen as desired by adjusting the excitation current and/or the bridge design.
4. A reference junction temperature or compensating device is not necessary.
5. The shape of the curve of output vs temperature can be controlled with the resistance sensor bridge design.
6. More electrical noise can be tolerated with resistance sensors and longer lead wires can be used.
7. Sensitivity to small temperature changes can be greater.
8. In moderate temperature ranges, absolute accuracy of calibration and stability of calibration for resistance elements can be better by a factor of 10 to 100.

The major problem with the use of thermocouples is the errors due to the spurious and parasitic emfs in the leads. This effect is primarily responsible for the difference in precision between thermocouples and resistance thermometers.

Small variations in the state or composition of a wire can produce Seebeck emfs anywhere a temperature gradient exists. These problems are most severe when the temperature gradient on the leads is changing. Annealing will reduce the causes of the effect but will not eliminate it. One should perform periodic recalibration of the thermocouples to offset the temperature distribution.

NONCONTACT SENSORS

If the temperatures are too high to allow a thermocouple or other contacting temperature sensing element to be used, noncontact methods are available. These devices rely on the fact that all hot bodies emit radiant energy with an intensity that is a function of the absolute temperature of the emitting surface.

Radiation pyrometry measures the radiated heat from a hot object. Practical radiation pyrometers are sensitive to a limited wave length band of radiant energy. The operation of these thermal radiation pyrometers is based on black-body radiation concepts.

The noncontacting instruments which respond to a relatively wide band of wave lengths operate according to the Stefan-Boltzmann law and are referred to as *total radiation pyrometers*. Another type of instrument uses narrow bands of wave length in the visible spectrum. Instruments of this type are referred to as optical pyrometers.

In a typical optical pyrometer, shown in Figure 6-10, a radiation source is viewed through a telescope system consisting of a lens and eyepiece. Inside the system is a small lamp and the current through the lamp is controlled. An optical filter is placed between the eye piece and lamp. Looking through the eye piece, the source is seen as a circle, square, or other shape, and in the center is the image of the lamp filament. The control resistance is adjusted until the brightness of the filament is equal to that of the radiation. The filament image then appears to merge into the radiation image and present a uniform picture to the eye. The control is calibrated directly in degrees. In recently developed instruments the operator has been replaced by electronic techniques to eliminate operator-to-operator differences.

In the total radiation instrument, shown in Figure 6-11, radiation from the source is focused onto a small disk of blackened platinum or similar black body. In some devices, a concave mirror is used for focusing. Fixed to the platinum disk are a number of thermocouples arranged in series as a thermopile; these are connected to the sensing element. The radiation heats the disk and generates an emf which is calibrated in degrees.

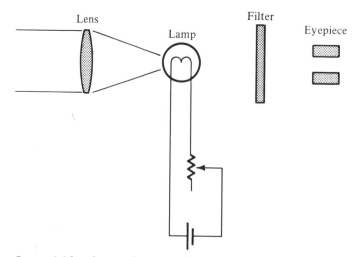

FIGURE 6-10   Optical Pyrometer Technique.

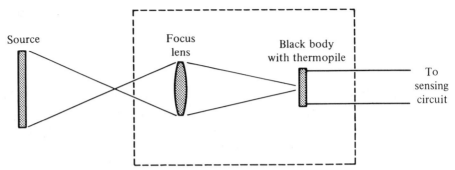

FIGURE 6-11   Total-Radiation Pyrometer Technique.

In some instruments the cold junction is quite close to the hot junction. The position of the hot and cold junctions ensures that both are nearly equally affected by any ambient temperature changes. The cold junction must be shielded, however, from radiation from the source.

The use of very fine thermocouple wire reduces conduction losses from the couple itself. If an excessive temperature rise of the pyrometer housing is likely to be encountered, the housing should be enclosed in an asbestos lined metal casing or in a water cooled jacket.

When cold junction compensation is required, a nickel resistance spool is sometimes connected as a shunt across the thermocouple leads at the cold junction end. The variation of this resistance with ambient temperature tends to compensate the thermocouple for the cold junction temperature change. Other designs use a bimetallic compensator to cut off part of the incoming radiation when the temperature of the housing increases.

The distance between the source of radiation and the pyrometer is important, since the radiation must be focused onto a receiving disk. In some designs, the position of the lens or mirror is adjusted so the radiation is always focused onto the same point, others use a fixed focus but require the radiation to completely fill the pyrometer opening.

Another type of instrument, referred to as a *partial radiation pyrometer*, uses photoelectric principles. It has some advantages over the total radiation pyrometer, which can be sensitive to smoke, water vapor, and CO_2.

The partial radiation pyrometer can be arranged to eliminate selectively part or all of the radiating effects of these constituents. It has been effectively used to measure the roof temperatures of steel melting furnaces, where a total radiation pyrometer is unsuitable due to the radiating properties of the water vapor and CO_2 present.

The use of a photocell has the advantage that variations in emissivity do not introduce as large an error as they do with other radiation pyrometers. Variations in ambient temperature can introduce a larger error, and in many cases partial radiation pyrometers are water jacketed.

SOURCES OF ERRORS IN TEMPERATURE SENSING SYSTEMS

The occasional user of temperature sensors may think that accurate measurements can be relatively easy. The use to which a temperature sensor is put can affect the accuracy in many ways. The successful application of the sensor depends on a knowledge of the errors involved. The following error sources are descriptive of the problem areas most commonly encountered. Most of these errors are common to all types of temperature sensors.

Calibration errors are due to the uncertainty of calibrating the instrument at a known temperature. These can be considered as a random error. Accuracies are a function of sensor configuration and calibration techniques and can range as low as 0.01°C at the ice point to 5°C at 1100°C.

Stability is usually a function of the environment. It generally refers to the long term changes in the calibration of the sensor. The output emf of a thermocouple tends to decrease with time at high temperatures, while the resistance of a wirewound resistance sensor increases. If enough data can be obtained on a sensor through periodic calibrations, the change can be considered as a systematic error and can be defined as:

$$\Delta T = (dT/dt)t$$

where

$$\Delta T = \text{temperature error}$$
$$dT/dt = \text{temperature drift rate}$$
$$t = \text{time between calibrations}$$

Repeatability is the ability of the sensor to repeat a calibration at a known temperature. This is a function of the sensor design since hysteresis affects the repeatability. Repeatability can be better than 0.25°C over 500°C for some platinum resistance sensors. Since the hysteresis effects are usually not well known, repeatability is considered as a random error.

Self-heating errors result from the I^2R or Joule heating in a resistance. This can cause the indicated temperature to be higher than the actual temperature. The self-heating error T can be expressed as:

$$\Delta T = Q (R_I + R_B)$$

where

Q = power dissipated by the element
R$_1$ = internal thermal resistance of the element
R$_B$ = thermal resistance of the boundary layer

A detailed analysis may require the solution of the steady state heat transfer problem using an internal heat source. It is usually easier to measure the self-heating effects directly. One way to measure the self-heating effect (if the temperature does not change rapidly) is to measure the resistance at several currents which result in measurable resistance changes. The I^{2}R power vs the change in resistance R or change in temperature T will be nearly linear. Self-heating effects can range from 0.01 mv/°C in some thermistors in still air to 500 mw/°C in platinum sensors in flowing fluids. This effect can be reduced by selecting a lower excitation current.

Stem conduction errors are steady state errors due to the heat transfer to or from the sensing element along the sheath and leads. This results in the temperature of the sensor being different from that of the fluid. Stem conduction effects can be measured directly by incrementally withdrawing the sensor and measuring the temperature at different positions.

A thermal lag error occurs when the temperature sensor does not respond fast enough to the environmental temperature changes. This error is a function of the heat transfer characteristics of the environment as well as the sensor. When the rate of the environmental temperature changes and the sensor response time are known, then the error due to thermal lag Tl can be calculated as follows:

$$Tl = (dT/dt) \, T_R$$

where

dT/dt = rate of temperature change
T$_R$ = sensor response time

Thermal radiation errors can occur when the sensor receives radiation from a source at a higher temperature than the environment around the sensor. The error is a function of the shape of the sensor with respect to the environment and the thermal radiation characteristics of the sensor. Thermal radiation errors may be reduced by the proper placement and shielding of the sensor.

Frictional heating can be an error source when a high viscosity fluid flows around the surface of a sensor at high velocities. Then the temperature indicated by a sensor in the stream may be higher than the actual stream temperature. The frictional heating effects are proportional to the square of the velocity and special shield geometries can be used to eliminate this effect.

Insulation becomes a source of errors in the presence of ionizing radiation, moisture, or high temperature increases. In some temperature sensors, the degraded insulation will act as a resistor in parallel with the sensing element. This tends to lower the indicated temperature. The insulation resistance of an isolated sensor can be made by measuring the resistance between the element and ground. Errors can be minimized by using better insulating materials.

Thermal emfs can be developed in resistance elements and lead wires. These can degrade sensor accuracy. In thermocouples, the errors may be 0.1°C or greater. They affect resistance sensors using dc excitation to a lesser degree. A 100-ohm platinum sensor with a sensitivity of 0.5 mv/deg C and 1 uv of thermal emf can be in error by 0.002°C.

Thermal emf errors can be reduced by maintaining the dissimilar metal joints at an equal temperature and the lead wires at the same temperature gradient. However, ac excitation can almost eliminate these effects.

Nuclear radiation effects can be caused by gamma rays, fast neutrons, and thermal neutrons. Radiation heating is the heat generated by the absorption of nuclear radiation. A sensor which can dissipate 100 mw/°C can be in error by 1°C with 100 mw of gamma heating. Fast neutrons can cause dislocations in metals which alter the sensor characteristics. The effects are worst at cryogenic temperatures. At room temperatures, the defects anneal out as quickly. The error for a platinum thermometer at −253°C is 1°C for a dose of 2×10^{17} nvt. Thermocouples may show errors about five times larger in a similar environment.

Thermal or slow neutrons alter the composition of materials by transmutation. Platinum tends to be transmuted to gold and mercury, altering its resistance/temperature characteristics. The change is dose dependent and irreversible. A platinum thermometer can shift by about 1°C after a dose of 10^{22} nv_0t. A Pt/Pt-10Rh thermocouple might shift about 10°C.

Bridge errors can also affect the sensor accuracy. Power supply errors can be introduced by the long term instability of the power supply. Resistor aging will not be a problem if stable wire wound resistors are used. A shift or less than ±0.01% in a five-year period can be obtained.

Potentiometer shifts and noise can be caused by shocks and vibration. The film type of potentiometer may tend to drift with age. Restricting the required adjustments can keep potentiometer effect errors to a minimum. Temperature coefficients should not affect bridge resistors unless subjected to outdoor environments. Resolution errors can occur when wire wound potentiometers are used.

Systematic errors have a direction and magnitude which may be predicted. Random errors have a magnitude and direction which can be predicted. Some amount of randomness may be found in systematic errors. This can be about three times the standard deviation of a group of measurements. If the individual errors can be estimated, the random error can be calculated by the square root of the sum of the squares of the individual errors.

A passive resistance bridge operating from a constant voltage supply is inherently nonlinear. This is due to the decrease in current that takes place when the sensor resistance increases. Employing a constant current supply is not completely successful because neither nickel nor platinum resistors are linear. The nonlinearity may be removed in scaling or conditioning or, in some control operations, it may not be significant. Nickel can be linearized by simple bridge but for platinum active circuitry must be used. The linearization can be done with a microcomputer. The millivolt bridge signal is digitized and fed into the microcomputer which computes the temperature. In a system involving a large number of temperatures with different ranges, even a high speed machine could become burdened. The different constants require some memory and each measurement requires the multiplication by at least two constants plus squaring the voltage and adding another constant. Another technique uses feedback from the sensor leg of the bridge to modulate the bridge power supply. By increasing the supply voltage as the temperature rises, the tendency for the output to decrease is overcome and a linear output is achieved. Temperature transmitters are also available which produce linear outputs for various resistance sensors.

EXERCISES

6-1. Describe the construction and operation of the resistance temperature detector.

6-2. In a resistance thermometer the resistance at 20°C is 100 ohms. It is 101 ohms at 25°C. Compute the resistance value at −100°C and at 150°C assuming a linear relationship.

6-3. Compute the resistance of the temperature sensitive element in Exercise 6-2 at 0°C assuming a linear relation within ±10% between resistance and temperature.

6-4. In the case of a platinum coil resistance thermometer, why is it better not to exceed a measuring temperature of 500°C?

6-5. Discuss the advantages and disadvantages of the thermocouple as compared to the resistance temperature detector. Consider such factors as distance, speed of response, multipoint installation, accuracy, operating range, computing possibilities, and relative expense and maintenance.

6-6. A platinum, platinum-rhodium thermocouple has an emf temperature relationship given as:

$$e = -3.28 \times 10^{-1} + 8.28 \times 10^{-3}T + 1.5 \times 10^{-6}T^2$$

where e is in millivolts and T is the temperature difference, in degrees Celsius, between the hot and cold junctions.

(a) Calculate, to three significant figures, the emf, if the cold junction is at 30°C and the hot junction is at 1200°C.

(b) Calculate the sensitivity, in microvolts, required to indicate a change of 1°C at 1200°C with the cold junction temperature maintained at 30°C.

6-7. Describe the operation and considerations of a transistor for measuring temperature.

6-8. Define the following: (a) Seebeck effect, (b) Peltier effect, (c) Thomson effect.

6-9. Discuss the importance of the use of various metals as applied to a thermocouple circuit.

6-10. Describe and explain, with the aid of a diagram, how a thermoelectric emf of the order of 0.001 volt can be balanced in a resistance temperature system.

6-11. Describe the use of multiple thermistors for temperature measurement.

6-12. It was found that a copper-constantan thermocouple followed the equation:

$$e = a(T_1 - T_2) + b(T_1^2 - T_2^2)$$

then

$$a = 3.25 \times 10^{-2} \text{ mv/C and } b = 4.75 \times 10^{-5} \text{ mv/C}.$$

$T_1 = 100°C$, and the cold junction T_2 is in ice. Compute the emf in millivolts.

6-13. Describe the operation of a typical integrated circuit thermometer used in industrial instrumentation.

6-14. List the possible errors due to the variation of resistance both internally and externally in thermocouple and resistance temperature measuring circuits.

6-15. The emf of an iron-constantan thermocouple was found to be 44.50 mv at 800°C with cold junction temperature of 0°C. If the cold junction temperature changes to 5°C, what is the corresponding emf? Assume a straight line relationship between temperature and emf with the variation of emf with temperature as 6.0×10^{-2} mv/°C.

6-17. Describe the construction and operation of a compensating loop used in a temperature measurement system.

6-18. Explain the importance of black-body radiation in high temperature measurement.

6-19. Describe with the aid of a diagram the operation of the total radiation pyrometer and a typical optical pyrometer.

6-20. Discuss the photocell type of partial radiation pyrometer and its advantages and disadvantages.

6-21. List some error sources in temperature measurement systems and the ways they may be minimized.

7

Pressure Transducers

PRESSURE TRANSDUCER CHARACTERISTICS

Pressure temperatures find many applications in aerospace, automotive, chemical, civil engineering, hydraulic, transportation, power generation, and medical areas. Frequently pressure transducers are used to measure altitude and water depth (Figure 7-1). Atmospheric pressure decreases with increasing altitude. The relationship is nonlinear since air density and temperature decrease with increasing altitude. The correct values for pressure vs altitude are listed in tables developed from measurements taken by satellites.

Pressure increases with water depth but it is affected by density variations due to temperature differences and by the salinity of seawater. If corrections for these variables are not required, then the average depth can be calculated using 0.445 psi per foot of depth for ocean water and 0.434 psi per foot of depth for fresh water.

Most electrical output pressure transducers detect pressure using a mechanical sensing element. These elements consist of thin walled elastic members, such as plates, shells, or tubes, which provide a surface area for the pressure to act upon (Figure 7-2). If the pressure is not balanced by an equal pressure acting on the opposite side of this surface, the element will deflect. This deflection is then used to produce an electrical output. If another pressure is allowed on the other side on this surface, then the transducer will measure differential pressure. When the other side of the surface is evacuated and sealed, absolute pressure is obtained. Gauge pressure is obtained when ambient pressure is allowed on the reference side (Figure 7-3). A typical calibration station for pressure transducers is shown in Figure 7-4.

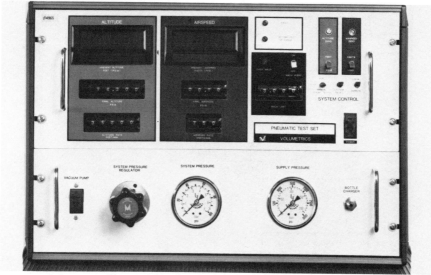

FIGURE 7-1 This air data test set can be used to measure altitude and air speed. (*Courtesy Volumetrics.*)

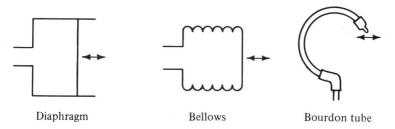

Diaphragm   Bellows   Bourdon tube

FIGURE 7-2 Some Pressure Sensing Elements.

INDUCTIVE TRANSDUCERS

Inductive transducers use the pressure to move a mechanical member which is used to change the inductance of a coil. The inductance is changed by the relative motion of a core and the inductive coil as shown in Figure 7-5. Inductive single coil transducers have been used in oscillator circuits where they were used to control the frequency. They suffer from difficulties in compensating for temperature effects, which requires a matching of the core and windings materials for temperature vs permeability characteristics.

Another type of inductive transducer uses the ratio of the reluctance of two coils. These are less sensitive to temperature effects than the single coil types.

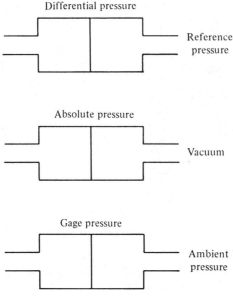

Figure 7-3   Pressure Reference Designation.

A small motion of about 0.003 inches will yield an ac output voltage of about 100 mV.

In diaphragm type, variable reluctance pressure transducers as shown in Figure 7-6, a diaphragm of magnetic material is supported between two symmetrical inductance assemblies. The diaphragm is deflected when there is a difference in pressure between the two input ports. This increases the gap in the magnetic flux path of one core and decreases the gap in the other, thus the reluctance varies with the gap. The overall effect is a change in inductance of two coils of the transducer. The inductance ratio L_1/L_2 can be measured in a bridge circuit to produce a voltage proportional to the pressure difference.

The measurement of force due to pressure is thus accomplished by the change in the inductance ratio of the coils. The force being measured changes the magnetic coupling path upon displacement of the core. The hysteresis errors are limited to the mechanical components. This advantage is diminished by the nature of the force summing members which are best suited to operate with the inductive principle.

The diaphragm is often used as part of the inductive loop. In this case the overall performance of the transducer may be less than optimum since the mechanical characteristics of the diaphragm must be compromised to improve the magnetic performance. The E core type of construction may be used to maintain good balance and low phase shift.

This type of transducer has the following advantages:

1. Responds to both static or dynamic measurements.
2. Continuous resolution.
3. High output.
4. Provides direct FM for telemetry.
5. High signal-to-noise ratio.

Disadvantages include:

1. Must be excited with ac.
2. Must be reactively and resistively balanced.
3. Frequency response normally limited by construction.
4. Large volumetric displacement and low frequency response.
5. Proximity to magnetic objects, or fields, causes erratic performance.
6. Mechanical friction may cause wear and errors.

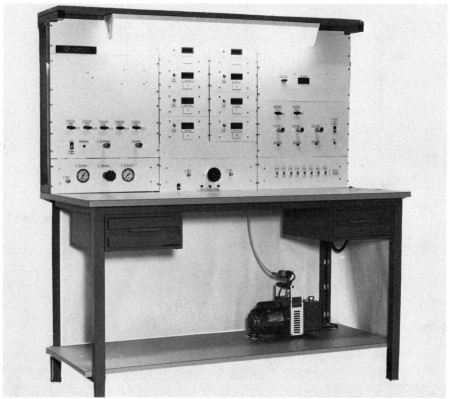

FIGURE 7-4 This calibration station can be used to calibrate differential, absolute or gage pressure transducers. (*Courtesy Volumetrics.*)

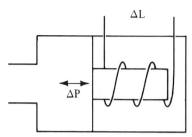

FIGURE 7-5   Single Coil Inductive Pressure Transducer.

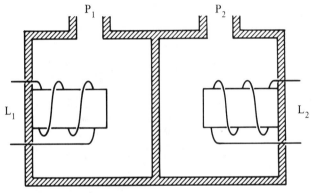

FIGURE 7-6   Differential Reluctive Pressure Transducer.

Some transducers contain dc-to-ac-to-dc conversion circuitry. Transducers with dc excitations of 28 and 5 V are available for absolute, gauge, and differential pressure measurements. Range is typically 1 inch of water to 12,000 psi.

Most of the ac transducers require a carrier frequency in the range of 60 Hz to 30 kHz. When dc conversion circuitry is used the internal carrier frequency can be much higher, allowing smaller coils and capacitors for a smaller transducer package.

The performance of these inductive transducers, with or without dc-to-dc conversion can compare in many ways to the best versions of other pressure transducers. Static error is typically ±0.5% with the nonlinearity accounting for the major portion. Errors due to hysteresis and nonrepeatability may be less than 0.2%. Proof pressure or overrange ratings of greater than six times normal range are available.

Temperature effects are minimized by using similar sensing element and coil materials. These errors can be 1% or 2% up to 100 °F. Pressure transducers without the dc conversion circuits may operate to 350 °F. The solid state components of the dc conversion circuits can limit the operating temperature of this type of transducer to less than ac types. The frequence response can range from 50 to 1000 Hz and is a function of the particular design.

Some constructions of these transducers, as with the diaphragm types, have a good tolerance to shock and vibration environments and good pressure overload capabilities in liquid as well as gas systems.

The linear variable differential transformer (LVDT) type of construction as shown in Figure 7-7 uses a sliding core which is connected to the pressure sensing element. LVDT transducers are also used widely for displacement and velocity measurement.

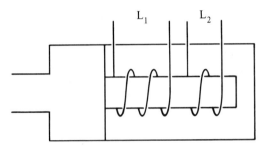

Figure 7-7   LVDT Pressure Transducer.

PIEZOELECTRIC TRANSDUCERS

Piezoelectric pressure transducers use a crystal which generates a charge or voltage when it is mechanically stressed. A diaphragm is normally used to react with pressure to produce the stress on the crystal. Piezoelectric pressure transducers have the ability to operate over a wide temperature range with relatively small errors due to temperature changes. Various crystal materials are used. Selected quartz crystals are used from those found in the natural state. Grown crystals include ADP (ammonium dihydrogen phosphate) which are grown in an aqueous solution. Various ceramic materials are also used.

The piezoelectric elements are cut from the crystal along existing crystallographic axes. Ceramic elements are made from powdered materials which are pressed into the required shape and fired at high temperatures. Piezoelectric characteristics result as they are polarized by an electric field during cooling. Ceramics can reach the Curie point when heated—the temperature at which the crystalline structure changes. Polarization is then lost, and the element is useless. Curie points vary from 300 °F to over 1000 °F. Repolarization of the element is possible. Quartz types can be used with temperatures ranging from −400 to 500 °F.

The output of a piezoelectric element can also be affected by the pyroelectric effect, which causes changes in the output proportional to the rate of change of temperature experienced by the crystal.

In the past some quartz crystal transducers have been used with amplifiers which permit static measurements. Most piezoelectric pressure transducers are used for the dynamic measurement of rapidly varying pressures. Sound pressure levels of up to over 180 db may be measured with an accuracy of 3% of full scale, including high frequency components. The measurement of dynamic pressures on aerodynamic loads on aircraft and spacecraft surfaces is a common application. Quartz units, mounted as sparkplugs, are used to measure pressures in internal combustion engines. Piezoelectric transducers also sense the pressure surges due to explosions and earthquake shocks. They have also been installed in turbines, pumps, and hydraulic equipment to measure dynamic stresses. The frequency response is in the range of 10 Hz to 50,000 Hz. The pressure ranges are 5 to 10,000 psi. Quartz and ADP types tend to have a higher natural frequency than the ceramic units. However, the ceramic crystals provide higher output levels. Since the piezoelectric sensor responds so well to high shock levels, this may be a limiting factor in some high level applications. The application of a high level shock could produce an overvoltage which would saturate the signal conditioning amplifiers for a period of 50 or more times the shock duration, causing loss of a data during this time. The normal output of a piezoelectric crystal is small and the impedance high so a high impedance amplifier is required for signal conditioning.

An operational amplifier can be used with capacitive feedback to compensate for the capacitance due to cabling. Integrated circuits allow the amplifier to be placed inside the transducer case.

CAPACITIVE PRESSURE SENSORS

Capacitive pressure transducers usually use a metal diaphragm with a metal plate positioned alongside the diaphragm as shown in Figure 7-8. Any movement of the diaphragm changes the capacitance between it and the fixed plate.

An ac signal across the plates is used to sense the change in capacitance, which occurs as a force applied to the diaphragm displaces one plate of the capacitor. This type of sensor may be used as part of an RC or LC network in an oscillator or it may be used as a reactive element in an ac bridge. When used in an oscillator circuit, the output detected may be ac, dc, digital, or a phase shift.

Capacitive transducers are small size with high frequency response and allow high temperature operation. It can measure both static and dynamic quan-

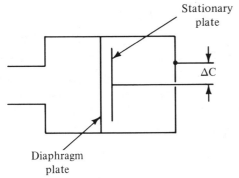

FIGURE 7-8   Capacitive Pressure Transducer.

tities. Range is .01 to 10,000 psi. Typical accuracy is .25% and units are available with accuracies of .05%.

Capacitance transducers should be reactively as well as resistively matched. Long lead lengths or loose leads can cause a variation in capacitance due to capacitive coupling. It is usually desirable to use a preamplifier close to the trans-ducer as well as matched cabling to minimize coupling effects.

Some capacitive sensor designs may use materials such as quartz for the ca-pacitor plates. One design uses two thin quartz disks with platinum electrodes on their inner surfaces. The disks are fused to form a small capsule and the electrodes separated by a 0.002 inch gap. As a vacuum is drawn inside the gap, the capaci-tance changes and absolute pressures of up to 30 psi can be measured. Gage pressures of up to 30 psi are measured with one port vented to the atmosphere. This device can provide an accuracy of ±0.5% as well as a 5V TTL output.

In general the major disadvantages of capacitive sensors are:

1. Motion of connecting cables or long lead length will cause distortion or erratic signals.
2. High impedance output must be reactively and resistively balanced.
3. Sensitive to temperature variations.
4. Receiving and conditioning circuitry can be complex.

The major advantages are:

1. The frequency response.
2. Inexpensive to produce.
3. Low shock response.
4. Can measure either static or dynamic phenomena.
5. Minimum diaphragm mass.
6. Small volume.

POTENTIOMETRIC TRANSDUCERS

A potentiometric transducer is an electromechanical device containing a resistance element which is contacted by a movable slider. Motion of the slider due to a pressure change results in a resistance change (Figure 7-9). Deposited carbon, platinum film, and other techniques may be used to provide the resistive element.

The potentiometer sensor is widely used despite its limitations. Its electrical efficiency is high and it provides an output to many control operations without further amplification.

The major advantages are:

1. High output.
2. Inexpensive.
3. May be excited with ac or dc.
4. Wide range of output functions available.
5. No amplification or impedance matching may be required for transmission.

The disadvantages are:

1. Large size.
2. The resolution finite in many devices.
3. High mechanical friction causes limited life.
4. Sensitive to vibration.
5. Can develop high noise levels with wear.
6. Requires large force because of friction.
7. Low frequency response.
8. Large displacements may be required from pressure sensing elements.

The potentiometric pressure transducer was first used in 1914, and still is used today due to low cost and connection simplicity. A wire wound or deposited conductive film may be used which is contacted by the movable slider or wiper connected to the mechanical sensing element. The output due to a pressure

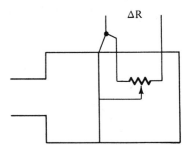

FIGURE 7-9   Potentiometric Pressure Transducer.

change is a function of the design of the potentiometric resistance curve which can be linear, sine, cosine, logarithmic, or exponential. These output can be used with ac or dc and no amplification or impedance matching may be required. A high output can be obtained with a high input voltage.

The more recent trends in potentiometric pressure transducers have taken two directions: (1) miniaturized devices with more relaxed tolerances, and (2) the use of a control to supplement the force of the diaphragm or capsule. These motor or force driven systems allow the use of low resolution multiturn potentiometers.

Typical potentiometric transducers have an accuracy of ±1%. A pressure range of 5 to 400 psi is available with some high pressure devices reaching 10,000 psi. Specialized devices may also be available with the following ranges: Resolution 0.2%, linearity ±.4%, hysteresis .5%, and temperature error ±.8%. Some advanced instruments may offer ±0.25% end point accuracy.

STRAIN GAGE PRESSURE TRANSDUCERS

Strain gage transducers use the force from a pressure change to cause a change in resistance due to mechanical strain. The pressure sensing element can be a diaphragm or even straight tube since the deflection required is small. The strain sensor gages may be mounted right on the tube sealed at one end. A pressure differential causes a slight expansion or contraction of the tube diameter. Other designs use a secondary sensing element or auxiliary member in the form of a beam or armature. Four or sometimes two arms of a Wheatstone bridge may be used for temperature compensation.

The force being measured displaces and changes the length of the member to which the strain gage is attached. The strain gage property known as a gage factor produces a change in resistance proportional to the change in length. The strain gages arranged in the form of a Wheatstone bridge circuit have one to four of the bridge legs active.

Strain gage transducers can be classed into two general categories: unbonded and bonded. The unbonded gage has one end fixed, the other end is movable and attached to the force member. The bonded gage is entirely attached by an adhesive to the member whose strain is to be measured (Figure 7-10). Strain

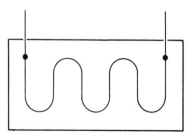

FIGURE 7-10   Wire Strain Gage, Bonded Type.

gages may be made from metal and metal alloys, semiconductor materials, and thin film materials.

The gage factor property which converts mechanical displacement is defined as the unit change in resistance per unit change in length.

Metal and metal alloys, in wire and foil form, are used in strain gages. Although all electrical conductors have a gage factor, only a few have the necessary properties to be useful as strain gages. For common metal strain gage materials, the gage factor is generally from 2.0 to 5.0. High gage factor materials tend to be more sensitive to temperature and less stable than low gage factor materials. The strain sensitive filaments must have stable elastic properties, high tensile strength, and corrosion resistance. Alloys possessing appropriate combinations of gage factor and thermal coefficient of resistivity will give optimum performance.

If the wire functions in the four elements of the bridge circuit, it is divided into four parts equal in resistance value. A resistance of 350 ohms is typical, but bridge resistances from 50 ohms to several thousand ohms can be provided by varying the diameter of the wire, the length of the filaments, and the number of loops of the resistance wire wound between the insulators. Because of this versatility, unbonded strain gage transducers have been used with a wide variety of systems and automated equipment.

In a typical unbonded configuration, a stationary frame is used with an armature. Four filaments of strain sensitive resistance wire are wound between rigid insulators mounted in the frame and armature. The filaments are of equal length and are arranged as shown in Figure 7-11.

The strain sensitive resistance wire must be wound to keep the filaments under some residual tension when the mounting posts are displaced to either extreme position. The resistance of the four strain sensitive elements is trimmed during assembly such that no signal appears in the output circuit when there is not an external force. If the bridge is balanced for zero output, the unbalanced electrical output of the bridge will have a linear relationship to a force applied.

In some devices the geometrical rearrangement is such that the travel of the armature is as small as ± 0.0003 in. This degree of movement is used in many miniature transducers. These transducers can provide excellent frequency response due to the low displacement and small mass used in the armature.

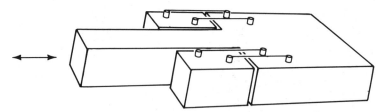

FIGURE 7-11   Strain Gage with Armature, Unbonded.

In the bonded strain gage, electrical insulation is provided by the adhesive and/or insulating backing materials on the strain gage. The force required to produce the displacement is larger than that required with an unbonded strain gage because of the additional stiffness. The bonded strain gage member will operate in a tension, compression, or bending mode.

Semiconductor material, usually silicon, is also used for strain gages. By controlling the amount and type of dopant, the strain gage properties can be optimized for a specific application. The silicon strain gage provides a higher gage factor than the metal strain gages, but also has a higher temperature coefficient. Gage factors of 50 to 200 are typical and they can be either positive or negative.

The available strain gage elements include:

1. Unbonded metal wire gages.
2. Bonded metal wire gages.
3. Bonded metal foil gages.
4. Thin film deposited gages.
5. Bonded semiconductor gages.
6. Integrally diffused semiconductor gages.

Unbonded wire elements as shown in Figure 7-11 are stretched and unsupported between a fixed and a moving end. They tend to have high sensitivity, but are sensitive to vibration. The unbonded strain gage may be used for pressures of less than 5 psi. Nichrome and platinum are common materials. The bonded elements are attached permanently over the active strain element which makes them much less sensitive to vibration. The cut or etched foil types allow a strong bond as well as more automated trimming. Foil, thin film, and semiconductor strain elements are normally bonded or deposited such that the semiconductor, thin film or foil, and a pressure diaphragm appear as a single part.

Thin film strain gages use manufacturing techniques similar to electronic microcircuitry. Vacuum deposition is used and by controlling the materials and deposition processes, the strain gage properties are varied to produce the desired characteristics. In these transducers a metal substrate provides the desired mechanical properties, then a ceramic film is vacuum deposited on the metal to provide the required electrical insulation. Four strain gages are then vacuum deposited on the insulator. They are connected into a bridge configuration by vacuum deposited leads. The multiple evaporations are made during a single pumpdown period using multiple sources and substrate masks. The lead wire is attached to the film by the microcircuit techniques of microwelding or thermocompression bonding of noble metal wire. The lead wire attachment is made directly to the film. The sensing element can have any configuration as the strain gages can be deposited on diaphragms, beams, columns, and other elements. The strain gage pattern can be designed to optimize the characteristics of a specific sensing element. A common pressure element is the flat plate diaphragm with four strain gages arranged in a bridge configuration.

As a force is applied normally to the plate, the plate is deformed causing the strain sensitive elements to elongate and increase in resistance. The change in resistance is proportional to the change in length and alters the balance of the bridge to produce an electrical output.

Diffused semiconductor strain gages are based on the technology of integrated circuits. A pressure sensing element is produced by diffusing a four arm strain gage bridge into the surface of a single crystal silicon diaphragm whose diameter and thickness is a function of the pressure range and application. The silicon has the proper pressure sensing mechanical properties being elastic and free from hysteresis.

These sensors have high gage factors and can produce relatively high outputs at low strain levels. The transducer is encapsulated into a transducer or transmitter housing using IC manufacturing techniques such as electrostatic or thermal compression bonding or electron beam welding.

The operational ratings such as shock, vibration, and overload are typical of other high quality microcircuit devices. Combined linearity and hysteresis effects are less than 0.06%. The low mass of the silicon diaphragm gives a fast response with minimum sensitivity to accelerations due to shock and vibration. In applications where the pressure media are not compatible with silicon, stainless steel, hastelloy, or other materials are used for isolating diaphragms.

The full scale output of a four element strain gage bridge for a pressure transducer using metal wire or foil gages is 50 to 60 millivolts for bonded gages and 60 to 80 millivolts for the unbonded gages, using 10 volt excitation. Compensating and adjusting resistors can reduce the output to about 20 to 30 millivolts for bonded and 30 to 40 millivolts for the unbonded gages. These resistors are used for zero and balance adjustments, full scale adjustment, thermal zero shift compensation, thermal sensitivity shift compensation, and also shunt calibration.

Semiconductor strain gage transducers provide a higher output of 200–400 millivolts for 5–7 mA excitation. Many transducers place a limitation on the excitation voltage. They produce a full scale output using internal circuitry to reduce the voltage to the level required for the proper bridge operation. They also use a constant current to provide some thermal compensation.

To provide TTL output levels, an amplifier must be used for all metal and many semiconductor strain gage transducers. Many integrated circuit transducers use integral amplifiers, which can also include some thermal compensation functions.

Strain gage transducers provide fast response times, good resolution, minimal mechanical motion, and good accuracy. Predictable compensation for temperature effects, low source impedance, and freedom from acceleration effects for the bonded type are other advantages. Obtaining zero output at zero pressure due to bridge unbalances can be a problem as well as the high vibration errors when unbonded types are used, especially for ranges lower than 15 PSI. The low output levels can cause noise problems and usually isolation of excitation ground from output ground is required in addition to other signal-conditioning requirements.

Temperature Compensation and Error Reduction

Ambient temperature variation is one of the major sources of errors in many pressure transducers. These errors can be of two kinds. The zero setting may shift with temperature. This could be due to unequal mechanical expansion of the instrument members. Also the calibration factor, span, or sensitivity, as it may be called, can change with the ambient temperature. This may be caused by the change in elasticity or spring constant of the gage members. Many metals have a temperature coefficient for Young's modulus of elasticity of about $-0.0007/$°C. To minimize the zero error with temperature, the differential expansion of any mechanical components must nearly balance, otherwise the armature of the transducer can be pulled off-center and the range of span becomes incorrect even if the zero shift is corrected. The differential expansion must be reduced until the total change over the ambient temperature range is a small fraction of full scale, then the remaining error can be compensated for electrically, as shown in Figure 7-12. Other variations are possible using circuits with compensating properties.

The compensating resistor should be at the same temperature as the transducer and preferably inside the case. The compensating resistors must not dissipate any appreciable heat unless an allowance is made for the increased resistance from this cause. The temperature coefficients of the compensating resistor's material can be determined from a test of the actual type used. Several factors affect the resistance of a resistor and the only positive method of obtaining the coefficient is to allow the test to duplicate the conditions of service.

The selection of the appropriate transducer is the first and most important step in obtaining accurate measurements. The transducer has the critical function of transforming the measurand from a physical quantity to a proportional signal. The accuracy of the final data can never be any better than the transducer capability.

In some cases, the sum of the individual accuracy factors can result in a total error which exceeds the desired system accuracy. It may be necessary to reduce

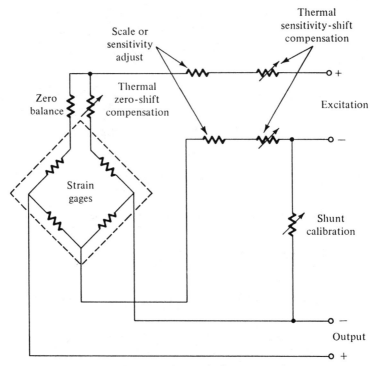

FIGURE 7-12   Strain Gage Bridge with Compensation.

the total error to the required accuracy limits by one or more of the following methods:

1. Specifying the accuracy over the total error band rather than on an individual parameter basis.
2. In system calibration using techniques with corrections performed by data reduction.
3. Monitoring the environmental changes and correcting the data accordingly.
4. Artificially controlling the transducer environment to minimize these errors.

Many of the individual errors may be random. The total error will usually be less than the algebraic addition of these individual errors. The use of an error band tends to simulate actual conditions because only the total accuracy deviation is considered, rather than the individual factors contributing to it. Some individual errors are predictable and can be calibrated out. When the system is calibrated, the calibration data can be used to correct the recorded data.

Environmental errors can be corrected by data reduction methods if the environmental effects are recorded simultaneously with the data. Then the data can be corrected using the environmental characteristics of the transducer.

These techniques may provide a significant increase in system accuracy. They are useful when individual errors are specified and when the error band specification is not used.

Another technique of improving the system accuracy is to control artificially the environment of the transducer. If the local environment can be kept unchanged, these errors will be reduced to zero. This type of control may require physically moving the transducer or providing the required isolation from the environment by a heater, isolation, or similar means.

The force balance technique is used in some transducers to reduce errors. In these pressure sensing systems, the output of the sensing device is fed to an amplifier, which in turn feeds back a restoring force equal to the applied variable. The sensing device is thus returned to the position it occupied prior to the application of the force being measured. The magnitude of the power fed back from the amplifier determines the output of the system. The repositioning of the sensing head is accomplished by motors or electromagnetic force. The force being measured can be sensed by almost any of the principles discussed. Many use a capacitive or differential transformer sensing element.

In some force systems optical techniques are used. The pressure sensor in one manometer system uses a hollow spiral quartz tube with two wire wound coils suspended from it. The coils are in a permanent magnetic field and a curved mirror welded to the tube transmits the tube motion to an optical detector.

Many other high accuracy systems for pressure measurement rely on the accuracy and readability of digital readouts. Many use displacement producing sensors such as bellows or diaphragms as the primary sensing device with analog-to-digital converters.

One technique uses a bourdon coil pressure sensing element and an optical encoder (Figure 7-13). These may be mounted on a common shaft which rotates in proportion to the applied pressure. A direct binary BCD or cyclic binary output is provided in serial or parallel form to displays. The digital format is compatible to microcomputers for error reduction processing, monitoring, and recording. Digital pressure instruments form an important part of pressure measuring equipment and their use is expected to grow. The high performance instruments which are produced are used to calibrate other pressure transducers or gages. Digital instruments are available which provide the pressure display in almost any pressure units (Figure 7-14).

FIGURE 7-13   This digital pressure transducer uses a bourdon tube which is tracked by an optical encoder. The output is parallel cyclic binary. (*Courtesy Siltran Digital.*)

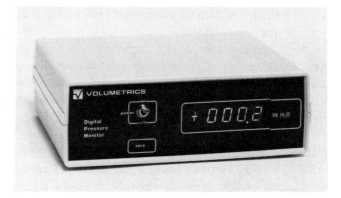

FIGURE 7-14   This digital pressure monitor can be calibrated to display pounds per square inch, inches of water and inches or millimeters of mercury. A parallel BCD output provides a simple interface to a microprocessor system. (*Courtesy Volumetrics.*)

EXERCISES

7-1. What pressure is obtained if the reference side of a sensor is evacuated?

7-2. Describe the advantages and other characteristics of a potentiometer pressure transducer.

7-3. What are the advantages of a capacitive pressure transducer over a potentiometer transducer in a microprocessor control system?

7-4. What are the advantages of a semiconductor strain gage over a metallic foil strain gage in a high volume, low cost product application?

7-5. What are the characteristics of piezoelectric pressure sensors that make them more suitable than other types in a low cost, low volume application? Discuss some typical applications.

7-6. Determine the pressure of liquid oxygen flowing through a high pressure pipe. Required accuracy is ±1%, range is 200 to 800 psi, time constant is 0.8 second; a proof pressure of 1,600 psig is required; flow rate is 80 to 150 gpm. Describe a monitor and alarm system for a series of 18 similar stations with outputs to a digital modem located at a maximum distance of 500 feet. Design the system for maximum safety and reliability.

7-7. A 300 ohm strain gage with a gage factor of 4.5 is to be measured using a 50 ohm cable in a bridge circuit having matched gage resistors. What current will flow for a 1,800 uin./in. strain when 10 V is applied to the circuit?

7-8. The circuit for a transducer is a complete and balanced Wheatstone bridge. The power supply is to be either alternating current or a carrier system. Discuss the choices and considerations involved.

7-9. Compare the advantages and disadvantages of the bonded and unbonded strain gages.

7-10. Describe the operation of a diffused diaphragm strain gage pressure transducer.

7-11. If a pressure of 6.0 inches of water acts on an effective area of 3.0 inches of a diaphragm, calculate the force, in pounds, acting on the push rod member.

7-12. Discuss the main physical differences and application of a foil and thin film strain gage elements.

7-13. If the approximate effective area of a diaphragm is given by $A = (R_a + R_b)/2^2$, calculate the effective force acting on a central push rod if a differential pressure of 3.0 psi is applied to the unit and $R_a = 1$ inch and $R_b = 1\frac{1}{4}$ inch.

7-14. What are some of the factors which determine the deflection range of strain gage pressure transducers?

7-15. Evaluate the important physical characteristics of crystals for pressure transduction.

7-16. The error band is defined as the band of allowable deviations of output values from a specified reference line or curve. These deviations are attributable to the transducer. How can the error band be derived?

7-17. A sensitivity adjustment controls the ratio of output signal to excitation voltage per unit measured. Describe some ways to accomplish this.

7-18. Describe the operation of the force balance technique.

7-19. Select a pressure transducer with the following advantages:
   a. High output
   b. Static or dynamic measurements can be made

    c. High accuracy

    d. High stability

    e. High resolution

    Justify your answer.

7-20. State some techniques for temperature compensation in strain gage pressure transducers.

8

Flow Sensors

DIFFERENTIAL PRESSURE FLOW TRANSDUCERS

Almost every type of process requires the measurement of flow. The media may be a fuel, gas, water, air, or processed liquids or gases which form the product itself (Figure 8-1). To specify a flow sensor for such a measurement, the designer must be familiar with the variety of devices available.

Some flow sensors respond directly to the flow rate of a fluid fall. These sensors can be grouped into the following three general types:

1. A restriction in a pipe or duct is used to produce a differential pressure which is proportional to the flow rate. This differential pressure is then measured using a pressure transducer system calibrated in flow units.
2. A mechanical member is used to sense the flow of a moving fluid by rotation or deflection in a tapered tube.
3. The fluid's physical characteristics are used to sense the flow.

The sensing elements in the first group are known as head meters since the differential pressure between two sensing ports may be equated to the head or the height of the liquid column. These elements are characterized by a constant area of flow passage. Venturi tubes, flow nozzles, orifice plates, and pitot tube are examples of this group. The pressure transducers used to measure flow with these sensing elements are described in Chapter 7.

The head type of flow measurement is simple, reliable, inexpensive, and is accurate. Since it is a direct measurement of flow rate, it is suited to automatic flow control systems.

Orifice plates have been the most popular restriction used for $\triangle P$ measurements in both liquid and gases. They are mounted in the flow line, perpendicular

FIGURE 8-1 This digital flow monitor is designed for the measurement of gas flow, a BCD output can be used to interface it to the microprocessor system. (*Courtesy Volumetrics.*)

to the flow. The flat orifice plate has an opening smaller than the inside diameter of the piping (Figure 8-2). Taps at two points in the line measure the pressure upstream from the orifice and downstream at the point of lowest pressure. The pressure difference between the two taps measures the rate of flow. They are not practical for slurries and dirty fluids because of accumulation and wear at the orifice. The sizing of an orifice is straightforward using the proper tables which may include compressibility factors, Reynolds curves, and thermal expansion factors. However, to have an accurate sensing unit for industrial applications, it is usual to depend on the orifice manufacturer to furnish the proper orifice from specifications for the applications.

The flow rangeability of a particular orifice plate is low, approximately of 3 or 4 to 1; range changes are effected by changing the plate size. There are limitations as to the minimum line size and minimum flow rates for accuracy when orifice plates are used. The quadrant edge orifice plate can be for low flow rates or for flows involving viscous fluids. The concentric thin plate square edge orifice has the widest use. Eccentric and segmental plates are used to a lesser extent. Accuracy of 1% is possible with careful installation.

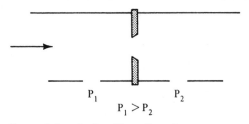

FIGURE 8-2 Orifice Plate Flow Sensor.

The advantages of orifice plates include:

1. A large amount of available coefficient data.
2. Low cost.
3. Capacity changes by switching plate size.
4. Applicable to wide range of temperatures and pressures.

The disadvantages are:

1. A straight approach of piping is required.
2. Upstream edge wear causes inaccuracies.
3. Low rangeability for a particular plate.
4. Unsuitable for slurry applications.
5. Unsuitable for low pressure systems.

A type of flow sensing element almost as well known as the orifice plate is the Venturi tube. The Venturi tube as shown in Figure 8-3 combines a short constricted portion or throat between two tapered sections and is usually installed between flanges in the line. It tends to accelerate the fluid and lower its pressure. Venturi tubes are usually used for liquids. They may be used for gas flow measurement when pressure recovery characteristics allow. While an orifice is a restriction at a point in the line, the Venturi spreads the restriction over a longer distance. Flow through the center section is at a higher velocity than at the end sections. Taps at the entrance and throat measure the pressure difference. An accuracy of 1% is normally achieved in many installations. Advantages of Venturi tubes include:

1. The use of fluids containing suspended solids.
2. Pressure recovery is better than for orifice plates.
3. High capacity.
4. Wear resistance due to abrasion.

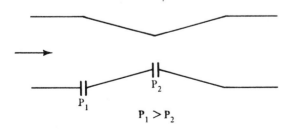

FIGURE 8-3   Venturi Tube Flow Sensor.

The disadvantages are:

1. Expensive.
2. Low rangeability of 3 to 4 to 1, and difficult to modify for major changes in flow range.
3. Produces a lower differential than an orifice for the same flow and throat size.
4. Larger sizes require considerable room.

Other tubes similar to the Venturi are the Dall and Foster.

The flow nozzle is a curved nozzle as shown in Figure 8-4, which is inserted between two sections of piping. The curvature of the contour must approach a gradual tangency to the throat without a sudden change of contour. The outlet end of the nozzle is beveled or recessed.

The advantages of flow nozzles are:

1. Suitable for fluids containing minor amounts of solids.
2. Capacity of 65% greater than for orifice plates of the same diameter.
3. More rugged than orifice plates and can be used for high velocity measurement.
4. Pressure recovery better than orifice plates.
5. More easily installed than Venturi tubes.
6. Not as susceptible to wear as orifice plates.
7. Produces higher differential pressures than Venturi tubes.

The pitot tube has limited importance in industrial systems. It is an effective laboratory tool and is sometimes used for spot checking flows. A typical tube is shown in Figure 8-5.

The tube has two pressure openings. One passage, which faces into the flowing fluid, is at the end of the tube towards the center of the pipe. This opening

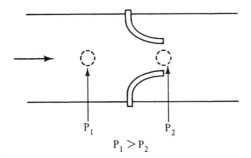

$P_1 > P_2$

FIGURE 8-4   Flow Nozzle Sensor.

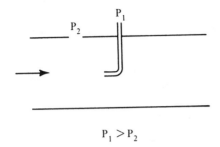

$$P_1 > P_2$$

FIGURE 8-5   Pitot Tube Flow Sensor.

intercepts a small portion of the flow and mainly reacts to the impact pressure. The other opening is perpendicular to the axis of flow and it measures the static pressure.

Other differential pressure sections are in use such as elbows, which can be useful for problem viscous fluids, and loops, which tend to react well to pulsating flows and slurries.

VARIABLE AREA METERS

The variable area meters may use a float in a tapered section of tubing (called *rotameters*), a spring restrained plug, or a spring restrained vane. The displacement of these elements causes the area of the flow passage to vary while the differential pressure or head remains constant. The displacement is measured to provide an output proportional to flow rate. Rotating elements include the turbine installed in a pipe section and the propeller installed in a flow stream. Both rotors turn at an angular speed proportional to the flow rate.

Rotameters measure a continuous stream flowing through a tapered vertical tube containing a float as shown in Figure 8-6. The float moves up or down in response to the rate of upward flow and stabilizes for a constant flow. Density and

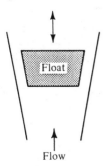

FIGURE 8-6   Rotameter (Variable Area) Flow Sensor.

viscosity of the liquid or gas being metered can affect float position. The area around the float increases as the float rises in proportion to the flow rate. If the rotameter tube is glass, the flow may be read by optical methods. With opaque tubes, magnetic techniques are used to monitor the float position.

Rotameter rangeability is 12 to 1 and accuracy to 1% is possible. Different sizes allow capacities from .6 cc to 5,000 gallons per minute. Temperatures to 1,000°F and pressures to 2,500 psi are allowed.

Rotameters produce a low pressure drop and allow some dirt in the fluids giving an average reading for pulsating flows. Their main drawback is vertical installation along with high cost for sizes larger than 2 inches.

The deflections of vanes and supports due to the force of flow are transducer by displacement sensors or strain gage bridges in a limited number of designs. The vane in these flowmeters is usually installed perpendicular to the flow. A wedge shaped vane used with strain gages is sometimes used for measuring air turbulence.

TURBINE FLOWMETERS

The turbine flowmeter uses the movement of the fluid to turn a turbine wheel (Figure 8-7). The speed of the rotor is a function of the flow rate.

Turbine flowmeters output the flow information as a precise number of pulses which depend on the volume of the fluid displaced between the rotor blades in a unit of time. The relationship is linear within the transducer limits for flow rate and viscocity. With a linearity of $\pm 0.5\%$ of rate, a repeatability of 0.02% of rate can be achieved.

Bearing friction, fluid, and magnetic drag and swirl in the fluid stream can cause errors in these transducers. These error causes can be minimized in the me-

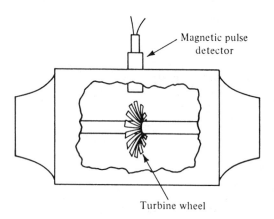

Magnetic pulse detector

FIGURE 8-7   Turbine Flowmeter.   Turbine wheel

chanical design. The swirl affects can be virtually eliminated by flow straighteners. Magnetic drag is not a concern except in miniature transducers. Thrust bearing friction is reduced using the hydrodynamic forces to balance the axial loads on the rotor.

The transducer case is made of a nonmagnetic metal and it is usually threaded for the detector coil. Since the amplitude of the output voltage from the coil is a function of the distance between the pole piece and turbine blade tips, the threaded coil can then be adjusted by turning.

Most turbine flowmeters use an electromagnetic oil of the permanent magnet type. The turbine blades are then made of a ferroelectric material. An accuracy of 0.5% can be achieved in the linear portion of the coefficient curve.

As each blade passes the coil's pole piece, a voltage is induced in the coil. The transducer output is an ac voltage with a frequency proportional to the flow rate. The number of cycles per revolution is given by the number of blades on the rotor. The rotor blades may be shaped to produce a sinusoidal output voltage with little harmonic distortion. The transducer output frequency band for a range of flow rates can coincide with a selected subcarrier frequency band for a telemetry system by selecting the number of rotor blades and their pitch.

This type of meter has become widely used in the process industries. Turbine flowmeters are selected because of their wide range and excellent performance. Turbine flowmeters are also used for gas flow measurement. Turbine meters should be selected with a capacity from 30 to 50% above the expected maximum flowrate. Operating the meter below maximum capacity provides a greater reliability. Turbine flowmeters are high performance instruments. Each meter is individually calibrated for one or a number of fluid conditions before shipment. They are among the most accurate flow monitoring devices in common use.

Turbine meters have been used successfully for flow rates from fractions of gallons per minute to tens of thousands of gallons per minute, with pressures to 7,000 psi, and temperatures from −400°F to 1,000°F. They are satisfactory with any reasonably clean fluid and have a rangeability of 20 to 1. A turbine meter installation complete with pickup electronics can be more expensive compared to a head meters type of installation.

The advantages of turbine meters include:

1. High pressure, high flow measurement capabilities over a wide temperature range.
2. Good flow range.
3. High accuracy and repeatability.
4. Short piping approaches required.
5. Good response times to flow changes.
6. Can be converted to mass flow with compensating hardware or software.

The disadvantages are:

1. Abrasive materials may wear out bearings.
2. High viscosities may affect measurements.
3. Pulsating flow or water hammer can cause damage.

FLUID CHARACTERISTICS SENSORS

A number of flow sensors use the fluid characteristics to measure flow. One type of sensor uses a heated wire in the form of a hot wire anemometer. The heated wire will transfer more of its heat as the flow velocity of the surrounding fluid increases. The resultant cooling of the wire causes its resistance to decrease. Another type of sensor uses a fluid containing a small amount of radioactivity to cause an increase in ionization current for an increase in flow velocity. When a slightly conductive fluid flows through a transverse magnetic field, it will produce an increasing voltage for an increasing flow velocity. If the boundary layer of a moving fluid is heated by a small heating element, the convective heat transfer to a temperature sensor located downstream from the heater will increase with an increase in flow velocity. These transducers which convert flow rate into a change of resistance all use the fluid characteristics. The methods used to obtain the resistive element response can be grouped into three transducer types: hot wire anemometers which are used primarily for air flow measurements, the thermal and boundary layer flowmeters, and the oscillating vortex flowmeters.

Hot wire anemometers use a wire element normal to the flowing stream that is heated electrically. Cooling due to flow changes affects the resistance of the wire. These resistance changes can be calibrated as a function of the flow velocity.

Some types maintain the wire temperature constant and measure the current required to maintain this temperature. The measured current is then a function of the flow.

Hot wire anemometers can measure mass flow so long as the product of the thermal conductivity, specific heat, and density remains constant. This is the case for many gases at low pressure. This type of transducer has the greatest application in the measurement of low flow rates of gases. They are also useful for gas velocity determinations. Anemometers are sensitive to flow changes but can be expensive.

The measurement of flow rate using heat transfer methods can be obtained by two types of instruments: the thermal flowmeter and the boundary layer flowmeter. Both types use an electrical heater to increase the heat of a portion of the moving fluid as shown in Figure 8-8. Temperature sensors, sometimes resistive, but usually thermoelectric, then measure the temperature of the fluid upstream and downstream from the heater. The heat transferred between the heater and the downstream sensor is obtained either by measuring the temperature differential

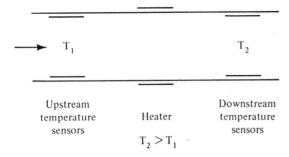

FIGURE 8-8   Heat-Transfer Technique for Flow Measurement.

between the two sensor areas at constant heat input or by measuring the change in heater current to keep this temperature differential constant.

The first thermal flowmeters used a heater immersed in the fluid. Since the entire core of the fluid was heated, the heater power required for some large designs was in the kilowatt region. This power requirement was a severe limiting factor in the application of these thermal flowmeters.

This disadvantage was overcome with the boundary layer flowmeter which heats only the thin boundary layer of the fluid adjacent to the pipe wall. The temperature sensors are flush with the inside pipe surface, and the heater is mounted around the outside of the pipe or embedded inside the wall. The heat input varies with mass flow rate and temperature differential; however, the relationship is more complex than for the internally heated thermal flowmeter. Heater power requirements are usually below 50 W in boundary layer instruments.

One type of thermal flowmeter uses a constant ratio bypass section to sample the gas flow in the main duct for flow ranges between 0 to 1.5 and 0 to 200 standard cfm. Ranges lower than these can be measured without the bypass section. The full scale output from the thermocouples is between 2 and 5 mV. Although this is not a boundary layer device, it has a relatively low power consumption. Power requirements for the transducer and readout are 15 W at 115 V ac, about 5 W are used by the transducer itself. A portable leak rate monitor which uses similar principles is shown in Figure 8-9.

The forced vortex meter, which is also known as the processing vortex meter or swirl meter, uses the vortex precession principle. Vortex precession occurs when a rotating body of fluid enters an enlargement as shown in Figure 8-10. Under certain conditions, a flowing fluid will oscillate with a well defined frequency that is proportional to flow. The output is a train of pulses whose frequency is proportional to flow rate. This train of pulses is created in the fluid by generating a hydrodynamic precession of the flow. Swirling can be imparted to the fluid by a fixed set of blades. Downstream of these blades is a Venturi-like contraction and expansion of the flow passage. In the region where the cross-sec-

FIGURE 8-9 This leak rate monitor uses three thermal mass flow meters which allow automatic corrections for temperature and pressure. (*Courtesy Volumetrics.*)

tion enlarges, the swirling flow precesses; it leaves the axial path of the meter center line and takes a helical path. The frequency of this precession is proportional to flow rate and is detected by a sensor. The oscillations of the fluid create variations in fluid temperature seen by either a platinum film sensor or a thermistor type resistive temperature sensor. The deswirl blades serve to straighten the flow as it leaves the flowmeter. Depending on operating conditions, the flowmeter has a possible linear range of 100:1 within $\pm 1\%$ of rate for an output of 10^{HZ} to 2 KHZ.

The vortex shedding flowmeter is a similar design that uses hydrodynamics. When a nonstreamlined obstruction is placed in a pipe, fluid does not flow smoothly. Eddies or vortices are formed which grow larger and are eventually detached from the obstruction. This detachment is called shedding and it occurs alternately at each side of the obstruction. The vortices form trails downstream of

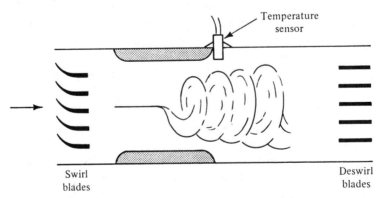

FIGURE 8-10 Forced Vortex Flowmeter.

the obstruction. The shedding frequency of the vortices is directly proportional to the flow rate.

The vortex shedding flowmeter has some features similar to other flowmeters. The obstruction is analogous to an orifice plate, and the output signal is similar to a turbine type flowmeter. The vortex shedding flowmeter senses flow velocity, thus is a volumetric measuring device. It can be a mass flowmeter if the fluid density is known. It has a wide range, universal calibration, and no moving parts. The vortex shedding flowmeter cannot be used in laminar flow and the Reynolds number of the fluid must be at least 3000. The vortex shedding phenomenon is applicable to both liquid and gaseous fluids.

Some types of sensors use radioisotopes and require the detection of nuclear radiation. One type uses a source mounted on the outside of the pipe upstream from the dectector (Figure 8-11). Neutrons from the source collide with the moving fluid and cause particle and electromagnetic radiation to be emitted. Most of this emitted radiation occurs at the location of the source, but some is emitted as the fluid passes the detector. The detected counts of radiation are a function of the flow rate of the fluid.

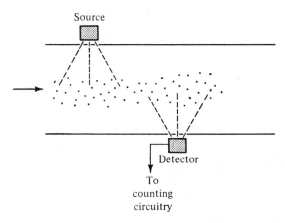

FIGURE 8-11   Nucleonic Flow Measurement.

Another type of nuclear flow sensing obtains the count by adding small amounts of a radioactive trace element to the fluid. These nucleonic flowmeters offer no obstructions to the fluid path. They are useful in the measurement of difficult fluids such as multiphase variable composition fluids, slurries, and suspensions.

ELECTROMAGNETIC FLOWMETERS

The electromagnetic flowmeters make use of Faraday's Law which states that a relative motion, at right angles between a conductor and a magnetic field, will in-

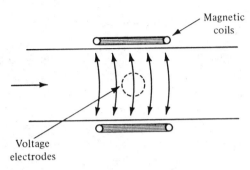

FIGURE 8-12   Electromagnetic Flowmeter.

duce a voltage in a conductor. This voltage is a function of the relative velocity of the conductor and the magnetic field.

An electromagnetic flowmeter as in Figure 8-12 is made from nonmagnetic materials and uses a conductive liquid. Magnetic coils are used to provide the magnetic field. As the liquid moves through the magnetic field, a voltage is generated proportional to the flow rate. Since electromagnetic flowmeters are calibrated to sense liquid velocity, the entire cross-sectional area of the pipe must be full. No gas bubbles should be carried by the liquid as they can cause measurement errors.

The electromagnetic meter's output ranges from microvolts to millivolts. The proper installation and grounding are required for accurate flow measurements.

These flowmeters are not affected by changes in liquid density or viscosity. Turbulence of the liquid and variations in piping also have limited effects. The conductivity of the fluid must be greater than about 10^{-8} mho/cm^3. Mixed phase fluids can cause conductivity variations and major measurement errors.

DC coil excitation should not be used since it may cause electrolysis problems. The linearity of output voltage with flow rate is a useful advantage in many applications.

PIEZOELECTRIC FLOWMETERS

The piezoelectric flowmeter was originally developed to measure fluid flows in aerospace applications during the 1950s. One type has two transducer pairs which establish upstream and downstream sonic paths diagonally across the fluid as shown in Figure 8-13. The difference in propagation velocity (the doppler effect) between the two paths can be calibrated as a function of the flow rate. A flow section is used to secure the transducers as shown in Figure 8-14.

Sonic and ultrasonic flowmeters can be used with liquids or sonically con-

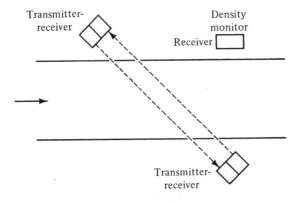

FIGURE 8-13   Doppler Ultrasonic Flowmeter.

ductive slurries through either closed pipes or open channels and are useful in water and waste plants, industrial process plants, power stations, and other industries. They can monitor bulk material flow on conveyors, chutes, and other machines.

Ultrasonic systems using level measurement, Doppler, or back-scatter Doppler have one common component: the transducer, which is perhaps the most critical part of the system. In level measurement systems, flow is monitored by measuring the rise and fall of a liquid flowing in an open flume using the sound

FIGURE 8-14   The flow section in this ultrasonic flowmeter contains a pair of sonic transducers to transmit and receive the sound waves. Pulse outputs can be used to interface to the microprocessor system. (*Courtesy Manning Technologies.*)

FIGURE 8-15   This ultrasonic flowmeter uses coupled transducers and is calibrated to measure energy flow in BTUs. (*Courtesy Controlotron Corporation.*)

transmission in air. Such transducers operate in the low ultrasonic portion of the spectrum.

Doppler systems are higher frequency devices, operating in the range of 100 to 1500 kHz. In some systems the transducers are coupled to the wall of the conduit through which the fluid flows as shown in Figure 8-15. Ultrasonic energy is transmitted through the wall and fluid at an oblique angle to the flow. A second transducer opposite the first receives the signal. Another transducer may be used in some systems to monitor the density variation as shown in Figure 8-13.

WEIRS AND FLUMES

Weirs and flumes are open channel devices which develop a liquid head used to measure flow rate. Often used for water flow in irrigation, waste, and sewage systems, they are used where flow is handled in open channels or in pipes and conduits that generally are not completely filled with liquid. Conditions must provide a tranquil surface for the head measurement. The liquid approach must be a straight run.

The weir is essentially a dam with a notched opening in the top through which the liquid flows. It is one of the simplest methods and it can be accurate for measuring the flow of water under the proper conditions. The rate of flow is determined by measuring the head of water above the lowest point of the weir opening through which the water flows. This height can be measured by a float installed in a box, called a stilling well, which is a part of the total weir structure. The float is located here so that it is not disturbed by the velocity of the flow, or by any turbulence in the stream. The V notch offers the widest range for a single size, since the small opening in the bottom of the V can accommodate small flows and the upper portion larger flows. It has the greatest head loss because of its shape. The rectangular notch is the oldest type and probably the most common because of its ease of construction. The trapezoid notch, also called the *Cipolletti*, is designed so that the trapezoidal sides produce a flow correction to allow the flow to be proportional to the length of the weir crest.

A weir must be cleaned periodically if the liquid being measured contains entrained material. Deposited materials can produce errors in the flow rate.

Parshall flumes are self-cleaning and operate with a small loss of head. The loss is about one-fourth that in weirs. These flumes can be used where sand, grit, or other heavy solids are present in the stream. Parshall flumes are designed for conditions in which flow velocities are moderate. They should be located so they are not affected by bends or other objects that can cause eddies, waves, or uneven flow patterns.

For the rectangular and trapezoidal weirs and the Parshall flumes, flow is proportional to the three-halves power of the measured head. For V notch weirs, flow is proportional to the five-halves power of the head. With the proper instrumentation, this can provide a wide range of flow measurement: 20 to 1 for rectangular and trapezoidal weirs and Parshall flumes, and up to 40 to 1 for the V notch weirs.

The flow rate through weirs and flumes measured by the level or head gives a direct relation to the volume flow. No correction is needed for the liquid density. Accuracy to 2% is attainable. Flow rates from a few gallons per minute to millions of gallons per minute have been measured.

POSITIVE DISPLACEMENT METERS

Positive displacement meters split flow into known volumes, based on the physical dimensions of the meter, and count them. They are mechanical meters with one or more moving parts to physically separate the fluid into increments. Energy to drive these parts comes from the flow stream and produces a pressure loss between the inlet and the outlet. Accuracy depends upon the clearances between the moving and stationary parts. Meter accuracy tends to increase as size increases. Positive displacement meters with well defined tolerances are within ± 1% over flow ranges to 20:1. They are used in batch processing and mixing or blending systems.

Since positive displacement meters require small clearances upon which their accuracy depends, the metered liquids must be clean.

Positive displacement meters do not require electric power or an air supply. They may be damaged by exceeding their capacity. Some types use electrical drive motors to deliver a given volume and electric pneumatics, or hydraulic power may be used for control purposes.

Piston pumps act as metering pumps to inject an exact amount of fluid into a flow line or a collecting vessel. The piston pump is generally the reciprocating type where a piston or plunger delivers a fixed volume on each stroke.

In the simple, single action unit, fluid is drawn into the piston cavity through an inlet valve. As the piston is moved back fluid is discharged into the flow line or vessel as the piston moves forward and fills the cavity. As the piston moves back and forth, it delivers a fixed volume in a pulsating flow.

Piston meters can be used to deliver controlled volumes at high pressures. If the pulsating flow cannot be tolerated, it can be averaged out using reservoirs. The amount of flow delivered may be changed by varying the length of the piston stroke or by changing the pumping speed. The stroke adjustment can be varied manually or automatically depending on the pump and the application.

The flow sensing elements discussed thus far are used for the measurement of volumetric flow rate. Mass circuit rate can be calculated from this measurement and a simultaneous measurement of density. A microcomputer may use density and volumetric flow rate inputs to provide an output of mass flow rate. Special sensing elements have been developed whose output is directly proportional to mass flow rate. Examples of such transducers are considered now.

MASS FLOWMETERS

A number of transducer designs are available for the direct measurement of mass flow to avoid the separate density and volumetric flow rate measurements necessary to compute mass flow rate.

The twin turbine mass flowmeter uses two bladed rotors with different blade angles, coupled by a spring to allow relative angular motion between them. As a result of the blade angle difference, the two turbines tend to rotate at different speeds but are restrained by the spring coupling. They assume an angular displacement with respect to each other which is proportional to flow momentum. The two rotors, considered as a unit, function as a volumetric flowmeter. The turbine speed is proportional to the average flow velocity. The angular displacement is measured as the phase angle between the outputs of the two transducer coils. The time interval in which the phase angle moves from a reference point is a measure of the mass flow rate. Digital counting circuitry is used to interpret the dual output.

Another twin rotor type uses the angular momentum of mass flow. This model incorporates two similar axial flow rotors. The upstream rotor is driven at a constant low speed by a synchronous motor. This rotor imparts a constant angular momentum to the fluid. The downstream rotor removes all angular momentum from the fluid. In so doing, a torque is exerted in accordance with Newton's Second Law. This torque is linearly proportional to the mass flow rate. The angular displacement of the turbine is proportional to the torque from a spring restraint. A reluctive transducer converts the angular displacement into an ac output voltage.

The gyroscopic mass flowmeter is based on the gyroscopic vector relationship. As shown in Figure 8-16, the liquid is caused to flow around a loop of pipe which is in a plane perpendicular to the input line. Following one cycle around the loop, the fluid is returned to the input axis. During the rotation, the fluid develops angular momentum similar to the rotor of a gyroscope. The loop is vibrated through a small angle about an axis in the plane of the loop. The vibration results in an alternating gyro-coupled torque about the orthogonal axis. The peak amplitude of the torque is proportional to mass flow rate. The torque acts against restraints to produce an angular displacement about the torque axis. Sensors convert the displacement to a proportional electric signal. The peak signal is proportional to the mass flow rate.

Flow monitoring will continue to expand because of the emphasis on energy conservation. Many manufacturers have lowered energy consumption with improved instrumentation. In some cases, power requirements have been drastically cut. New applications of physical principles, such as hydrodynamic oscillation, are being applied to new flowmeters to increase reliability, extend applications, and reduce costs.

One new meter uses a self-excited oscillating prism to produce a signal over the range from 4 to 200 Hz. The output is linear with flow and is relatively unaffected by fouling within the meter. A prism-shaped oscillator, with a pendulum support, is excited by the flowing fluid. The frequency of the oscillation is linearly proportional to the flow. A magnetic inductive system picks up the vibration and

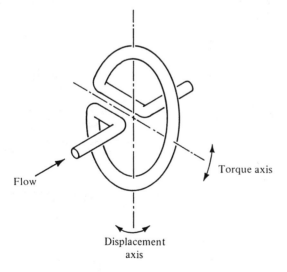

FIGURE 8-16 Gyroscopic Mass Flowmeter Principle.

converts it into an electrical signal in one of three different ways: by a frequency converter into an analog current proportional to the flow, into a digital signal by counting the number of oscillations in a given time period, or into a count representing the total amount of fluid that has passed through the pipe.

In one type of fluidic flowmeter the meter body acts as a fluidic oscillator with a frequency that is linear with volume flow rate. The oscillations are detected by a sensor and amplified to provide a digital pulse output. The meter body is shaped so that the flowing stream attaches itself to one of the side walls using the Coanda effect. A portion of the flow is diverted back through a feedback passage to a control port. The feedback flow acts on the main flow and diverts the main flow to the opposite wall where the same feedback action is repeated. This results in continuous self-induced oscillations. The oscillation frequency is a linear function of the liquid velocity.

Accurate measurements under the conditions that are also compatible with the control system are always important. Flowmeters exist for all situations for measurement under almost any condition. Most of these produce an analog output that must be converted to a digital format to interface the meter to a digital system. Some types of meters, such as the turbine and vortex, can be used to produce a digital frequency output.

EXERCISES

8-1. Discuss how the Venturi tube can be used for liquid flow measurement with a monitoring system for a local microprocessor alarm control system located within 30 feet of the pressure duct.

8-2. Compare the operation, construction, advantages, and disadvantages of the Venturi tube and orifice plate.

8-3. List the merits and limitations of the industrial type of Pitot tube.

8-4. Water is flowing down a vertical pipe 8 feet long. The pipe has a diameter of 4 inches. If the quantity of water is 380 gallons/min. nominal with expected variations of ±30, describe a flow monitoring and pumping control system using a low cost microprocessor. Select a transducer as well as a microprocessor.

8-5. Describe the operations required to measure liquid flow in an open channel by means of a simple microprocessor system. Select a microprocessor and show its operation with a flow chart.

8-6. Discuss the importance of a temperature sensor with a short time constant in the oscillating vortex flowmeter. Explain what methods you could use to determine the average velocity of airflow through a large circular pipe and a large rectangular air duct.

8-7. Describe the operation of a mass flowmeter that uses gyroscopic motion.

8-8. What is the function of the downstream rotor in a twin turbine flowmeter?

8-9. Costs are one of the major items to be considered in selecting a flow element. Select several types of flow meters and discuss how line size as a measure of capacity can form a basis for making cost comparisons.

8-10. The various head meters possess unique features which must be compared when a decision must be made as to the type to be installed. Discuss these features as related to the application criteria. Summarize your results in a head meter selection chart.

8-11. Discuss the use of each flow sensor listed here for a flow of dilute sulfuric acid in a stainless steel pipe of 2.5 inches diameter. The flow is medium speed with no turbulence.

    a. Electromagnetic.
    b. Ultrasonic.
    c. Vortex shedding.
    d. Turbine.

8-12. Discuss how each of the flow sensors listed here might possibly give usable flow information for a flow that is mainly liquid, with some undissolved solid matter (flow is fairly rapid and in a 3 inch diameter plastic pipe):

    a. Electromagnetic.
    b. Ultrasonic.
    c. Vortex shedding.
    d. Turbine.
    e. Heat transfer type.

8-13. Select a flowmeter to be used for the following applications and discuss your answers:

    a. Gaseous flow.
    b. Homogenous nonconductive fluid flow.
    c. Sea water flows.
    d. Conductive fluid flows.
    e. Flows of small particle solid materials.

8-14. Discuss the use of either an ultrasonic or turbine flowmeter for the following:

    a. High velocity of nonconductive fluid.
    b. Medium flows of conductive fluid.
    c. Very low flow rates.
    d. The output must be a sinewave, with constant amplitude and variable frequency.
    e. The output must be a pulse train of variable pulse repetition rate.

8-15. There are several ways of obtaining mass flow rate from conventional weighing devices. Weighing belts can be provided with a belt speed sensor; the signal from this sensor is combined with the weight signal to provide a mass flow rate signal. Show how this can be accomplished with a low cost microprocessor. Draw a flow chart for the computations.

8-16. Mass flow can be inferred from measurements made with an orifice plate or any of the head meters by additional equipment used to measure, compensate, and make computations for specific gravity, flow temperature, and pressure. Show how to do this with a low cost microprocessor. Write a short routine for the computations in assembly language.

9

Viscosity, Moisture, and Humidity

VISCOSITY MEASUREMENT

Viscosity must be measured at flow rates low enough for the fluid to move in layers. If the flow is turbulent, the measurement must be corrected. The unit of viscosity is the *poise,* which is defined as the force required to move one of two parallel planes, each one square centimeter in area, a distance of one centimeter at a velocity of one centimeter per second, when the planes are separated by one centimeter of fluid. Molasses, water, and air have the approximate values of 2000, 0.01, and 0.00018 poises. As most liquids have low viscosities, the centipoise (0.01 poise) is a more convenient unit. It also has the advantage in that the viscosity of water at 68.4°F is 1 centipoise.

Some basic methods of measuring viscosity use the dropping of a ball bearing through a column of viscous material in a clear glass or plastic tube, then measuring the time it takes for the ball to reach the bottom. It is also possible to estimate the viscosity of a fluid by sending air bubbles up through a column and timing the rise of bubbles through the fluid. These techniques are methods of viscosity measurement that involve Stoke's law and are not valid except for flows at very low Reynolds numbers; these are encountered in viscous oils or gases of high hydrogen content. The accuracy is always questionable unless calibration comparisons are made. Viscosity can be measured on an intermittent basis using samples or continuously in a system. Usually, intermittent measurements are made by laboratory instruments.

When viscosity is to be controlled under dynamic conditions, a continuous flow viscometer is used. Selection is based on the range of viscosity and whether or not the medium is a Newtonian fluid. A Newtonian substance, when subjected to shear stress, undergoes a deformation such that the ratio of the shear rate or flow to the shear stress of force exerted is a constant. A non-Newtonian substance does not have a constant ratio of flow or force.

PRESSURE DROP TECHNIQUES

Some viscosity measuring systems for continuous control use pressure drop techniques. The liquid being measured is pumped through a friction tube or orifice plate at a constant rate and temperature. The viscosity is measured by a pneumatic differential transmitter or by a viscosity sensitive rotameter calibrated in viscosity units. Other methods that are used include ultrasonically vibrated probes, torque elements, and constant volume pumps.

In the use of an orifice plate to obtain a pressure drop to measure viscosity, the Reynolds number is used with respect to viscosity. The Reynolds number is an index of the ratio of inertia to viscous forces. Inertia forces dominate when Reynolds numbers are high, and the coefficient of the Venturi tube, the flow nozzle, or orifice plate becomes a constant for practical purposes. If the viscosity is variable, the variation in the restriction device coefficient complicates the flow measurements for low Reynolds numbers. The volume V passing through the pipe per unit time is proportional to the difference of pressure, to the fourth power of the radius, and to the reciprocal of the length of the tube, L and the coefficient of viscosity n:

$$V = \frac{r^4 (P_1 - P_2)}{8_n L}$$

where r is the radius and P_1 and P_2 are the pressures. The continuous capillary viscometer uses this relation to convert differential pressure readings into viscosity values.

The continuous capillary viscometer as shown in Figure 9-1 uses a constant sample flow rate of approximately 1GHP. The measurement span is determined by the bore and length of the capillary. Measuring a large variety of viscosity ranges is possible.

Most continuous capillary viscometers are designed to measure the viscosity of Newtonian liquids. Since pressure transducers are used to transmit the measured viscosity, it is adaptable to the automatic computer control of processes. This type of viscometer has been successful for viscosities up to 15,000 poises and at temperatures of up to 900°F. Since viscosity depends on the temperature of the fluid, the measuring system must be temperature controlled with a resistance thermometer being used either to measure the temperature at which the viscosity is measured or to control the temperature in the measured fluid to maintain a constant viscosity.

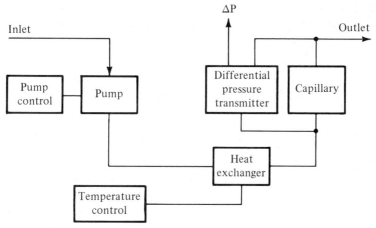

FIGURE 9-1   Continuous Capillary Viscometer.

A number of viscometer types use a rotameter type of float. The single float viscometer as shown in Figure 9-2 is a direct reading instrument for the continuous measurement of viscosity. As illustrated, a positive displacement pump provides the constant sample flow rate through the instrument. The single float viscometer may be used with non-Newtonian fluids with viscosities less than 400 centipoises and Newtonian fluids up to 10,000 centipoises, with a maximum span of 6:1 and a minimum of 3:1. Accuracy is ±4% of indication, and reproducibility is ±1%.

The two float viscometer can be used for local indication and viscosity signal transmission where automatic control is not possible. It incorporates two floats.

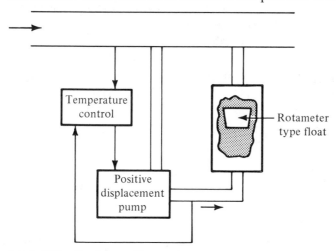

FIGURE 9-2   Float-Type Viscometer.

The fluid flow rate is manually adjusted to a predetermined value as indicated by the position of the upper float. By maintaining a constant reference flow rate as indicated by the flow rate float, the position of the other float will indicate the viscosity of the fluid.

The two float viscometer can be used for Newtonian fluids with viscosities from 0.3 to 250 centipoises and with a span range of 10:1. Accuracy is within ±4% for viscosities higher than 25 centipoises, and ±2% for lower viscosities. Reproducibility is ±1%.

The concentric viscometer uses a differential pressure regulator to maintain a constant pressure drop and a variable area flowmeter with a viscosity sensitive float. As the fluid enters the unit, it splits into two streams. The fluid that flows upward around the differential pressure float controls the pressure drop. The upper end of the differential pressure float is used as a control valve, and as flow rate changes, it throttles the flow to maintain a constant pressure drop which is determined by the weight of the float. The portion of fluid that flows downward enters the viscometer tube by an orifice and then passes the viscosity sensitive float. Constant flow rate through the tube is maintained as the fluid flows through the orifice at a fixed pressure drop. Measurement of the viscosity is made under the constant flow rate. An extension attached to the float transmits its movement through magnetic coupling.

OSCILLATION TECHNIQUES

The ultrasonic viscometer as shown in Figure 9-3 consists of a probe and associated circuitry connected by a coaxial cable. The electronic section sends out pulses of current to a coil inside the probe and around the thin blade. The resulting field excites the magnetostrictive member and causes the blade to vibrate at its natural frequency which is determined by the length of the strip. The probe acts as the transducer, measuring the damping effect or viscous drag of the liquid. The damping effect on the ultrasonic vibration is a function of the liquid density as well as of the viscosity, thus compensation may be required. This type of viscometer is a more recent development, but it is already widely used by the chemical, petroleum, paper, textile, rubber, plastic, and paint industries.

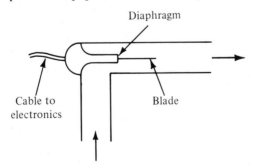

FIGURE 9-3 Ultrasonic Viscometer.

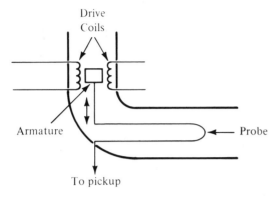

FIGURE 9-4   Vibrating Reed Viscometer.

Vibrating reed type viscometers of the type shown in Figure 9-4 consist of a frequency generator, vibrating spring rod, probe, and a pick-up unit. The principle of operation is that the amplitude of probe vibration depends upon the viscosity of the media. The resistance to shearing action caused by the probe vibration increases with an increase in the media viscosity. The viscometer is installed in a process loop where temperature, pressure, and flow rate are controlled in order to maintain laminar flow.

TORQUE AND WEIGHT TECHNIQUES

Rotating cone viscometers as shown in Figure 9-5 measure viscosity by sensing the torque required to rotate a spindle in a liquid. In a process application, the sample is continuously replaced and is subjected to a constant shear rate. The measurement of non-Newtonian apparent viscosity is possible, as well as the absolute viscosity of Newtonian fluids. A synchronous induction motor drives a cage coupled through a calibrated spring to a spindle arm which supports the spindle or cylinder in the fluid being measured. During a measurement, the spring tends to wind until its force equals the viscous drag on the spindle. The cage and spindle now rotate at the same speed but with an angular relationship to each other which is proportional to the torque on the spring. In this instrument a variable capacitor is used. The measured capacitance is proportional to the angular relationship between the cage and spindle. A variation of this type uses a potentiometer to sense the angular displacement. The torque element viscometer is a flowthrough system with the flow upward or vertically into the measuring system which is housed in a stainless steel flowthrough body. The total system consists of a housing, a measuring cell made up of a measuring cup, a magnetic coupler, and a torque meter made up of a three-speed gearbox, synchronous motor, potentiometer, and torque spring. It provides an electrical output for control or recording.

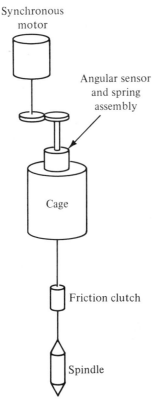

Synchronous
motor

Angular sensor
and spring
assembly

Cage

Friction clutch

Spindle

FIGURE 9-5   Rotating Cone Viscometer.

The agitator power viscometer also operates in a similar manner as the rotating cone viscometer except that the torque exerted is measured by a transmitting wattmeter or thermal converter. It measures the power consumed in driving an agitator in the mixing tank. It is widely used in the paper industry to control and measure the consistency of paper pulp slurries. It is self-cleaning and uses an agitating design, which is ideal for materials that have a tendency to cling to parts or to settle out from suspensions.

The falling piston viscometer design which is shown in Figure 9-6 has an excellent reproducibility which allows it to be used to measure Newtonian and non-Newtonian viscosities for in-process measurements.

When the piston is raised, a sample of liquid to be measured is drawn in through openings in the sides of the tube, and this liquid fills the tube as the piston is withdrawn. During the measurement, the piston is allowed to fall by gravity, forcing the sample out of the tube through the same route as it entered. The time of fall is a measure of the viscosity using the clearance between the piston and the wall of the tube as an orifice. The timed interval is displayed or recorded as viscos-

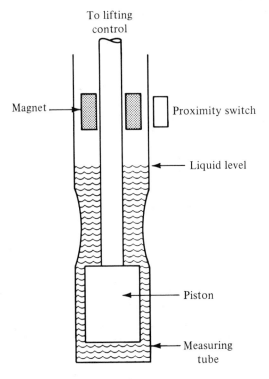

To lifting
control

Magnet

Proximity switch

Liquid level

Piston

Measuring
tube

FIGURE 9-6   Falling Piston Viscometer.

ity. This instrument should not be used when response times of less than one minute are required.

The automatic efflux cup type of viscometer shown in Figure 9-7 is a low cost on-line device used where accuracy is not critical and intermittent measurements with major time lags between measurements can be tolerated. Fillings, efflux timing, and solvent washing operations are controlled by the cycle time programmer control. For the best results, the liquid should flow through the calibrated orifice in the bottom of the cup in approximately 20 to 40 seconds. Repeatability of $\pm 2\%$ is possible under these conditions.

Selection of the right viscometer for the application is not always possible and therefore it may have to be customized. Viscometer vendors or consultants can assist in selection and modification. Many plants that use on-line viscosity measurement as the main method of controlling process operations use customized devices.

MOISTURE AND HUMIDITY MEASUREMENT

The moisture content of the atmosphere is vital in many industrial processes, such as the manufacture of textiles, tobacco, paper, soap, chemical solvents, fertilizers,

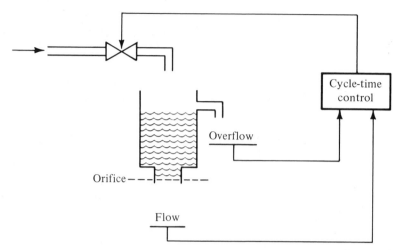

FIGURE 9-7   Automatic Efflux-Cup.

and wood products. Humidity measurement and control are necessary during processes like the drying of chemicals. Another application of humidity measurement and control is heating and air conditioning systems. Psychrometry is the study of the dynamics of the atmosphere using the measurement of a number of variables. One of these variables, humidity, is a measure of the water vapor present in the air. Relative humidity is defined as either (1) the ratio of the moisture content of the air to that which the air would have at the same temperature if saturated or (2) the ratio of actual water vapor pressure in the air to the water vapor pressure in saturated air at the same temperature. The dew point is the temperature at which the water vapor present in the air saturates the air and begins to condense; thus, dew begins to form. As the temperature of the air is reduced below the dew point, the air remains saturated and the partial pressure of the water vapor decreases because of condensation.

Many of the more common humidity sensing elements also perform the transduction, for example, the resistive element, which is actually a combination sensing/transduction element.

The elements for humidity and moisture transducers can be grouped into three categories on the basis of their measuring techniques. Hygrometers measure humidity directly. Among these, the resistive hygrometer is widely used. The use of displacement producing elements has been decreasing since reliable resistive elements became available. Psychrometers measure humidity indirectly. Resistive temperature transducers are frequently used for "wet bulb" and "dry bulb" measurements. Dew point elements give a direct indication of the dew point from which the humidity can be derived. In this category the cold mirror devices are the most common.

Hygrometer Techniques

Resistive hygrometer elements are among the most popular sensing elements used. A variation of relative humidity will produce a variation in the resistance of certain materials such as hygroscopic salts and carbon powder. These materials are applied as a film over an insulating substrate and terminated with metal electrodes (Figure 9-8). Hygroscopic salt films are ionizable materials with a resistance that varies as a function of the vapor pressure of water in the air. The best known hygroscopic salt is lithium chloride. The element uses a film consisting of an aqueous solution of less than 5% of lithium chloride in a plastic binder. The lithium chloride element is sometimes referred to as the *Dunmore element* or *Dunmore hygrometer.*

In this hygrometer technique, because of the steep resistance to relative humidity change that occurs, the element spacing or the resistance properties of the film for specific humidity ranges must be varied. This results in several resistance elements to cover a standard range. Systematic calibration is essential since the resistance grid varies with time and contamination as well as with exposure to temperature and humidity extremes. Other materials besides lithium chloride are used. The general classification of resistance sensors includes carbon strips, the aluminum oxide hygrometer, electrolytic conductive elements, ceramic elements, polyvinyl chloride elements, and certain crystals.

In the aluminum oxide hygrometer element, the electrical properties of anodized aluminum are used for humidity measurement in small strip, needle, or rod elements. When the aluminum surface is anodized, a thin layer of aluminum oxide is formed. A thin metal coating of aluminum or gold is deposited on the outside surface of the aluminum oxide layer. It acts as an electrode and the aluminum base acts as the other electrode, as shown in Figure 9-9. The number of water molecules absorbed on the oxide structure is a function of the change in impedance of an element in the equivalent circuit of the hygrometer structure. The transduction is both capacitive and resistive. The impedance change is measured

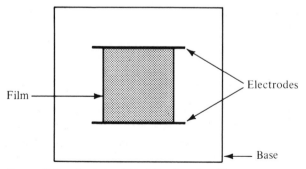

FIGURE 9-8   Resistive Humidity Sensing Element.

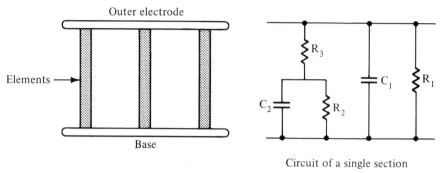

FIGURE 9-9   Aluminum Oxide Humidity Sensor.

to indicate the humidity. A thin film strip aluminum oxide element is shown in Figure 9-10.

A hygroscopic film is used in certain capacitive humidity sensing elements where it acts as a humidity sensitive dielectric. Changes in the dielectric cause a change in capacitance with the humidity. The capacitance of the element changes with the amount of water vapor in the dielectric. One instrument uses a thin film polymer capacitor with the following specifications:

> *Range:* 0 to 100% R.H.
> *Response Time:* 1 sec. to 90% humidity change at 20°C
> *Hysteresis:* 1% for humidity excursion of 5 to 80 to 0%; 2% for humidity excursion of 0 to 100 to 0%
> *Temperature Co-efficient:* 0.05% R.H./°C
> *Drift (7 hours):* 1% from 0 to 60% R.H.; 2% at 80% R.H.; 5% above 90% R.H.
> *Accuracy:* ±3% from 5 to 80% R.H.; ±5 to 6% from 80 to 100% R.H.
> *Sensitivity:* 0.2% R.H.
> *Input Power:* 3.6 ± 0.01 VDC
> *Power Consumption:* 11 mA
> *Output:* 0 to 100 mV
> *Operating Temperature Range:* −40° to 120°F (−40° to 50°C)

Oscillating crystal hygrometers obtain the measurement of humidity by using a quartz crystal with a hygroscopic coating operating in an oscillator circuit. The mass of the crystal changes with the amount of water in and on the coating. One type of hygrometer compares the changes in frequency of two hygroscopically coated quartz crystal oscillators. As the mass of the crystal changes due to the absorption of water vapor, its frequency changes. The oscillatory reading is related to the moisture content of the incoming air. Commercial versions use an integral system to produce a dry reference gas and frequency change is corrected for flow and the type of gas. Many instruments use lithium chloride as the hygro-

(a)   (b)

FIGURE 9-10a   This aluminum-oxide sensor is composed of three layers. The middle layer is the aluminum-oxide porous surface; this is covered with a thin-layer of gold and mounted on an aluminum base. (*Courtesy Panametrics.*)

FIGURE 9-10b   The sensor is then covered with a porous protective cone to complete the probe assembly. (*Courtesy Panametrics.*)

scopic coating on the crystal. The base frequency is generally 9 megahertz. If two crystals are used, while one is in service the other is being dried by the reference gas system. This type of instrument is relatively expensive. Its flow sensitivity, susceptibility to damage by contact with water, and calibration dependence on carrier gas have made it a difficult instrument to use in process control applications.

Spectroscopic hygrometers measure humidity based on the analysis of absorption bands of water vapor in a gas sample. These instruments use a sensing path as the sensing element, such as a cylinder, through which a beam of radiation passes (infrared, microwave, ultraviolet, or visible). The absorption characteristics are determined in the gas sample which extends over this sensing path in the cylinder. The unit requires an energy source, energy detector, an optical system for isolating wave lengths in the spectral region of interest, and a measurement system for determining the attenuation of radiant energy caused by the water vapor present in the optical path. In one type a sample cell contains the gas being sampled and provides the means for calibrating the instrument. Separate calibration gas bottles are used to provide a "zero" and "span." Sampling lines, flowmeters, and pressure regulators are also needed in the system.

Mechanical hygrometers use various materials which change their dimension with absorption and desorption of water from the air. Organic materials have been used for humidity sensing elements considerably more than others. The first and most popular material is hair, particularly human. The second is animal membrane. Other materials, such as paper, wood, textiles, and plastics are less satisfactory. Human hair as well as animal membrane changes length with increasing humidity. This dimensional change has been used for the transduction in a number of different designs. The change in length of human hair over the range 0% to 100% RH is quite small and nonlinear. Sensitivity decreases from approximately 0.0004 in. of hair length per % RH to about 0.00005 in. per % RH at 85% RH at one gram tensile force. These materials also tend to have broad hysteresis effects and are generally unreliable below 32 °F. The time response is generally inadequate for monitoring a changing process.

PSYCHROMETRIC TECHNIQUES

Psychrometer elements measure humidity by the "wet and dry bulb" method using temperature sensors. Psychrometric sensing elements are dual elements with separate outputs. One element, the dry bulb, measures the temperature of the ambient air. The other element, the wet bulb, is enclosed by a wick that is saturated with distilled water. When the measurement is made, the ambient air ventilates over the wick and cools the sensing element below the ambient temperature by causing an evaporation of water from the wick. The evaporation of moisture from the wick depends on the vapor pressure or moisture content of the ambient air. The temperature sensing elements are usually resistive, but are sometimes thermoelectric thermistors. The wick may be cotton or another textile material. It can also be a porous ceramic sleeve, fitted over the resistive element. This is a steady state, open loop, nonequilibrium process which is dependent on the purity of the water, cleanliness of the wick, radiation effects, accuracy of the temperature sensor and density, viscosity, and thermal conductivity of the gas. Many are also functions of the pressure, temperature, and type of gas. The existing systems and corrections that are used are primarily valid for wet bulb temperatures above 32°F (0°C). When the bulb is encrusted with ice, other correction factors must be used. When properly used with atmospheric pressures and gases, it is a calibration standard. In process use, it is susceptible to operator error and contamination.

DEW POINT SENSORS

Dew point sensing elements measure a discrete temperature at which liquid water and water vapor, or ice and water vapor, are in equilibrium. At this temperature only one value of saturation vapor pressure exists for water vapor. The absolute humidity can be determmined from this temperature and knowledge of the pressure. To determine the dew point at any given air or gas temperature, the temperature of a surface is lowered until dew (or frost) first condenses on it. As this point is reached, the temperature of the surface is measured. A dew point sensing element must perform the function of temperature sensing as well as sensing the change from vapor to liquid (or solid) phase. The temperature of the surface must be measured at the instant when condensation first occurs since the characteristics of the condensate will not change appreciably as the surface continues to be cooled below the dew point. Dew point sensing elements sense relative humidity as do psychrometric sensing elements. In both cases, tables are used from which the relative humidity values are calculated on the basis of the measurement obtained and the ambient temperature. Relative humidity is determined from dew point data by the use of saturation vapor pressure tables. These tables can be stored in semiconductor memory and used by a microprocessor in instruments such as shown in Figure 9-11.

The instant of condensation function is usually provided by a thin disc or plate with a smooth surface closely coupled thermally to a cooling element and a condensation detector. Thermoelectric coolers have proved satisfactory. Conden-

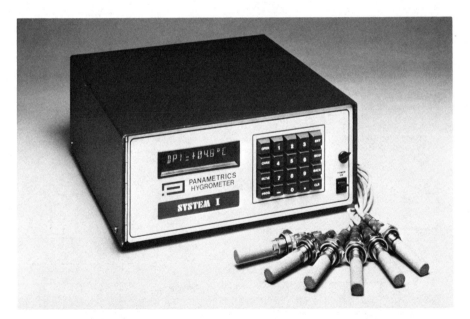

FIGURE 9-11   This microprocessor based moisture analyzer has six channels for the independent calculation of moisture in parts per million by weight or volume, relative humidity, dew and frost points and pressure and temperature. The main operating program is stored in EPROM and the analyzer is easily interfaced to the system control microprocessor. (*Courtesy Panametrics.*)

sation detectors include: photoelectric, which uses a mirror as the condensation surface, a light source to illuminate the surface and one or more light sensors which detect light reflections from the surface; resistive, which uses a metal inlaid in the condensation surface, in which a change in surface resistance occurs when the condensation forms; and nucleonic, in which an alpha or beta particle radiation source is located within the condensation surface and a radiation detector senses the drop in radiation when condensation forms over the source. Common versions of these condensation detectors are shown in Figure 9-12. The temperature at which condensations first occurs upon cooling of the surface is sensed by resistive or thermoelectric sensing elements. Platinum wire resistance temperature sensors have been preferred over thermistors or thermocouples but all three types have been used.

An actual thermoelectrically cooled, optical dew point hygrometer uses a continuous sample of the atmosphere gas over the mirror. The mirror is illuminated by a light source, and observed by a photodetector bridge network. The change in reflectance is detected by a reduction in the reflected light due to the light scattering effect of the individual dew molecules. This light reduction forces the optical bridge toward a balance point and reduces the input error signal to an amplifier which proportionally controls the power supply to the thermoelectric

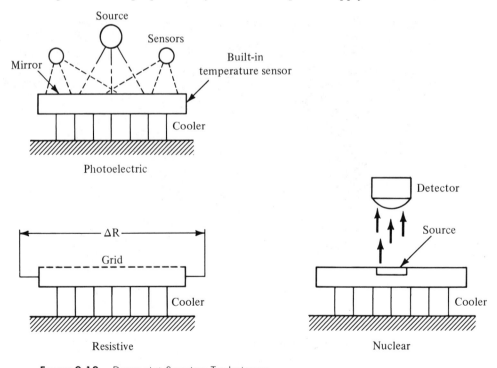

FIGURE 9-12   Dewpoint Sensing Techniques.

cooler. This maintains the mirror at a temperature at which a constant thickness dew layer is retained. Embedded within the mirror, a temperature sensor measures the dew point temperature. This hygrometer is continuous measuring, direct reading, and is suited for process monitoring and control. Its sophistication and accurate measuring technique make it a relatively expensive instrument.

In another design a lithium chloride element acts as a self-regulating heater. The element heats itself by the vapor pressure of a lithium chloride solution, which is in equilibrium with the pressure of the ambient atmosphere. A resistive or thermoelectric temperature sensing element is contained within the solution. The output is a function of the dew point since the vapor pressures for saturated lithium chloride are well established.

ERROR CONSIDERATIONS

Of all the factors that must be considered in humidity measurement, one of the most important is the sampling system, which in most cases can be a probe sample cell (Figure 9-13). The leakage, pressure, and temperature gradients and moisture absorption/desorption characteristics of the sampling system must be considered in the measurement system.

The problems due to leakage are relative to the ambient conditions. If the dew point being measured is close to the ambient room dew point, leakage into the system may not change the reading substantially. If the system is pressurized above atmospheric so as to create a leakage out of the system, the error introduced will be even less. The degree of error due to leakage also depends on the actual dew point being measured.

FIGURE 9-13   A moisture probe sample cell such as this provides a simple bypass loop, which is adequate for many measurements. The probe also allows servicing of the probe without affecting most measurements. (*Courtesy Panametrics.*)

The temperature stability of the sampling system components may be more important. For a given equilibrium sampling condition, a specific amount of moisture will be absorbed onto the sampling system's wetted surfaces. Any factors which upset this equilibrium, such as a change in sample concentration of the process or an ambient temperature fluctuation of the sampling system, will cause a new equilibrium condition to be established before the true dew point can be measured. Thus, any control of the sample line temperatures should be noncyclic, e.g., a constant or proportional control should be used instead of on-off control. A portable calibration system like that shown in Figure 9-14 is useful when equilibrium factors may change.

Similar errors can be caused by the material absorption/desorption characteristics on the overall system response. Stainless steel, glass, and nickel alloy tubing are the most useful nonhygroscopic materials for low dew point applications (0°F to −100°F). Teflon begins reducing system response due to desorption at the lower dew points. Copper, aluminum alloys, and stabilized polypropylene tubing are useful above the −20°F dew point. Most plastic and rubber tubing is unacceptable in all ranges.

Unless it is attacked by the sample, the effect of the more hygroscopic materials is not that of a contaminating nature, but one of introducing a severe lag into the system during the establishment of an equilibrium condition. Some plastics such as nylon are not used at low dew points since the equilibrium condition may take days to be completed.

Another factor is the effect of absolute pressure on the humidity measurement. Dew point analyzers determine the dew point at the actual total pressure within the sensor. If a mixture of gas and water vapor is subjected to a change in pressure and there is no precipitation or addition of water to the system, Dalton's law states that the partial pressure of the water vapor must change in the same ratio as the total pressure. This results in a dew point change for the change in total pressure.

System contamination effects on dew point measurement can be due to condensibles or noncondensibles. Since the optical dew point analyzer measures the dew point of any substance that condenses on the mirror surface, contamination constituents in a sample will not condense unless their dew point temperature is reached.

Condensibles may be soluble or insoluble. If insoluble with a dew point at or above that of the constituent being measured, the relative concentration level will determine the effect on the measured dew point. If the concentration level of the contaminant is low, then it has a low partial pressure compared to the water vapor. Then it can be removed by heating the mirror surface.

For high concentration levels the dew point analyzer may measure the dew point of the contaminant rather than the water vapor dew point. This problem is

FIGURE 9-14   A calibration system such as this can generate a repeatable concentration of water vapor in a carrier gas stream. It is based on saturation/dilution techniques. (*Courtesy Panametrics.*)

reduced due to the high light attenuation characteristics of dew or frost compared to many common contaminants.

If the contaminant is soluble in the constituent being measured, it will modify the vapor pressure and, thus, the dew point of the degree of solubility.

HUMIDITY AND MOISTURE CONTROL

One can use dry and wet bulb temperatures to estimate relative or absolute humidity or dew point. By measuring the relative humidity and dry bulb temperature, the absolute humidity can be determined and thus the dew point. But when a process must be controlled, the choice of the measurements which best represent the state of the process is critical.

When air is to be held at one point on the psychrometric chart, two variables must be controlled. To minimize the interactions between these two control loops, the choice of measurements must relate closely to the manipulated variable.

One might control air conditions using both a heating coil and a water spray. The heat will affect the dry bulb and wet bulb temperatures, as well as the relative humidity. It will not change the dew point. Spraying water into the air lowers the dry bulb temperature and raises the relative and absolute humidities as well as the dew point, but the wet bulb temperature will remain steady.

Optimal control might have a dew point controller for water flow, and a wet bulb controller for adjusting the heater. This combination of loops will not interact. With other controlled variables such as steam injection or fresh air input, other measurement choices would be used.

To provide air of a certain moisture content at a constant barometer but variable temperature, dew point control could be used. Variations in pressure would have to be taken into account. If the operating pressure were a variable, a composition analyzer may be required to determine the partial pressures.

Many times problems from condensation can create corrosive conditions. In these cases, controlling the temperature difference between dew point and dry bulb can provide the best solution.

Some materials, such as paper and wood, change their dimensions with relative humidity. To control the dimensional stability, measure the dimension of a similar material like hair. The hair hygrometer is a useful humidity measuring device for this type of application.

Drying solid materials presents other problems. If air is heated at constant humidity and then blown across the moist material, adiabatic evaporation occurs. As a result, the air's wet bulb temperature is the same before and after contact with the product material. Usually the temperature of the material is similar to the wet bulb temperature of the air, except when extreme dryness is present. The wet bulb wick acts as a model of the solid being dried.

The evaporation rate is proportional to the temperature difference between the air and the product; this is the difference between the dry and wet bulb temperatures. For these reasons, relative humidity and dew point are not useful variables for dryer control.

EXERCISES

9-1. Define a poise using a diagram.

9-2. Describe the way in which the viscosity of liquids and gases can change with respect to temperature and pressure changes.

9-3. Describe, using diagrams, the apparatus for some basic methods of determining the viscosity of water or some other liquid.

9-4. What is the difference between Newtonian and non-Newtonian substances?

9-5. Describe, with the aid of a diagram, the following methods of determining the viscosity of a liquid: (1) continuous capillary method, (2) float method, (3) vibration reed method, and (4) torque method.

9-6. Show that the following statement is true:

$$\text{Kinematic viscosity (centistokes)} = \frac{\text{absolute viscosity (centipoise)}}{\text{mass density}}$$

at the same temperature as the viscosity reading.

9-7. Describe the operation of continuous reading viscometer monitored by a microprocessor controlled display. Draw a flow diagram for the microprocessor.

9-8. When a sphere of radius r cm sinks in a viscous liquid at a constant velocity of v cm/sec, and u is the absolute viscosity of the liquid, in poises, the resistance to the motion of the sphere is found to be

$$R = 6urv \text{ dynes.}$$

Compute the viscosity of the liquid in which a sphere of diameter 0.0622 in. sinks 20 cm in 21.8 sec. The density of the liquid is 0.98 g/cm^3 and that of the sphere is 7.8 g/cm^3.

9-9. An oil with a kinematic viscosity of 0.005 ft^2/sec flows through a .0-in.-ID pipe with a velocity of 1 ft/sec. If the oil weighs 58 lb/ft3, calculate the friction drop, in pounds per square inch, over 3,000 ft. of pipe length.

9-10. Define the relationship among the following: (1) dew point, (2) relative humidity, (3) specific humidity, (4) partial water vapor pressure, and (5) dry and wet bulb temperatures.

9-11. What is the difference between hygrometers and psychrometers?

9-12. Describe the operation of an aluminum oxide hygrometer. How would you interface it with a microprocessor?

9-13. How could an oscillating crystal hygrometer be connected as a control input to a microprocessor system?

9-14. What are the advantages and disadvantages of the spectroscopic hygrometer?

9-15. How can the properties of mechanical hygrometers be used to the best advantage?

9-16. Describe the types of condensation detectors available for optical dew point sensors.

9-17. Why is it important to know both the process pressure and the pressure in the dew point sensor? Is this true of all gas analysis instrumentation?

9-18. Why does the area of leakage become important in humidity measurement systems operating below ambient pressure?

9-19. Discuss the effects on dew point measurement of contaminants which are noncondensibles, which could be solubles, primarily salts. How could a microprocessor be used to reduce or minimize these effects?

9-20. How should a combination of any two measurements of humidity and temperature be used to control the condition of air? Show with the aid of a flow chart how a microprocessor could be used to optimize the control system.

10

Displacement, Proximity, Velocity, and Acceleration

DISPLACEMENT AND PROXIMITY MEASUREMENT

Sensors for the measurement of displacement and proximity use resistive, capacitive, inductive, or optical methods. Displacement sensors measure the linear or angular position and are connected mechanically between the point or object being sensed and a reference or fixed point or object. Proximity sensors also measure linear or angular motion but without any mechanical linkage. The output from a displacement or proximity sensor may be an analog or digital equivalent of the absolute distance being sensed or it may be a function of the distance from a given starting point. Many of the displacement and proximity techniques have been encountered as primary sensors in other transducers such as many of the pressure types.

RESISTANCE SENSORS

Variable resistors can be used as voltage or current dividers to provide displacement information. The wiper or movable arm of the resistor slides over the resistance element. Sensors are available for linear and angular measurement, including fractional and multiturn operation.

The potentiometric displacement transducer is one of the cheapest devices available, however it is not as accurate as other devices due to its mechanical design.

Resistance elements can be wire wound, carbon ribbon, or deposited conductive film. Excitation can be either ac or dc. No amplifiers are required. The ouput can be made to conform to a variety of functional characteristics: linear, sine, cosine, logarithmic, or exponential. Disadvantages are finite resolution (wire wound type), friction, limited life due to wire wear, increasing electrical noise due to wear, and sensitivity to shock. Typical characteristics are: resolution of 0.2%, linearity of ± 1%, repeatability of ± 0.4%, hysteresis of 0.5–1%, and temperature error of ± 0.8%. Precision instruments are available with total errors in the range of ± .5%.

CAPACITIVE SENSORS

Capacitive sensors are generally used for linear rather than angular proximity measurement. Either the dielectric or one of the capacitor plates is movable for angular or linear displacement measurement. Capacitive proximity sensors use the measured object as one plate, while the sensor contains the other plate. The capacitance changes as a function of the area of the plates, the dielectric, or the distance between the plates. Capacitive transducers are available with packaged signal conversion circuitry for dc output operation. Accuracy for small displacements can be near the order of .25%, with accuracies of up to .05% available at a higher cost.

Capacitive devices are accurate, relatively small devices with excellent frequency response. Their greatest weakness is probably their sensitivity to temperature or the need for additional electronics to produce a usable output. A typical capacitor transducer circuit is shown in Figure 10-1. An ac voltage is applied across the plates to detect changes. The capacitor can also be made part of an oscillator circuit causing an output frequency to vary.

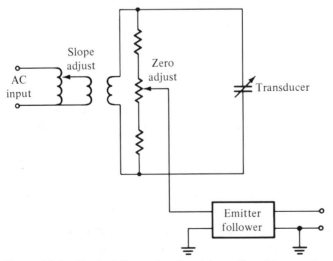

FIGURE 10-1   Typical Capacitor Transducer Conditioning Circuit.

These devices have good linearity and good output resolution. Disadvantages are the temperature and cable sensitivity which require the amplification circuitry to be located close to the transducer.

INDUCTIVE SENSORS

Inductive sensors consist of single coil units which use a change in the self-inductance of the coil and multiple coil units which rely on the change in magnetic coupling or reluctance between coils. Single coil displacement sensors use a movable core to change the self-inductance while single coil proximity sensors use the magnetic properties of the object itself to modify the self-inductance as shown in Figure 10-2. The change in inductance is usually sensed with a bridge circuit or oscillator.

Multiple coil inductive sensors use the differential transformer technique and its variations. The linear variable differential transformer (LVDT) uses three windings and a movable core to sense linear displacement.

A typical LVDT configuration is shown in Figure 10-3. The transformer's

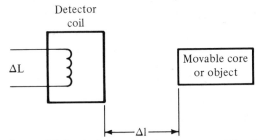

FIGURE 10-2   Inductive Displacement and Proximity Sensors.

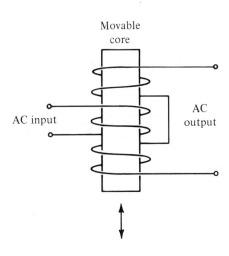

FIGURE 10-3
LVDT Displacement Transducer.

secondary windings are wound to produce opposing voltages and are connected in series. With the core in the neutral or zero position, voltages induced in the secondary windings are equal and opposite and the net output is a minimum. Displacement of the core increases the magnetic coupling between the primary coil and one of the secondary coils and decreases the coupling between the primary coil and the other secondary coil. The net voltage increases as the core is moved away from the center position and the phase angle increases or decreases as a function of the direction.

A demodulator circuit is used to produce a dc output from this winding configuration. Input frequency ranges from 60–30 kHz. Devices are available with built-in conversion circuitry for dc input and output.

Differential transformers are also available for angular measurement in which the core rotates about a fixed axis. Variations in the winding configurations are used in synchros, resolvers, and microsyns. Inductance bridge sensors use two coils with a moving core to change the inductance of the coils which form one-half of an ac bridge. These sensors are available in both linear and angular configurations.

The temperature error for variable reluctance transducers is typically 1–2% for a 100°F. change. Transducers without the conversion circuitry can operate up to 350°F., while transducers with the conversion option are limited to −65° to +200°F. Typical linearity is ±0.5%. Repeatability and hysteresis can be less than 0.2%.

Variable reluctance transducers can offer good shock and vibration characteristics along with successful dynamic response, but the ac conditioning circuitry required adds to a system's cost and error budget.

DIGITAL TECHNIQUES

Another class of transducer is becoming prominent since it simplifies interfacing to displays and data acquisition equipment; these transducers are purely digital. They have either a digital frequency or a digital coded output which is a function of either displacement or proximity. Types with a frequency output use a controlled pulse technique with control being a function of sensing movement. Transducers with a digital coded output detect position and convert this deflection into a digitally coded word. Besides allowing easy interfacing with other digital components, the digital signal allows more accurate processing, resulting in greater system accuracy, and is more suited to transmission for remote monitoring.

Digital outputs or pulses can be produced in displacement and proximity sensors using changes in electrical conduction, induction, or photoelectric conduction.

Conducting encoders use brushes or wipers to detect the position of a coded disc or plate. If a single track is employed, a number of pulses is produced as the disc or plate is moved. Direction sensing is performed by adding another track which is offset to produce sequence logic. Electronic counting circuitry is used to count the number of pulses and perform the conversion to angular or linear measurement. Multiple track encoders provide a digital or binary coded output which is a function of the absolute angular or linear position.

Rotary encoders have been in existence for many years; they evolved from complex rotary switches which produced multiple outputs or combinations of outputs as the switch position was rotated. It was logical to adapt the switch outputs to fit a coded pattern which would depict the position of the shaft. Thus shaft encoders using wipers or brushes and conductive disks were developed. The disks were plated with precious metal alloys for good conduction. Wipers were also plated with precious metal alloys to reduce arcing and material migration. These features were costly, and the contacting encoder, with its limited life, was used primarily in applications where replacement was easy and cost was not critical.

Magnetic displacement sensors use gears of ferromagnetic material to produce pulses from a change in linear or angular position as shown in Figure 10-4. Direction sensing is obtained by shaping the gear teeth in an asymmetrical pattern to modify the output waveform. Photoelectric encoders use a light source and the detector with discs or plates of transparent and opaque windows. Operation is similar to conducting encoders, except that switching is accomplished by breaking the path of the light beam between the source and detector.

Optical encoders offered improvement over mechanical types since the switching was done with a light beam. The plate or disk uses transparent and opaque windows to depict position.

The early light source was a hot filament bulb with or without a lens, and the

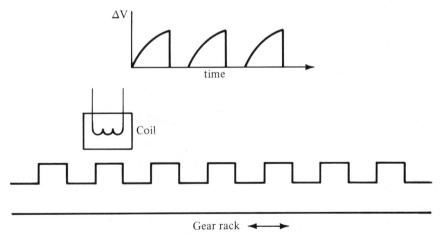

FIGURE 10-4   Magnetic Displacement Sensor with Pulse Output.

light detector was an open semiconductor chip. The hot filament bulb was the weak point in this design, and was subject to premature failure due to vibration and voltage surges. These units were designed for easy bulb replacement and were used for noncritical applications. Accuracy was achieved by bulbs with high brightness, which pushed up the size and power, but reduced reliability.

The development of light emitting diodes (LEDs) provided a light source which was not damaged by vibration, had a life in excess of several hundred thousand hours, consumed little power, and was small. These features allowed linear encoders to be developed as an alternative to rotary shaft encoders used with rack and pinion gearing (Figure 10-5).

Encoders may be of the incremental type, which use one or two tracks of equal spaces of transparent and opaque windows and produce a series of pulses as a function of position. The direction is detected by comparing the two outputs.

The absolute position encoder requires no signal processing to determine linear position and uses a multiple track and arrays of emitters and detectors as shown in Figure 10-6.

In the incremental encoder the position pulses are created when the light path is broken and must be counted and subtracted from the zero setting to obtain the relative position at the rest point. Counting and tracking the position pulses is done with electronics either internal or external to the transducer enclosure.

The laser interferometer uses a laser beam which is directed at a reflector on the measured object. Changes in the linear displacement of the object produce interference fringes which are counted by electronic circuitry. The system is very accurate, but expensive.

TRANSDUCER SELECTION

Since a variety of displacement and proximity sensors are available, selection usually begins with the technical and economic requirements. The primary cri-

FIGURE 10-5   This small absolute-position encoder produces an 8-bit output for interfacing with the microprocessor system. (*Courtesy Siltran Digital.*)

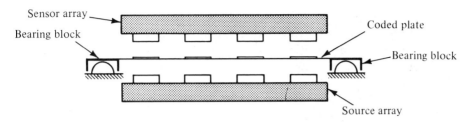

Sensor array

Bearing block

Coded plate

Bearing block

Source array

FIGURE 10-6   Linear Optical Encoder.

teria are the characteristics of the system in which the sensor is to be used—this determines the type of output. The second most important criteria governing selection are range and overall system accuracy, which tend to reduce the number of choices. Then the physical requirements of the system must be considered for optimum sensor installation. Should a coupled or noncontacting transducer be used? Additional factors for consideration include availability, and the manufacturer's capabilities and reputation.

In dc systems potentiometric transducers are simple to apply and can be used with output levels to 50 V, or higher. Devices are available for displacements to 24 inches or from 5 and 3600 degrees. Resolution is approximately 0.5% of full scale for the smaller displacements and can improve slightly with increasing displacements when a greater number of wire turns with lower resistance can be used. Accuracy and friction error improve for the longer range devices.

Reluctive transducers with dc-to-dc conversion circuitry have become available in recent years. Typical displacement spans are between 0.01 and 120 inches and between 0.05 and 90 degrees. Full scale output is typically adjusted to 5V; however, lower and higher outputs are available.

Capacitive and inductive proximity sensors as well as photoelectric sensors are used for the measurement of small displacements. These noncontacting sensors can detect displacement changes as small as one micro-inch.

Typical applications of noncontacting proximity sensors include the measurement of shaft eccentricity, bearing film thickness, rolled sheet thickness, and machined parts.

In ac systems multiple coil inductive sensors have been used more than all others. Displacements are the same as for the dc output versions. The output is proportional to the excitation voltage and to the excitation frequency which may range from 50 to 10,000 Hz. Single coil inductive as well as potentiometric sensors are also used in ac measuring systems.

Digital control systems can obtain the maximum accuracy with incremental and absolute digital displacement sensors along with units such as interferometers. Shaft angle encoders are popular and can provide a system accuracy of less than 5 seconds of arc. The total cost of a digital system can be greater than that of an analog system, but a digital system can adapt to digital computation or display with less software. They are the logical choice for most microprocessor position control systems. Many designs specify digital encoders in applications from pulp and paper manufacturing to automatic plotting and drafting machines. The military use units for control stick monitoring in advanced aircraft and flight simulators.

LINEAR VELOCITY TECHNIQUES

Linear velocity transducers are usually the electromagnetic type, in which a change in flux induces an electromotive force in a conductor. The flux change results from the relative motion between a coil and a permanent magnet. Some designs use a fixed magnet and the moving coil is the sensing element. In others the coil is fixed and the moving magnet is the sensing element.

The basic electromagnetic linear velocity transducer consists of a coil in a steel housing and a cylindrical permanent magnet or core with a threaded end (Figure 10-7). The core moves freely with the motion of the object to which the shaft is connected. The output from the coil is proportional to the velocity of the core movement.

In another fixed coil design the permanent magnet is supported between two springs. Bearing rings are pressed on the ends of the cylindrical magnet to allow its motion within a stainless steel sleeve with a minimum of friction. The mechanical assembly is sealed by threaded retainers in the ends of the transducer case. Relative motion between coil and magnet causes the flux change necessary to obtain an output whose magnitude is proportional to velocity.

In a typical moving coil design the coil is part of an armature pivoted at bearings at the end of the transducer assembly. It moves within a magnetic field established by the pole pieces of a fixed magnet.

Linear velocity may also be obtained with a gear rack and electromagnetic sensor as shown in Figure 10-4.

TACHOMETERS

Angular speed sensing elements use a tachometer technique. The sensing shaft can be a solid cylinder, or it may be a radial hole: splined, serrated, square, slotted, threaded, or conical. In a few noncontacting angular speed tachometers a rotating member other than a shaft acts as the sensing element.

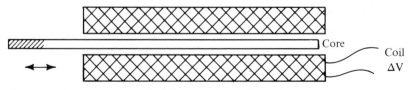

FIGURE 10-7   Electromagnetic Linear Velocity Transducer.

The electromagnetic angular speed transducers consist of a group of widely used tachometers whose output is a varying dc voltage or an ac voltage varying in amplitude or frequency. The dc tachometer uses either a permanent dc magnet or a separately excited winding as its stator and a conventional generator with a commutator equipped rotor. The output of the stator winding type is between 10 and 20 V per 1000 rpm. The brushes required by the commutator require maintenance, but an advantage for some applications is the indication of direction of the shaft rotation since the output is dependent on it. This feature makes the dc tachometer an angular velocity transducer. The ac induction tachometer operates as a variable coupled transformer in which the coupling coefficient is proportional to the speed. When the input winding is excited by an ac voltage, an ac voltage at the output terminals is provided by the secondary winding. The output amplitude varies with rotor speed. A squirrel-cage rotor is used in some devices. Others may use a cup shaped rotor made of a high conductance metal. The shaft rotation produces a shift of flux distribution on which the operation is based. Ac permanent magnet tachometers use the magnetic interaction of a permanent magnet rotor and a stator winding to provide an ac output. The amplitude as well as the frequency of this ac magneto are proportional to angular speed. In some applications circuitry is used to convert the output into a dc voltage. One advantage of operating on the frequency, rather than the amplitude, is the relative freedom from the effects of loading, temperature variations, and armature misalignments caused by vibrations.

Toothed rotor tachometers use a ferromagnetic toothed rotor with a transduction coil around a permanent magnet. These are the most common group of frequency output, angular speed transducers. The technique is illustrated in Figure 10-8. As the magnetic steel tooth on the rotating shaft passes in the proximity

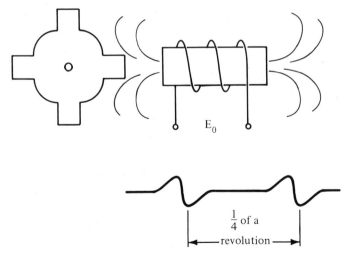

$\frac{1}{4}$ of a
revolution

FIGURE 10-8   Electromagnetic Frequency Tachometer.

of the permanent magnet, the lines of flux of the magnet cut across the coil winding inducing an electromotive force in the coil. An output is created once per each tooth. The time between pulses is inversely proportional to the shaft speed. With increasing speed the number of pulses increases and can be displayed on an events per unit time (EPUT) scale.

Rotors with continuous teeth are used more frequently than single tooth rotors because they produce a higher output frequency as well as a greater voltage amplitude at low shaft speeds. At a shaft speed of 600 rpm, the output frequency of the transducer shown in Figure 10-9, which has four teeth, will be 40 pulses per second. The rms output voltage increases with decreasing clearance between the coil assembly and the teeth, with increasing shaft speed and increasing tooth size.

An optical encoder with a noncoded disk may also be used as a tachometer. Self-contained units or kits which are separately packaged can be used to provide a square or a sinusoidal wave shape of constant amplitude typically between 2 and 8 V zero to peak. The output frequency is determined by the shaft speed and the number of transparent sectors in the disc. These devices are usually preferred for microprocessor control applications.

Reluctive angular speed transducers use toothed rotors along with a reluctive, rather than an electromagnetic, transduction element. Some designs use a C-core transformer with an excitation winding on one end of the core and an output winding on the other end. A toothed rotor is placed in the gap of the core so that the reluctance path is varied by the movement of ferromagnetic material in the gap. A differential transformer design uses an E-core with the excitation primary winding on the center portion and two secondary windings on the outer legs of the core. A toothed rotor in one of the two gaps varies the reluctance path between the primary and one secondary winding. This type of transducer is rarely produced in the United States.

Strain gage, angular speed transducers use strain gages with a deflecting beam and an eccentric disc attached to the sensing shaft, as shown in Figure 10-9. A sinusoidal output can be obtained from the strain gage bridge circuit, two arms of which are on the bending beam. Transducers of this type have been mostly manufactured in countries other than the United States.

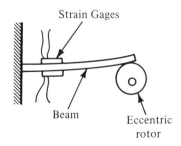

FIGURE 10-9   Strain-Gage Tachometer Technique.

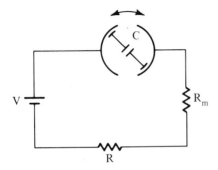

FIGURE 10-10   Switch Type Differentiating Tachometer.

Some switch type, angular speed transducers use rotary switches as fre-quency output tachometers, but the more common designs incorporate a dif-ferentiating circuit so that a low cost damped milliammeter can be used. In the schematic of Figure 10-10, rotating contacts on each end of the capacitor C allow the capacitor to change to one polarity, then to the opposite polarity. R is the in-ternal resistance of the milliammeter and R is the resistance in interconnecting wiring if the meter is located remotely. The meter is sufficiently damped to re-spond to the average of the individual current pulses. The average current is kCVw, where k is a constant and w is the angular speed. One transducer uses a double pole throw reed switch for reversal of capacitor polarity. An alternate de-sign obtains a switching revolution by an eccentric pin attached to the sensing shaft, which moves a yoke up and down to actuate the polarity contacts. Shunt resistors across the indicating milliammeter are used to adjust the meter range.

ACCELERATION TRANSDUCER CHARACTERISTICS

All acceleration transducers use a seismic or proof mass, which is restrained by a spring, and whose motion is usually damped (Figure 10-11). When an accelera-tion is applied to the transducer, the mass moves relative to the transducer case. As the acceleration ends, the spring returns the mass to its original position, when acceleration is applied to the transducer in the opposite direction, the spring is compressed. Under steady state acceleration conditions, the displacement of the

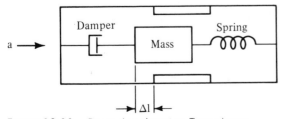

FIGURE 10-11   Basic Acceleration Transducer.

mass is given by the acceleration multiplied by the ratio of the mass M to the spring constant k. Under dynamic or varying acceleration conditions the damping constant enters into a modified version of this equation.

The seismic mass in a linear accelerometer is usually a circular or rectangular section. It can be linked to the case by slides or bars, but it is always restrained from motion in any but the sensing axis. The seismic mass of an angular accelerometer may be a disc pivoted at its center and restrained by a spiral spring, which responds to angular acceleration with an angular displacement.

Capacitive acceleration transducers use a change of capacitance in response to acceleration. Some transducer designs use a fixed stator plate and a diaphragm to which a disc shaped seismic mass is attached. The diaphragm acts as the restraining spring as well as the moving electrode of the capacitor. Acceleration acting on the mass causes the diaphragm to deflect and its capacitance to the stator to change proportionally.

Piezoelectric Accelerometers

Piezoelectric accelerometers use the force on a crystal to measure acceleration. The crystal may be bonded to the mass which also may act as the elastic member. Another design uses an annular crystal bonded to a center post on its inside surface and to an annular mass on its outside surface. The upward or downward deflection of the mass causes shear stresses across the thickness of the crystal. Typical compression designs use a stacked arrangement to increase the low output of quartz crystals. One transducer uses seven crystals which are stacked and connected for output multiplication as indicated in Figure 10-12. A coaxial connector is typically used for all piezoelectric acceleration transducers. The base and case of many piezoelectric accelerometers are connected to one crystal electrode or set of electrodes. The case is sealed to prevent moisture damage. The impedance across the crystal electrodes is high and, in general, the larger the accelerometer, the higher its sensitivity and the lower its resonant frequency.

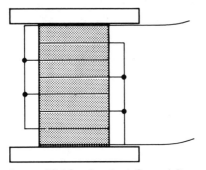

FIGURE 10-12   Stacked Crystal Piezoelectric Accelerometer.

The higher the resonant frequency, the lower the capacitance or sensitivity and the more difficult it is to provide mechanical damping. The amplification factor is defined as the ratio of the voltage sensitivity at its resonant frequency to the voltage sensitivity in the band of frequencies in which the sensitivity is independent of frequency. This ratio depends on the amount of damping in the seismic system, and it decreases with increased damping. Some units with resonant frequencies below 20,000 Hz use silicon oil damping but most piezoelectric devices are essentially undamped and have amplification ratios from 5 to 50.

The total measurement system capacitance includes the cable capacitance, so a maximum cable length can be used for a maximum frequency and voltage. Source followers can be used for impedance matching; this will solve most of the problems associated with high impedance voltage amplifiers. Calculations for a particular source follower can be used to determine the frequency and output voltage for different values of cable length, or the maximum output voltage and/or frequency for a given cable length. The problem of matching with a follower can be minimized by the use of a charge amplifier system.

Charge Amplifiers

A charge amplifier is an operational amplifier with capacitive feedback as shown in Figure 10-13. It allows E_{out} to be proportional to the charge produced by the accelerometer. Since the amplifier detects charge rather than voltage, the system is independent of shunt capacitance and of changes in shunt capacitance. This means that: (1) cable length can be ignored in system output calculations, and (2) the system characteristics of the transducer are not affected by changes in capacitance of the transducer or the cable. Other advantages of the charge amplifier are (1) the lower dynamic input impedance of less than 1 MΩ, which reduces the effects of humidity and cable connector contamination and noise, and (2) flat frequency response to 10 Hz or less.

A typical charge measuring system provides a voltage closely corresponding to the actual motion of the accelerometer at the amplifier output. The amplitude of this signal is a function of the transducer charge sensitivity, the actual vibration

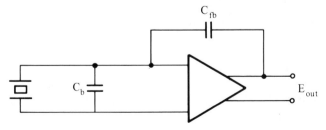

FIGURE 10-13   Charge Amplifier.

amplitude, and the amplifier gain setting. The wave form of the signal is controlled by the amplifier frequency response.

Charge amplifier temperature response is a function of the charge characteristics of the accelerometer. Compensation can be accomplished by a series capacitor to flatten the charge-versus-temperature curve. However, the gain of the charge amplifier then changes.

Accelerometer Calibration

The sensitivity of an accelerometer can be determined by mounting it on a vibration table. The table is set to a desired set peak-to-peak displacement like 0.040 in. at 100 Hz. The displacement is measured optically and the frequency monitored by a counter. The acceleration level can be calculated to an accuracy of $\pm 1\%$ using

$$a = 0.1023 \, d_o f^2$$

where

$$d_o = \text{peak displacement}$$
$$f = \text{frequency}$$
$$a = \text{resultant acceleration in peak g}$$

The charge sensitivity can be calculated from:

$$S_q = \frac{\sqrt{2} \, eCt}{Aa}$$

where

S_q = charge sensitivity in coulombs per graviation unit

e = system output in volts rms

a = acceleration level in peak g

A = follower gain

Ct = total capacitance, of the accelerometer, cable, and the follower input capacitance, in picofarads

The open circuit voltage sensitivity can then be calculated from:

$$S_{eo} = \frac{S_q 10^3}{Ca}$$

where

S_{eo} = open circuit voltage sensitivity

Ca = accelerometer capacitance in picofarads

Cross axis sensitivity or traverse response can be found by mounting the accelerometer on the sensitive axis and recording the output at a measured frequency and displacement. The accelerometer is then mounted perpendicular to the sensitive axis and rotated 360° while vibrating at the same g level to determine the maximum output. The cross-axis sensitivity is the maximum output perpendicular to the sensitive axis divided by the output on the sensitive axis at the fixed vibration level.

The frequency response can be determined using a calibrated shaker. The drive of the table is set using an NBS certified standard, then the response is read on a working standard. The certified standard is now replaced by the accelerometer to be calibrated. The data from these standards are averaged, while the working standard is used as the table drive control during the calibration.

The natural frequency can be found by mounting the accelerometer on a plate with at least ten times the effective mass of the accelerometer. This system is suspended and a variable frequency oscillator is used to drive the accelerometer. The accelerometer current is held constant as the frequency is changed until the output across a resistor is a maximum with a 90° phase shift. This is the natural frequency of the accelerometer plate system as well as the frequency of minimum impedance.

The temperature response can be determined by measuring the accelerometer output at some desired frequency and level like 100 Hz and 2 g at ambient temperature and then at the temperature extremes.

EXERCISES

10-1. What is the difference between a displacement sensor and a proximity sensor? Give some examples.

10-2. A linear potentiometer is used as a displacement sensor. It has 5 volts excitation across it and is composed of 1,800 turns of wire of 0.0008 inch diameter. What is the expected noise voltage amplitude?

10-3. An angular position transducer uses a rotary potentiometer with a conductive film element. The contact resistance changes from .2 ohms. The external load is 1 k ohms. The transducer is a 10 ohm unit with 5 volts across it. What is the noise voltage generated by the potentiometer?

10-4. Discuss the operation of a capacitive sensing circuit. What are the critical parameters?

10-5. A differential transformer has a sensitivity of 5 volts per inch. If this transducer senses a movement of 0.0001 inch, what will the output be?

10-6. Discuss the effects of the excitation frequency on differential transformer performance.

10-7. Draw the plate pattern for a 4-bit linear binary conducting encoder. What would be the advantage of using a cyclic binary code or a BDC coded plate?

10-8. Show the operation of an optical encoder using a diagram. What are the critical parameters that affect the output level?

10-9. What is the difference between an absolute and an incremental encoder? How does this affect the interface to a microprocessor?

10-10. The thickness of plywood sheets must be monitored during manufacturing. Select the transducers and show the complete system block diagram of a microprocessor monitor and control system for the thickness and pressure controls required. The microprocessor will control the hydraulic pressure to the plywood pressure rollers.

10-11. Discuss basic techniques used to measure linear velocity.

10-12. What are the critical parameters when a gear and electromagnetic sensor are used to measure velocity?

10-13. Describe the differences between dc and ac tachometers.

10-14. What are the advantages of an optical encoder tachometer over a reluctive or strain gage tachometer in a microprocessor control system?

10-15. Describe the characteristics common to all accelerometers.

10-16. What are some of the characteristics of the piezoelectric accelerometer? Write an equation defining the amplification factor.

10-17. Discuss the use and advantages of a charge amplifier in accelerometer measurement circuits.

10-18. Write an expression defining the cross-axis sensitivity for an accelerometer.

10-19. The temperature response of piezoelectric accelerometers must be measured as a quality control test. Write a test procedure and select a temperature monitoring method. The data is to be collected and analyzed by a microcomputer system. Define the essential system elements required.

10-20. Discuss a control system for automating the determination of the natural frequency of piezoelectric accelerometers. The accelerometers weigh four ounces and 200 must be tested per hour.

11

Stress, Strain, Force, and Torque

STRESS AND STRAIN MEASUREMENT

Stress and strain are generated in parts, and systems by weight, temperature, pressure, vibration, or displacement forces. The most popular method of making these measurements is by the use of strain gages. Other methods include stress coating and photoelastics.

When a load or stress is imposed on an object, the object expands or contracts or is subjected to shear. If a grid of wire or foil with the desired resistive characteristics is bonded to the object, it will stretch or be compressed as does the surface to which it is attached. The metallic resistance strain gage is based on the principle that, as a conductor is subjected to a tensile or compressive strain, it exhibits a change in resistance. The magnitude of this change is proportional to the magnitude of the applied strain. This strain is defined as:

$$\text{Strain} = \frac{\text{Change in length}}{\text{Original length}}$$

In a strain gage, use is made of a constant known as the gage factor (GF). This constant ranges from 2 to 6 for most strain gage alloys. The gage factor is based on the change that occurs in the total resistance as related to the change of length in the conductor with respect to its total length, or:

$$\text{GF} = \frac{\triangle R/R}{\triangle L/L}$$

The primary ambient temperature performance characteristic of all strain gages is the gage factor. It is usually shown as a nominal value with tolerances. The complete specifications should include the strain range over which the linearity, hysteresis, and creep tolerances apply, as well as the allowable overload ac strain limit which does not cause output errors. Linearity often applies only to a specified partial range. Hysteresis and creep are usually expressed in units of strain. Drift in strain gages refers to the changes in gage resistance at a constance extreme temperature with no applied strain. Since many strain gage applications require a bridge configuration, the electrical characteristics of the strain gage element and the degree of matching required from the other elements, as well as the internal geometry of the combination, become important. Active/dummy combinations may be used by both types of elements.

Dimensions are usually more critical for strain gages than many other types of transducers. The gage length and width are usually defined. The tolerances must be commensurate with the magnitude of the dimensions.

When the gage is furnished with leads, the diameter or thickness of the leads, lead material insulation, and length and spacing between leads may be important. When the gage is bonded to a carrier or is encapsulated, the base length, width, thickness, and material should be known. Some gages are surface-transferable and the size and material of the strippable carrier become important. The material of the grid wire diameter or foil thickness are important considerations.

The recommended methods for the bonding or other attachment of the strain gage to the measured surface or its reaction to specific materials should be known. This information includes the type of cement, material compatibilities, cure time, temperature and pressure, insulation resistance and changes at temperature extremes, resistance to humidity, and moisture proofing.

Metal Strain Gages

Resistive metal wire strain gages are manufactured in bondable, surface transferable, and weldable configurations. The unbonded strain gage, consisting of a wire stretched between two posts and unsupported between its ends, is used in some pressure transducers but rarely for strain measurements.

The bondable strain gage is also called a *bonded* strain gage since it is usually bonded, with cement, to the measured surface. It is the most widely used strain gage. A basic flat wire grid bondable strain gage consists of a thin wire or filament arranged in a winding or pattern and cemented to a base. Lead wires are soldered or welded to the ends of the filament for the external electrical connections. The length of the grid is the gage length. A cover plate is sometimes cemented over the wire grid.

The choice of filament material is usually based on five criteria: (1) The material should provide as high a gage factor as possible so that its resistance change with strain is large, (2) The temperature coefficient of resistance should be low to minimize temperature errors, (3) The resistivity of the material should be high to allow winding within the smallest possible area, (4) The filament should have a high mechanical strength to allow high stresses to be applied to the filament, and (5) The material should generate the lowest possible thermoelectric potential at the junctions with the leads.

Gages with multiple grids, i.e., rosette gages, can be used for the simultaneous measurement of strain in different directions. Multiple wire gages use a sandwich construction with the grids laid over each other and insulated by cement. Weldable strain gages are mounted within small flat or tubular envelopes made preferably of the same metal as the measured surface. The base of the envelope is slightly larger than the main body of the gage for welding to the surface. Since the base is thin and the edge is narrow, microwelding techniques are used for mounting weldable gages. Such gages have been used successfully on the inside and outside surfaces of tanks containing cryogenic liquids. Strain gages for strain measurements in concrete are encapsulated in a waterproof container to protect the filament during installation when pouring the concrete.

Metal foil gages use photoetching techniques similar to mass produced printed circuit boards to allow better use of a given area. Foil gages can be made smaller than wire grid gages and they can be used to measure larger strains than wire types. They are not damaged as easily. For many applications, foil gages are superior in all essential respects. The foil grid consists of a thin (3 to 8 microns thick) layer of foil, some of which is removed by etching to the desired grid shape. The grid can be cemented to a base.

The versatility of the etched foil technique allows a number of special designs. Dual and triple element rosettes are available, as are biaxial gages. One type of biaxial gage has two elements spaced 90 degrees from each other and spaced at 45 degree angles from an identifiable center line. This spacing is useful for torsional strain determinations in torque measurements. Triaxial rosettes with elements spaced at 60 degrees from each other are also available. A spiral foil gage is usable for measuring tangential strain in a diaphragm. The foil pattern is bifilar to facilitate its construction and to cancel inductive effects. Full bridge rosettes may use opposing biaxial combinations for structural applications or an in-line configuration for bending beam applications.

A typical stress strain gage uses two uniaxial strain sensing elements oriented at 90 degrees to each other. The elements have a common connection so that either of them may be observed independently for conventional strain measurement or the series combination of the two elements can produce an output

Principal axis

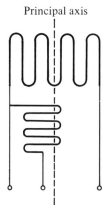

FIGURE 11-1   Two Element Stress Strain Gage Technique.

proportional to stress along the principal axis as shown in Figure 11-1. The latter is achieved by making the ratio between the resistances of one element to the other element equal to Poisson's ratio of the material to which the gage combination is to be applied. If the stress strain gage is to be applied to aluminum, with Poisson's ratio 0.33, the resistances of the elements are 115 and 345 ohms. Strain in either the principal axis or the transverse axis is measured in the usual manner using the gage factors for each element. Stress is determined by using a manufacturer supplied stress gage factor.

Thin film techniques have been used in the manufacture of strain gages in which a thin film is applied directly to the measured surface. Such gages can be made smaller than other equivalent metal strain gages. The surface is first coated with an insulating substrate upon which the gage or rosette is formed by evaporative or bombardment methods. Applications of thin film metal strain gages have been mainly to the diaphragms of pressure transducers and can be expected to be extended to other transducer sensing elements.

Flame sprayed strain gages have been applied to structures which are exposed to hostile environments. In these applications they have been an alternative to welded strain gages although the application techniques are most costly. They are thinner than other strain gages, which is an advantage in some applications, such as surfaces which may be subject to aerodynamic heating. A strain sensitive metal grid is applied to an insulating ceramic substrate on the measured surface. The substrate is formed by a process in which the end of a ceramic rod is heated and the molten particles projected onto the surface. Substrates 0.05 to 0.1 mm thick have been used on aircraft engines with satisfactory results. A surface transferable metal gage is then applied to the substrate and permanently bonded to the surface by a flame sprayed ceramic coating. Temperatures to 2200°F can be used for flame sprayed strain sensors.

Semiconductor Strain Gages

Semiconductor strain gages use the piezoresistive effect which is much larger in semiconductors than in conductors. The gage factors of semiconductor strain gages are between 50 and 200, while those of metal strain gages are no greater than 6 and are frequently around 2. However, semiconductor gages tend to be more difficult to apply to measured surfaces; their strain ranges are usually limited along with their operating temperature range and they require more temperature compensation. The gage factor is dependent on resistance change due to dimensional change as well as the change of resistivity with strain. It is also nonlinear with applied strain. This nonlinearity is usually minimized by using a strain gage bridge circuit. One method is to use a pair of active gages, matched to each other, as adjacent bridge arms. Other methods use cancellation of nonlinearity by an opposite nonlinearity in a bridge circuit or by using a four active arm bridge with two gages in tension and two in compression with gages that are matched, or by shunting a single active arm. Another technique is prestressing the gage in tension during mounting to shift its operation into a more linear portion. The amount of doping can also be increased to obtain better linearity with a reduction in gage factor. The linearity of highly doped gages approaches most metal gages.

Semiconductor gages are affected by temperature changes more than metal gages. The temperature coefficient of resistance of the semiconductors used is 60 to 100 times greater than that of constantan. The variation of gage factor with temperature is three to five times greater, and the Seebeck coefficient or thermoelectric potential generated at connections can be ten to twenty times as large. The differential thermal expansion between the semiconductor and the measured surface can also be greater. The linear expansion coefficient of a semiconductor gage is about one-half that of a metal gage.

Temperature compensation can be achieved by similar techniques. A dummy gage can be connected to one bridge arm. Resistors (or thermistors) with controlled thermal resistance changes can be connected in series or parallel with the active semiconductor gage bridge arms or with one or both of the excitation leads. A p type semiconductor with a positive gage factor and temperature coefficient can be connected in an active bridge arm when using an active n type gage. The p and n type gages can be manufactured as a dual element combination. Thermocouples can be so used and connected to provide an opposite voltage. Gages can be selected such that the thermal expansion and temperature coefficient of resistance tend to balance each other. Typical semiconductor strain gage configurations are shown in Figure 11-2. The gages can be surface transferable or encapsulated. Gage lengths vary from 0.01 to 0.25 inches. Lead materials are gold, copper, or silver wire. Silicon is used almost exclusively as the semiconductor material. Some configurations fuse the dopant directly into portions of silicon blocks

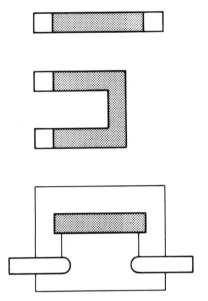

FIGURE 11-2   Semiconductor Strain Gages.

or diaphragms. These sensing elements with integral diffused strain sensors have been used in pressure and force transducers.

Strain Gage Application

Electrical characteristics are comprised of the gage resistance, power rating, or maximum current which depends on the heat sink provided by the measured surface and the bond; for gages bonded to a carrier or encapsulated, the insulation resistance is important for weldable gages since no insulation is provided by any bonding cement. Gage resistances typically are 60, 120, 240, 350, 500, and 1000 ohms and for semiconductor types, 5000 and 10,000 ohms.

A mounted strain gage senses the strain in its plane which is separated from the measured surface by cement and the mounting surface. When axial strains are being sensed, the strain path lies on the surface and is transmitted through the cement and mounting surface to the gage. However, when the specimen is under a bending stress, the strain at a surface is defined by the distance of the surface from the neutral axis and the radius of curvature of the axis. Figure 11-3 shows the gage lies farther from the neutral axis than the measured surface. The gage then senses a strain larger than that occurring on the specimen. This is called the *offset error*, which is minimized by placing the gage as close to the specimen as possible. The center plane of a wire grid gage may be approximately 0.004 inches above the specimen surface while the center plane of a foil gage is approximately 0.0025

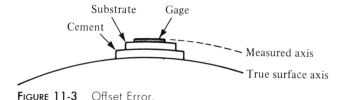

FIGURE 11-3   Offset Error.

inches above. Thus a foil gage has a lower offset error than a wire gage. In many applications, offset error can be eliminated by proper calibration.

The measurement of strain with a resistance strain gage requires the accurate measurement of a small change in resistance. The most common circuit for this application is the Wheatstone bridge. Using a constant supply, the voltage across the output is zero for a balanced bridge and has a predictable and measurable amplitude when one arm of the bridge changes. Since it is possible to relate the resistance change ratio to the applied strain, the output voltage can also be related to strain. The relationship between voltage and strain is kept linear by proper choice of operation limits. The output voltage is a few millivolts, which makes it necessary to use high sensitivity amplifiers to obtain suitable signal levels. With a well designed system, the signal-to-noise ratio can be controlled to give a reliable signal output. A circuit is shown in Figure 11-4 for a dc excited gage system. Only one active gage is shown in the figure, but two, three, or four gages can be active; unstrained gages called *dummy gages* can be used to obtain compensation. Gages can be used in half bridge or full bridge configurations. Figure 11-5 illustrates the use of two active gage circuits; Figure 11-6 illustrates the use of four active gage circuits. These configurations are used mainly for static strain measurements.

Strain Gage Calibration

Several techniques can be used to calibrate strain gages and strain gage systems. One method is to place known loads on the gages and measure the output. In

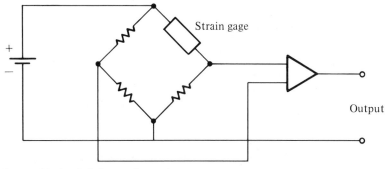

FIGURE 11-4   DC Strain Gage System.

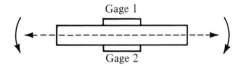

Gage 1

Gage 2

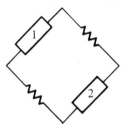

For axial load with bending compensation

FIGURE 11-5   Double Active Bridge.

many applications this method cannot be used, because the system cannot be exercised to the calibration values. If one calibrates the gages to eliminate all the errors that have been discussed, a computer model would be required. Errors exist as a result of (1) the temperature coefficient of the materials, (2) sensitivity changes with temperature variations and the methods of correction, (3) calibration errors caused by series or shunt resistances, and (4) resistance of transmission lines and the configuration. For most purposes, when use is made of the half-bridge configuration, the following equation gives a R_{cal} value which can be used:

$$R_{cal} = \frac{Rg}{\epsilon K} - Rg$$

When the signal leads are of a length to have a significant resistance this equation can be modified to:

$$R_{cal} = R_T/\epsilon K - R_T$$

where

R_{cal} = calibration resistance in ohms
R_T = total resistance of gage and leads in ohms
Rg = gage resistance
K = gage factor
ϵ = strain value in microinches per inch

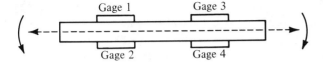

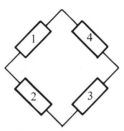

For bending load with axial compensation

FIGURE 11-6   Quad Active Bridge.

VISUAL TECHNIQUES

Photoelasticity is a visual technique for measuring the strain and stresses in parts and structures. When a photoelastic material is subjected to forces and viewed under polarized light, the resulting strains are seen as fringe patterns. The pattern reveals the overall strain distribution, and quantitative measurements can be made of the strain directions and magnitudes. The three major photoelastic techniques are (1) two dimensional model analysis, (2) three dimensional model analysis, and (3) photoelastic coating analysis. Of these, the photoelastic coating method is the most widely used since it allows the use of photoelasticity on actual parts of any size and shape operating under actual or simulated service conditions. Two dimensional analysis is used to determine stress concentration factors; the three dimensional analysis is used to provide a complete stress analysis both inside and outside the part.

Brittle coatings are lacquers sprayed on parts or structures. After drying, the part is tested, and the coating will exhibit fine cracks, revealing the location of maximum strain and the direction of principal strains. This method is not suited for detailed quantitative analysis like photoelasticity. Brittle coatings are primarily used to determine in advance the exact areas for strain gage location and orientation.

Moire fringe analysis can be used to determine the components of displacement or strain. On a part subjected to analysis, a grid of equidistant lines is deposited. When stressed, deformation of the part and of the grid applied to it occurs.

The deformed grid is then superimposed and compared to an undeformed grid and this superposition produces an optical effect known as *Moire fringes*. The technique is primarily applied to the measurement of strains that cannot be resolved easily, accurately, or economically by other techniques. Applications include high temperature strain measurements, measurement of large elastic and plastic strains, and absolute measurements of strain to establish material properties.

FORCE AND TORQUE MEASUREMENT

Most force and torque transducers use a separate sensing element which converts the measured quantity into a small mechanical displacement, usually a deformation of an elastic element. An elastic deformation used to sense force is strain or deflection. A maximum of each occurs at some location in the sensing element but not necessarily at the same location. This maximum of either strain or deflection is applied to the force or torque element. The element is made of a homogeneous material and is usually manufactured to close tolerances. Various types of steel are commonly used as element materials.

Beams are the simplest force sensing elements. The maximum deflection of a beam occurs at the point of force application; the maximum deflection of a cantilever beam always occurs at its free end. The point of maximum strain in a simple cantilever beam is at the fixed end. Cantilever beams can be of the constant strength type in which they have a triangular or parabolic tapered shape which is narrowest at the point of force application. In a constant strength beam the strain is constant along the bottom of the beam. In a simply supported beam the point of maximum strain is at the point of force application.

Diaphragms are circular plates which are used as force-sensing elements. Maximum strain and deflection occur at the center of the diaphragm where the force is always applied.

Proving rings and the related flat proving rings or frames are a frequently used type of force sensing element (Figure 11-7). Maximum deflection as well as maximum strain occur at the point of force application. However, strain has an almost equal magnitude at points 90 degrees in either direction from the point of force application, and strain sensing is more convenient at these points.

Column force sensing elements usually have the point of maximum deflection at their vertical center at one-half of the column height. Their characteristics depend primarily on the height-to-width ratio and on the wall thickness for hollow cylinders. Compression and tension in a column are normally sensed as strain and sometimes as changes in magnetic characteristics or natural vibration frequency, but rarely as displacement changes.

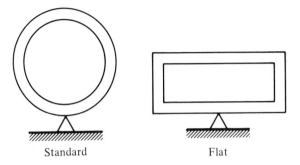

Standard   Flat

FIGURE 11-7   Proving Ring Force Elements.

Piezoelectric Force Transducers

Several types of elements have been successful in force and torque sensing. Force transducer designs normally use either the strain or the deflections of sensing elements. Although piezoelectric force transducers can be used for a measurement of quasi-static response when they are used with charge amplifiers, their application is mostly in dynamic force measurement. Column type sensing elements are normally used with the ceramic or quartz transduction element located at the center of the cylindrical column. One design uses a pair of disc shaped piezoelectric crystals separated by a thin electrode. The electrode is insulated from the case and is connected to a coaxial cable. The crystal assembly is placed between two thick metal discs with each threaded for mounting. Another design is the washer type

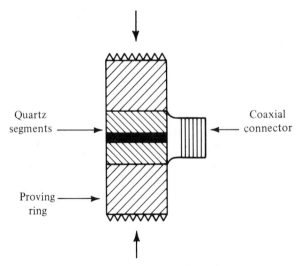

FIGURE 11-8   Piezoelectric Force Transducer.

transducer, or force washer, illustrated in Figure 11-8. The sensing element is an annular column of sandwich construction. The central hole can be used with a bolt or stud. The force washer is often used in threaded fastener mountings of machinery and other equipment. The piezoelectric transducer responds to compression forces only. It can be preloaded so that a force is continuously exerted to produce a static level. If a bidirectional dynamic force is applied, the force will provide an output corresponding to the alternate tension and compression.

Piezoelectric force transducers do not require an excitation but they do have a very high output impedance. Voltage amplifiers, emitter followers, and charge amplifiers are used in these transducers. This equipment has been discussed in the chapters dealing with piezoelectric acceleration and pressure transducers.

Reluctive Force Sensors

Reluctive force transducers using the deflection of a sensing element are widely used. Typical transducers use differential transformers as illustrated in Chapter 10 Figure 10-3. Deflection of a proving ring or diaphragm causes a relative motion between a core and a coil assembly. This changes the amplitude and phase of an ac voltage in the secondary windings. The coil assembly is normally encapsulated and shielded. The excitation is usually between 2 and 10 V at between 50 and 10,000 Hz. Sensitivity improves with increasing excitation frequency. When the secondary windings are connected in the most common configuration, the output voltage increases equally from the core center null point for sensing deflection due to compression or tension, and the output phase changes from 90 towards zero degrees in one direction and towards 180 degrees in the other direction. A phase sensitive detector can improve output accuracy.

Capacitive Force Transducers

Capacitive force transducers convert force into a change of capacitance. These transducers require more complex circuitry than strain gage or even reluctive transducers. They are used primarily for dynamic force measurements where a wide frequency response is desired and in some specialized applications, such as measurements in low pressure and high temperature environments. Grounded beams and diaphragms have been used as rotors in many capacitive transducers. The capacitance changes can be converted into changes of a dc output voltage by signal-conditioning. Transducer operation tends to be more efficient at higher excitation frequencies as larger changes in reactance are observed. The variable capacitor represented by the transducer may be one leg of an ac bridge or a feedback element in an ac amplifier, a shunting capacitor across an amplifier

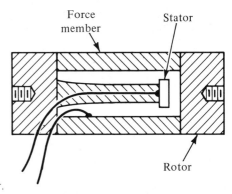

FIGURE 11-9   Capacitive Force Transducer.

input, or an element of a tuned L-C circuit. In the transducer shown in Figure 11-9, the capacitive element uses a coil to form a tuned tank circuit. One end of the moving capacitor rotor is grounded. The rotor forms one end of a hollow cylinder force sensing element. The transducer base forms the other end. The end caps are provided with threaded mounting holes. A voltage in the megahertz range is coupled to the tank through a matching coil. Changes in capacitance are detected as variations in the impedance of the matching coil due to changes in tuning the tank circuit.

Vibrating wire force transducers use a wire between a pair of permanent magnets which is stretched between two points of a deflecting force sensing element. As the wire is caused to vibrate at its resonant frequency in a feedback oscillator, the oscillator frequency will change with wire tension. This principle has been applied in force transducers with frequency modulated output, designed primarily in countries other than the United States. When a second wire, not affected by sensing element deflection, is made to oscillate at the same frequency as the sensing element wire when no force is applied, the beat frequency (frequency difference) between the two frequencies of oscillation will be proportional to tension in the sensing wire only.

Torque Sensors

In-line torque sensing elements for use with rotating shafts are special sensing shafts inserted between the mechanical power source and its load. If torque is applied to one end while the other end is fixed, a line on the shaft surface originally parallel to the axis of rotation becomes a helix or part of a helix. This same twisting, but of a much lower magnitude, occurs during normal use of the shaft. It can be sensed as surface strain as shown in Figure 11-10. Torque transducers for shaft torque measurements use surface strains or stress patterns of the shaft to a greater

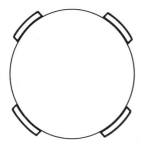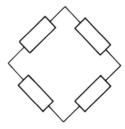

FIGURE 11-10   *Active Strain Gage Bridge Mounted on Rotating Shaft.*

extent than deflection. Torque transducers can be used to measure torque regardless of the direction of shaft rotation. Force transducer may respond equally or unequally to tension and compression and some designs can be used for one force direction only. More strain is produced by machining a section of the diameter of the shaft and sensing the strain on the surface of this section only. A square notch provides the advantage of easier strain gage mounting, with electrical connections in the low stress area near the corners. A rodlinked notch is used with special transduction elements. One design contains four rods of square cross-section. This torque sensing element offers some performance improvements over other types when used with strain gages.

Encoder type torque transducers and torque measuring systems use the discs of incremental digital angular displacement transducers. Two discs are mechanically attached to the sensing shaft so that the shaft deflection due to torque causes a relative angular displacement between the discs. If toothed ferromagnetic discs are used, then the phase difference between two electromagnetic or reluctive output reading devices is proportional to the shaft deflection. The same technique is used with optically coded discs using alternating transparent and opaque portions with photoelectric reading devices.

Another photoelectric technique uses a single light beam through both discs. The two discs are first aligned so that the combination of the opaque and transparent sectors through which the light beam passes produces a null output at the light detector. As torque is applied to the shaft, the discs will pass more or less light, depending on the null point selected. The output of the detector will be proportional to torque exerted on the shaft.

The basic design parameters of force and torque sensing elements include the relative size and shape, material, modulus of elasticity, sensitivity in terms of strain and deflection, dynamic response, and the effects of loading by the transducer on the measured system. Upon deciding if a universal tension or compression, force transducer, or a shaft torque or reaction torque transducer is to be used and after selecting the required measurement principle, the following characteristics should be considered: case, mounting dimensions and tolerances, torque and force connecting provisions, electrical connections, and optional devices.

EXERCISES

11-1. What is the importance of the gage factor?

11-2. If a circuit using a 350Ω strain gage, having a 4.0 gage factor, is to be measured by means of a 50Ω cable in a bridge circuit having matched gage resistors, what current will flow for a 1,500 uin./in. strain when 10 V is applied to the circuit?

11-3. Assuming a gage resistance of 350Ω, a lead resistance of 1.0Ω, a gage factor of 4, and a strain of 2000 uin./in., find the R_{cal} value.

11-4. What is the R_{cal} value of a 120Ω gage with a 1.0 lead resistance and a gage factor of 4 for 4,000 uin./in. strain application?

11-5. Show that the smaller the resistance of a strain gage, the greater the error a given lead resistance produces.

11-6. Describe what happens if all the bridge components of a strain gage bridge are not simple resistors.

11-7. What are the advantages and disadvantages of a semiconductor strain gage?

11-8. Describe the possible use of photoelasticity, brittle coatings, and Moire fringes in automated strain analysis using a microprocessor.

11-9. Describe why the construction of the foil gage grid and methods of manufacturing the grid material should make it easier to control the properties of a foil than a wire. How does this result in a greater degree of reproducibility among gages and better stability of the individual elements?

11-10. Show using a block diagram how a microcomputer could be used to reduce strain gage errors during calibration.

11-11. Draw a strain gage bridge configuration for measuring the shear in a steel panel which compensates for bending forces.

11-12. The vibration sensitivity of a strain gage can be defined as the peak instantaneous change in output at a given vibration level. It is usually expressed as a percentage of full scale output per g level over a given frequency range. It can also be specified as the total error in percentage of full scale output for a given acceleration level. Define a test sequence for determining the vibration sensitivity of an unbonded strain gage. Use a microprocessor for control and sensitivity calculation. Write a flow chart for the microprocessor.

11-13. Describe how a column could be used to measure force using strain gages to provide a readout to a small portable microprocessor based instrument. How could the instrument be used to calibrate itself?

11-14. A reluctive force transducer is to be used with a proving ring for measuring force. Describe with the aid of a diagram the physical and electrical connections for the system. The maximum sensivity and accuracy are desired. Describe any special considerations.

11-15. Describe a self-contained capacitive force transducer. What are the interface requirements for connection to a microprocessor bus?

11-16. Describe with the aid of diagrams two techniques of using photoelectric encoding for torque measurement. Which of these could provide digital information in the format required by a microprocessor? What additional circuitry or equipment would be required?

11-17. List the basic specifications required to specify a force or torque sensing element.

Bibliography

Abramson, N., and **Kuo, F. F.** Eds. *Computer–communications networks.* Englewood Cliffs: Prentice-Hall, 1973.

Adams, L. F. *Engineering measurements and instrumentation.* London: English Universities Press, 1975.

Ahrons, J. W., and **Gardner, R. D.** Interaction of technology and performance in COS-MOS integrated circuits. *IEEE Journal of Solid State Circuits,* 1970, *SC-5.*

Allison, D. R. A design philosophy for microcomputer architectures. *Computer,* 1977, *10,* 2.

Altman, L. Charge coupled devices move in on memories and analog signal processing. *Electronics,* 1974, *47.*

Altman, L., Ed. *Microprocessors.* New York: Electronic Magazine Book Series, 1975.

Amelio, G. E. et al. Experimental verification of the charge coupled device concept. *Bell System Technical Journal,* 1970, *49.*

Amelio, G. E. et al. Charge coupled imaging devices—design considerations. *IEEE Trans. on Electronic Device,* 1971, *V.ED-18.*

American Micro-Systems, Inc. *MOS integrated circuits.* New York: Van Nostrand Reinhold, 1972.

Anderson, D. A. Design of self-checking digital networks using code techniques, PHD Thesis, Report R 527, Urbana, IL: University of Illinois, October, 1971.

Andrews, M. *Principles of firmware engineering in microprogram control.* London: Computer Science Press, 1980.

Avizienis, A. Fault-tolerant computing—an overview. *Computer,* 1971, *4.* January–February.

Baker, W. D. Oxide isolation brings high density to production bipolar memories. *Electronics,* 1973, *46,* March.

Barna, A., and **Porat, D. I.** *Integrated circuits in digital electronics.* New York: John Wiley & Sons, Inc., 1973.

Barrett, J. C., Bergh, A., Horank, T., and **Price, J. E.** Design considerations for a high speed bipolar read only memory. *IEEE Journal of Solid State Circuits,* 1970, *V. SE-5,* No. 5.

Bartogiak, G. Guide to thermocouples. *Instruments and Control Systems,* November, 1978.

Baum, A., and **Senzig, D.** Hardware considerations in a microcomputer multiprocessor system. San Francisco: COMCON Paper, February, 1975.

Baumgart, G. E., and **Jones, D. W.** Modular approach speeds DC drive system build-up. *Control engineering,* 1978, *25,* No. 1, January.

Bazjanac, V. Architectural design theory: Models of the design process in *Basic Questions of Design Theory,* W. R. Spillers, ed. New York: American Elsevier Publishing Co., Inc., 1974.

Bell, C., and **Newell, A.** *Computer structures.* New York: McGraw-Hill, 1970.

Benedict, R. P. *Fundamentals of temperature, pressure and flow measurements.* New York: Wiley, 1969.

Berlekamp, E. R. *Algebraic coding theory.* New York: McGraw-Hill, 1968.

Bernhard, R. "Bubbles take on disks," IEEE Spectrum, May, 1980.

Blasso, L. Flow measurement under any conditions. *Instruments and Control Systems,* February, 1975.

Breuer, M. A., and **Griedman, A. D.** *Diagnosis and reliable design of digital systems.* Woodland Hills, CA: Computer Science Press, 1976.

Brooks, F. P. An overview of microcomputer architecture and software. *Micro Architecture, EUROMICRO 1976 Proceedings.*

Brooks, F. P. *The mythical man-month, essays on software engineering.* Reading, MA: Addision-Wesley, 1975.

Burroughs Corporation, *Digital computer principles.* New York: McGraw-Hill, 1969.

Burton, D. P., and **Dexter, A. L.** Handle microcomputer I/O efficiently. *Electronic Design* 13, June 21, 1978.

Burzio, G. Operating systems enhance uCs. *Electronic Design.* June 21, 1978.

Camenzind, H. R. *Electronic integrated system design.* New York: Van Nostrand Reinhold, 1972.

Chandy, K. M., and **Reiser, M.** (Ed.). *Computer performance.* Amsterdam, Netherlands: North-Holland Publishers, 1977.

Childs, R. E. Multiple microprocessor systems: Goals, limitations and alternatives. Digest of Papers, COMPCON, Spring, 1979.

Chu, Y. *Computer organization and microprogramming.* Englewood Cliffs, N.J.: Prentice-Hall, 1972.

Coffee, M. B. Common-made rejection techniques for low-level data acquisition. *Instrumentation Technology,* July, 1977.

Cohen, T. Structured Flowcharts for Multiprocessing. *Computer Languages, 13,* 1978.

Cragon, H. G. The elements of single-chip microcomputer architecture. *Computer,* October, 1980.

Crick, A. Scheduling and controlling I/O operations. *Data Processing,* May–June, 1974.

Cushman, R. H. The Intel 8080: First of the second-generation microprocessors. *EDN, 19,* 9, May 1, 1974.

Cutler, H. Linear velocity ramp speeds stepper and servo positioning. *Control Engineering.* Vol. 24, *5,* May, 1977.

Dal Cin, M. Performance evaluation of self-diagnosing multiprocessor systems. Conference of Fault-Tolerant Computing, Toulouse, France, June, 1978.

Davis, S. Selection and application of semiconductor memories. *Computer Design, 13,* 1, January, 1974.

De Forest, W. S. *Photoresist materials and processes.* New York: McGraw-Hill, 1975.

Dijkstra, E. W. *A discipline of programming.* Englewood Cliffs, N.J.: Prentice-Hall, 1976.

Doebelin, E. O. *Measurement system—application and design.* New York: McGraw-Hill, 1975.

Doherty, D. W., and **Wells, E. J.** Digital power drive dynamic are characterized by ROMs. *Control Engineering, 25,* 1, January, 1978.

Donovan, J. J. *Systems programming.* New York: McGraw-Hill, 1972.

Dowsing, R. D. Processor management in a multiprocessor system. *Electronic Letters,* November, 1976.

Eckhouse, R. H., Jr. *Minicomputer systems.* Englewood Cliffs, N.J.: Prentice-Hall, 1975.

Elliott, T. C. Temperature, pressure, level, flow-key measurements in power and process. *Power,* September, 1975.

Enslow, P. H., Jr. (Ed.). *Multiprocessors and parallel processing.* New York: John Wiley, 1974.

Farnbach, W. A. Bring up your uP bit-by-bit. *Electronic Design, 24,* 15, July 19, 1976.

Foskett, R. Torque measuring transducers. *Instruments and Control Systems,* November, 1968.

Foster, C. C. *Computer architecture.* New York: Van Nostrand Reinhold Company, 1970.

Frankenberg, R. J. *Designer's guide to semiconductor memories.* Boston, MA: Cahners, 1975.

Franklin, M. A., Kahn, S. A., and **Stucki, M. J.** Design issues in the development of a modular multiprocessor communications network. Sixth Ann. Symp. *Computer Architecture,* April, 23–25, 1979.

Freedman, M. D. *Principles of digital computer operation.* New York: John Wiley & Sons, Inc., 1972.

Friedman, A. D., and **Memon, P. R.** *Fault detection in digital circuits.* Englewood Cliffs, N.J.: Prentice-Hall, 1971.

Fung, K. T., and **Torng.** On the analysis of memory conflicts and bus contentions in a multiple-microprocessor system. *IEEE Trans. Computers, C-27,* 1, January 1979.

Garland, H. *Introduction to microprocessor system design.* New York: McGraw-Hill, 1979.

Gear, C. W. *Computer organization and programming.* New York: McGraw-Hill, 1974.

Gonzalez, M. J., and **Ramamoorthy, C. V.** Parallel task execution in a decentralized system. *IEEE Transactions on Computers,* December, 1972.

Greene, R., and **House, D.** Designing with Intel PROMs and ROMs. Intel Application Note AP-6, Intel Corporation, Santa Clara, CA., 1975.

Grimsdale, R. L., and **Johnson, D. M.** A modular executive for multiprocessor systems. Sheffield, England: *Trends in On-Line Computer Control Systems.* April, 1972.

Gutzwiller, F. W. (Ed.). *SCR manual.* General Electric, Syracuse, N.Y. 1967.

Hall, J. Flowmeters—Matching applications and devices. *Instruments & Control Systems,* February, 1978.

Hall, J. Solving tough flow monitoring problems. *Instruments & Control Systems,* February, 1980.

Hamilton, M., and **Zeldin, S.** Higher order software—A methodology for defining software. *IEEE transactions on software engineering, SE-2,* 1, March, 1976.

Harris, J. A., and **Smith, D. R.** Hierarchical multiprocessor organizations. Fourth Ann. Symp., *Computer Architecture,* March 23–25, 1977.

Hill, F. J., and **Peterson, G. R.** *Digital systems: Hardware organization and design.* New York: John Wiley, 1973.

Hodges, D. A. Alternative trends for advanced memory systems. *Computer,* September, 1973.

Hodges, D. A. *Semiconductor memories.* New York: IEEE Press, 1972.

Hordeski, M. F. Digital control of microprocessors. *Electronic Design,* December 6, 1975.

Hordeski, M. F. Digital sensors simplify digital measurements. *Measurements and Data,* May–June, 1976.

Hordeski, M. F. When should you use pneumatics, when electronics? *Instruments & Control Systems,* November, 1976.

Hordeski, M. F. Guide to digital instrumentation for temperature, pressure instruments. *Oil, Gas and Petrochem Equipment,* November, 1976.

Hordeski, M. F. Digital instrumentation for pressure, temperature/pressure, readout instruments. *Oil, Gas and Petrochem Equipment,* December, 1976.

Hordeski, M. F. Innovative design: Microprocessors. *Digital Design,* December, 1976.

Hordeski, M. F. Passive sensors for temperature measurement. *Instrumentation Technology,* February, 1977.

Hordeski, M. F. Adapting electric actuators to digital control. *Instrumentation Technology,* March, 1977.

Hordeski, M. F. Fundamentals of digital control loops and factors in choosing pneumatic or electronic instruments, presentation at the SCMA Instrumentation Short Course, Los Angeles, April 6, 1977.

Hordeski, M. F. Balancing microprocessor-interface tradeoffs. *Digital Design,* April, 1977.

Hordeski, M. F. Digital position encoders for linear applications. *Measurements and Control,* July–August, 1977.

Hordeski, M. F. Future microprocessor software. *Digital Design,* August, 1977.

Hordeski, M. F. Radiation and stored data. *Digital Design,* September, 1977.

Hordeski, M. F. Microprocessor chips. *Instrumentation technology,* September, 1977.

Hordeski, M. F. Process controls are evolving fast. *Electronic Design,* November 22, 1977.

Hordeski, M. F. Fundamentals of digital control loops. *Measurements and Control,* February, 1978.

Hordeski, M. F. Using microprocessors. *Measurements and Control,* June, 1978.

Hordeski, M. F. *Illustrated dictionary of micro computer terminology.* Blue Ridge Summit, PA: Tab, 1978.

Hordeski, M. F. *Microprocessor cookbook.* Blue Ridge Summit, PA: Tab, 1979.

Hordeski, M. F. Selecting test strategies for microprocessor system. ATE Seminar Proceedings, Pasadena, CA, January, 1982. New York: Morgan-Gramprian.

Hordeski, M. F. Selection of a test strategy for MPU systems. *Electronics Test,* February, 1982.

Hordeski, M. F. Trends in displacement sensors. Pasadena, CA: *Sensors and Systems Conference Proceedings,* May, 1982 (Network Exhibitions, Campbell, CA.)

Hordeski, M. F. The impact of 16-bit microprocessors. Las Vegas: *International Instrumentation Symposium Proceedings,* May, 1982. (Research Triangle Park, N.C.: Instrument Society of America.)

Hordeski, M. F. Diagnostic strategies for microprocessor systems. *ATE Seminar Proceedings,* Anaheim, CA., January, 1983. (New York: Morgan-Gramprian.)

Hnatek, Eugene R. *A user's handbook of semiconductor memories.* New York: John Wiley, 1977.

Hnatek, Eugene R. Current semiconductor memories. *Computer Design,* April, 1978.

Hougen, J. O. *Measurements and control applications.* Research Triangle Park, N.C.: Instrument Society of America, 1979.

Husson, S. S. *Microprogramming: Principles and practices.* Englewood Cliffs, N.J.: Prentice-Hall, 1970.

Intel Corp. *8086 user's guide.* Santa Clara, CA: Intel Corp., 1976.

Intel Corp. *4004/4040 assembly language programming manual.* Santa Clara, CA: Intel Corp., 1974.

Intel Corp. *8080 user's manual.* Santa Clara, CA: Intel Corp., 1975.

Intel Corp. *8080 assembly language programming manual.* Santa Clara, CA: Intel Corp., 1975.

Jackson, M. A. *Principles of program design.* New York: Academic Press, 1975.

Janki, C. What's new in motors and motor controls? *Instruments and Control Systems,* November, 1979.

Jones, J. C. *Design methods.* New York: Wiley-Interscience, 1970.

Jutila, J. M. Temperature instrumentation. *Instrumentation Technology,* February, 1980.

Kaufman, A. B. Monitor acceleration, velocity or displacement. *Instruments and Control Systems,* October, 1979.

Klingman, E. E. *Microprocessor systems design.* Englewood Cliffs, N.J.: Prentice-Hall, 1977.

Klipec, B. How to avoid noise pickup on wire and cables. *Instruments & Control Systems,* December, 1977.

Knuth, D. E. *The art of computer programming.* Reading, MA: Addison-Wesley, 1973.

Kohonen, T. *Digital circuits and devices.* Englewood Cliffs, N.J.: Prentice-Hall, 1972.

Kolk, W. R. PFM—Control candidate in energy limited systems. *Control Engineering, 24,* 10, October, 1977.

Kuck, D. J. *The structure of computers and computations.* Volume 1. New York: Wiley, 1978.

Leventhal, L. V. *Microprocessors: Software, hardware, programming.* Englewood Cliffs, N.J.: Prentice-Hall, 1978.

Liptak, B. G. (Ed.). *Instrument engineers' handbook on process measurement.* Radnor, PA: Chilton, 1980.

Liptak, B. G. Ultrasonic instruments. *Instrumentation Technology,* September, 1974.

Lomas, D. J. Selecting the right flowmeter. *Instrumentation Technology,* 1977.

Lorin, H. *Parallelism in hardware and software.* Englewood Cliffs, N.J.: Prentice-Hall, 1972.

Madnick, S. E., and **Donovan, J. L.** *Operating systems.* New York: McGraw-Hill, 1974.

Mano, M. M. *Computer logic design.* Englewood Cliffs, N.J.: Prentice-Hall, 1972.

Martin, Donald P. *Microcomputer design.* Chicago: Martin Research Ltd., 1975.

Mazur, T. Microprocessor basics: Part 4: the Motorola 6800. *Electronic Design,* July 19, 1976.

McGlynn, D. R. *Microprocessors.* New York: Wiley, 1976.

McKenzie, K., and **Nichols, A. J.** Build a compact microcomputer. *Electronic Design, 24,* 10, May 10, 1976.

Medlock, R. S. Vortex shedding meters. London: Liquefied Gas Symposium, 1978.

Moss, D. Multiprocessing adds muscle to uPs. *Electronic Design 11,* May 24, 1978.

Motorola Semiconductor. *MC68000 microprocessor user's manual.* Austin, TX: Motorola, 1979.

Motorola Semiconductor. *M6800 microprocessor applications manual.* Phoenix, Ariz.: Motorola, 1975.

Motorola Semiconductor. *M6800 programming manual.* Phoenix, Ariz.: Motorola, 1975.

Motorola Semiconductor. *MECL integrated circuits data book.* Phoenix, Ariz.: Motorola, 1972.

Murphy, H. N. Flow measurement by insertion turbine meters. Delaware: Instrument Society of America Symposium, 1979.

Muth, S. (Ed.). *Synchro conversion handbook.* Bohemia, N.Y.: ILC Data, 1974.

Myers, G. J. *Reliable software through composite design.* New York: Petrocelli Charter, 1975.

Myers, G. J. *Advances in computer architecture,* New York: Wiley, 1978.

National Semiconductor. *The NS16000 family of 16-bit microprocessors.* Santa Clara, CA: National Semiconductor.

National Semiconductor. *Digital integrated circuits.* Santa Clara, CA.: National Semiconductor, 1973.

Newton, R. S. *An exercise in multiprocessor operating system design.* Agard Conference on Real-Time Computer-based Systems, Athens, Greece: NATO Advisory Group on Aerospace R & D, May, 1974.

Nick, J. R. Using Schottky 3-state outputs in bus-organized systems. *Electronic Design News, 19,* 23, December 5, 1974.

Norton, H. N. *Handbook of transducers for electronic measuring systems.* Englewood Cliffs, N.J.: Prentice-Hall, 1969.

Noyce, R. N., and **Hoff, M. E., Jr.** A history of microprocessor development at Intel. *IEEE Micro,* February, 1981.

Oppenheim, A. V., and **Schafer, S.** *Digital signal processing.* Englewood Cliffs, N.J.: Prentice-Hall, 1975.

Palmer, R. Nonlinear feedforward can reduce servo settling time. *Control Engineering, 26,* 3, March 1978.

Park, R. M. Applying the systems concept to thermocouples. *Instrumentation Technology,* August, 1973.

Patel, J. H. Processor-memory interconnections for multiprocessors. *Proceedings of Sixth Annual Symposium Computer Architecture,* April 23–25, 1979.

Patterson, D. A., and **Seguin, C. H.** Design considerations for single-chip computers of the future. *IEEE Trans. Comp.,* C-29, February, 1980.

Peters, L. J., and **Tripp, L. L.** Is software design wicked? *Datamation, 22,* 6, June, 1976.

Peuto, B. L., and **Shustek, L. J.** Current issues in the architecture of microprocessors. *Computer,* February, 1977.

Plumb, H. H. *Temperature: Its measurement and control in science and industry.* Research Triangle Park, NC: Instrument Society of America, 1972.

Richman, P. *MOS field effect transistors and integrated circuits.* New York: Wiley-Interscience, 1973.

Schaeffer, E. J., and **Williams, T. J.** *An analysis of fault detection correction and prevention in industrial computer systems.* Purdue University: Purdue Laboratory for Applied Industrial Control, October, 1977.

Sheingold, D. H. *Analog-digital conversion handbook.* Norwood, MA: Analog Devices, 1972.

Shepherd, M., Jr. *Distributed computing power: Opportunities and challenges.* National Computer Conference, 1977.

Sigma Instruments. *Stepping motor handbook.* Braintree, MA: Sigma Instruments, 1972.

Skrokov, M. R. (Ed.). *Mini and microcomputer control in industrial processes.* New York: Van Nostrand Reinhold, 1980.

Slomiana, M. Selecting differential pressure instruments. *Instrumentation Technology,* August, 1979.

Sobel, H. S. *Introduction to digital computer design.* Reading, MA: Addison Wesley, 1970.

Soisson, H. E. *Instrumentation in industry.* New York: Wiley, 1975.

Soucek, B. *Microprocessors and microcomputers.* New York: Wiley, 1976.

Stevens, W. P., Myers, G. J., and **Constantine, L. L.** *Structural design.* New York: Yourdon, 1975.

Stone, H. S. *Introduction to computer architecture.* New York: McGraw-Hill, 1975.

Tasar, O., and **Tasar, V.** A study of intermittent faults in digital computers. *AFIPS Conference Proceedings,* Vol. 46, Montvalue, N.J.: AFIPS Press, 1977.

Texas Instruments. *TMS 9900 microprocessor data manual.* Dallas: Texas Instruments.

Texas Instruments. "The TTL Data Book for Design Engineers," Texas Instruments, Dallas, 1973.

Texas Instruments. *The microprocessor handbook.* Houston: Texas Instruments, 1975.

Thomas, T. B., and **Arbuckle, W. L.** *Multiprocessor software: Two approaches.* Conference on the Use of Digital Computers in Process Control, Baton Rouge, LA, February, 1971.

Thurber, K. J., and **Masson, G. M.** *Distributed processor communication architecture.* Lexington, MA: Lexington, 1979.

Tippie, J. W., and **Kulaga, J. E.** Design considerations for a multiprocessor-based data acquisition system. *IEEE Transactions on Nuclear Science,* August, 1979.

Toong, H. D., and **Gupta, A.** An architectural comparison of contemporary 16-bit microprocessors. *IEEE Micro,* May, 1981.

Torrero, E. A. Focus on microprocessors. *Electronic Design,* September 1, 1974.

Torrero, E. A. (Ed.). *Microprocessors: New directions for designers.* Rochelle Park, N.J.: Hayden Book Co., 1975.

Vacroux, A. G. Explore microcomputer I/O capabilities. *Electronic Design,* May 10, 1975.

Wegner, W. (Ed.). *Research directions in software technology.* Cambridge, MA: MIT Press, 1978.

Yourdon, E., and **Constantine, L. L.** *Structured design.* New York: Yourdon, 1975.

Zaks, R. *Microprocessors.* Berkeley, CA: Sybex, 1980.

Zaks, R., and **Lesea, A.** *Microprocessor interfacing techniques.* Berkeley, CA: Sybex, 1979.

Zilog Corp. Z8000 user's guide. Cupertino, CA: Zilog Corp.

Index